T0281023

Lecture Notes in Computer Science 14532

Founding Editors

Gerhard Goos
Juris Hartmanis

The series Lecture Notes in Computer Science (LNCS), including its subseries Lecture Notes in Artificial Intelligence (LNAI) and Lecture Notes in Bioinformatics (LNBI), has established itself as a medium for the publication of new developments in computer science and information technology research, teaching, and education.

LNCS enjoys close cooperation with the computer science R & D community, the series counts many renowned academics among its volume editors and paper authors, and collaborates with prestigious societies. Its mission is to serve this international community by providing an invaluable service, mainly focused on the publication of conference and workshop proceedings and postproceedings. LNCS commenced publication in 1973.

Bong Jun Choi · Dhananjay Singh ·
Uma Shanker Tiwary · Wan-Young Chung
Editors

Intelligent Human Computer Interaction

15th International Conference, IHCI 2023
Daegu, South Korea, November 8–10, 2023
Revised Selected Papers, Part II

 Springer

Editors
Bong Jun Choi 🅾
Soongsil University
Seoul, Korea (Republic of)

Dhananjay Singh 🅾
Saint Louis University
St. Louis, MO, USA

Uma Shanker Tiwary 🅾
Indian Institute of Information Technology
Allahabad, India

Wan-Young Chung 🅾
Pukyong National University
Busan, Korea (Republic of)

ISSN 0302-9743   ISSN 1611-3349 (electronic)
Lecture Notes in Computer Science
ISBN 978-3-031-53829-2   ISBN 978-3-031-53830-8 (eBook)
https://doi.org/10.1007/978-3-031-53830-8

This Springer imprint is published by the registered company Springer Nature Switzerland AG
The registered company address is: Gewerbestrasse 11, 6330 Cham, Switzerland

Paper in this product is recyclable.

Preface

IHCI is an annual international conference in the field of Human-Computer Interaction, dedicated to exploring research challenges arising from the intricate interaction between machine intelligence and human intelligence. The fifteenth edition of this event focuses on the theme of "AI-Powered IHCI," delving into the dynamic intersection of artificial intelligence and human-computer interaction.

We are pleased to introduce the proceedings of the 15th International Conference on Intelligent Human-Computer Interaction (IHCI 2023), hosted both on-site and online by EXCO Daegu, South Korea, from November 8 to 10, 2023. The event took place in Daegu, South Korea, and featured 4 special sessions, 2 workshops, 2 tutorial sessions, and 2 panel discussions, all aligned with the conference's central theme. Out of 139 submitted papers, the program committee accepted 71 for oral presentation and publication, following the recommendations of at least 2 expert reviewers in each case.

The 15th IHCI conference included keynote and invited talks facilitated by expert session chairs with substantial experience in both industry and academia. It drew more than 200 participants from over 25 countries.

IHCI has become the primary global gathering for academic researchers, graduate students, leading research think tanks, and industry technology developers in the field. We believe that participating in IHCI not only contributes to personal and professional growth but also fosters success in the business realm, ultimately benefiting society as a whole.

Our sincere gratitude goes to all the authors who submitted their work to IHCI 2023. The Microsoft CMT portal conference system played a crucial role during the submission, review, and editing stages, and we extend our thanks to the technical program committee (TPC) and organizing committee for their invaluable efforts in ensuring the conference's success. Finally, we express our appreciation to the speakers, authors, and participants whose contributions made IHCI 2023 a stimulating and productive conference. The continued success of the IHCI conference series relies on their unwavering support in the future.

November 2023

Bong Jun Choi
Dhananjay Singh
Uma Shanker Tiwary
Wan-Young Chung

Organization

General Chairs

Wan-Young Chung	Pukyong National University, South Korea
Uma Shanker Tiwary	Indian Institute of Information Technology, Allahabad, India

Advisory Chairs

Ajay Gupta	Western Michigan University, USA
Dae-Ki Kang	Dongseo University, South Korea
Jan Treur	Vrije Universiteit Amsterdam, The Netherlands
Venkatasubramanian Ganesan	National Institute of Mental Health & Neurosciences, India

Program Chairs

Madhusudan Singh	Oregon Institute of Technology, USA
Sang-Joong Jung	Dongseo University, South Korea
Jong-Hoon Kim	Kent State University, USA

Technical Program Chairs

David (Bong Jun) Choi	Soongsil University, South Korea
Dhananjay Singh	Saint Louis University, USA
Uma Shanker Tiwary	Indian Institute of Information Technology, Allahabad, India
Wan-Young Chung	Pukyong National University, South Korea

Tutorial Chairs

Hanumant Singh Shekhawat	Indian Institute of Technology, Guwahati, India
Ikechi Ukaegbu	Nazarbayev University, Kazakhstan
Annu Sible Prabhakar	University of Cincinnati, USA

Workshop Chairs

Nagamani Molakatala	University of Hyderabad, India
Ajit Kumar	Soongsil University, South Korea
Rajiv Singh	Banasthali Vidyapith, India

Session Chairs

Jong-Hoon Kim	Kent State University, USA
Ikechi Ukaegbu	Nazarbayev University, Kazakhstan
Irish Singh	Oregon Institute of Technology, USA
Annamaria Szakonyi	Saint Louis University, USA
Young Sil Lee	Dongseo University, South Korea
Sang-Joong Lee	Dongseo University, South Korea
Swati Nigam	Banasthali Vidyapith, India
Jan-Willem van't Klooster	University of Twente, The Netherlands
Ajit Kumar	Soongsil University, South Korea
Tatiana Cardona	Saint Louis University, USA
Bong Jun Choi	Soongsil University, South Korea
Manoj Kumar Singh	Banaras Hindu University, India
Aditi Singh	Cleveland State University, USA
Naagmani Molakatala	University of Hyderabad, India
Maria Weber	Saint Louis University, USA
Madhusudan Singh	Oregon Institute of Technology, USA
Saifuddin Mahmud	Bradley University, USA

Industrial Chairs

Mario José Diván	Intel Corporation, USA
Garima Bajpai	Canada DevOps Community of Practice, Canada
Sandeep Pandey	Samsung, India

Publicity Chairs

Zongyang Gong	Vanier College, Canada
Mohd Helmy Bin Abd Wahab	University Tun Hussein Onn Malaysia, Malaysia
Sanjay Singh	Motilal National Institute of Technology Allahabad, India

Local Organizing Chairs

Sang-Joong Jung	Dongseo University, Korea
David (Bong Jun) Choi	Soongsil University, Korea
Jong-Ha Lee	Keimyung University, Korea
Dhananjay Singh	Saint Louis University, USA

Technical Support Chairs

Swati Nigam	Banasthali Vidyapith, India
Irish Singh	Oregon Institute of Technology, USA
Aishvarya Garg	Banasthali Vidyapith, India

Plenary Speakers

Wan-Young Chung	Pukyong National University, South Korea
Uma Shanker Tiwary	Indian Institute of Information Technology, Allahabad, India
Dhananjay Singh	Saint Louis University, USA

Keynote Speakers

Uichin Lee	Korea Advanced Institute of Science and Technology, South Korea
Mary Czerwinski	Microsoft Research Lab – Redmond, USA
Henry Leung	University of Calgary, Canada
KC Santosh	University of South Dakota, USA
Shiho Kim	Yonsei University, South Korea

Invited Speakers

Yashbir Singh	Mayo Clinic, USA
Hyunggu Jung	University of Seoul, Korea
Hyungjoo Song	Soongsil University, Korea
Gaurav Tripathi	Bharath Electronics Limited, India
Jan-Willem van't Klooster	University of Twente, The Netherlands
K. S. Kuppusamy	Pondicherry University, India
Siba K. Udgata	University of Hyderabad, India

Jungyoon Kim   Kent State University, USA
Anshuman Shastri   Banasthali Vidyapith, India
Madhusudan Singh   Oregon Institute of Technology, USA

Tutorial Speakers

Urjaswala Vora   Pennsylvania State University, USA
Shael Brown   McGill University, Canada
Colleen M. Farrelly   Staticlysm LLC, USA
Yashbir Singh   Mayo Clinic, USA

Panel Discussion Speakers

Jong-Hoon Kim   Kent State University, USA
Maria Weber   Saint Louis University, USA
Tatiana Cardona   Saint Louis University, USA
Annamaria Szakonyi   Saint Louis University, USA
Urjaswala Vora   Pennsylvania State University, USA
Wan-Young Chung   Pukyong National University, South Korea
Uma Shanker Tiwary   Indian Institute of Information Technology,
                             Allahabad, India
Dhananjay Singh   Saint Louis University, USA

Workshop Speakers

Shiho Kim   Yonsei University, South Korea
Ashutosh Mishra   Yonsei University, South Korea
Dimple Malhotra   Imatter Institute of Counselling and Behavioral
                             Sciences, USA

Technical Program Committee

Abhay Kumar Rai   Central University of Rajasthan, India
Aditi Singh   Cleveland State University, USA
Ahmed Imteaj   Southern Illinois University, USA
Aishvarya Garg   Banasthali Vidyapith, India
Ajit Kumar   Soongsil University, South Korea
Andres Navarro- Newball   Pontificia Universidad Javeriana, Colombia

Ankit Agrawal	ADGITM, India
Anupam Agrawal	IIIT Allahabad, India
Arvind W. Kiwelekar	Dr. Babasaheb Ambedkar Technological University, India
Awadhesh Kumar	Banaras Hindu University, India
Bernardo Nugroho Yahya	HUFS, South Korea
Bhawana Tyagi	Banasthali Vidyapith, India
Wan-Young Chung	Pukyong National University, South Korea
David (Bong Jun) Choi	Soongsil University, South Korea
Dhananjay Singh	Saint Louis University, USA
Irina Dolzhikova	Nazarbayev University, Kazakhstan
Irish Singh	Oregon Institute of Technology, USA
Jeetashree Aparajeeta	VIT Chennai, India
Jong-Hoon Kim	Kent State University, USA
Jungyoon Kim	Kent State University, USA
Kanike Sreenivasulu	DRDO Hyderabad, India
Lei Xu	Kent State University, USA
Manju Khari	JNU, India
Manoj Kumar Singh	Banaras Hindu University, India
Manoj Mishra	Banaras Hindu University, India
Madhusudan Singh	Oregon Institute of Technology, USA
Mohammad Asif	Indian Institute of Information Technology, Allahabad, India
Nagamani Molakatala	University of Hyderabad, India
Nidhi Srivastav	SKIT Jaipur, India
Pooja Gupta	Banasthali Vidyapith, India
Pooja Khanna	Amity University, Lucknow, India
Rajiv Singh	Banasthali Vidyapith, India
Raveendra Babu Ponnuru	Cleveland State University, USA
Ravindra Hegadi	Central University of Karnataka, India
Renu Dalal	Guru Govind Singh Indraprastha University, India
Ritu Chauhan	Amity University, Noida, India
Ruchilekha	Banaras Hindu University, India
Sachchida Nanad Chaurasia	Banaras Hindu University, India
Saifuddin Mahmud	Bradley University, USA
Sakshi Indolia	Narsee Monjee Institute of Management Studies, India
Sarvesh Pandey	Banaras Hindu University, India
Shakti Sharma	Bennett University, India
Siddharth Singh	University of Lucknow, India
Sonam Seth	Banasthali Vidyapith, India
Sudha Morwal	Banasthali Vidyapith, India

Contents – Part II

Mobile Computing and Ubiquitous Interactions

Social Computing and Interactive Elements

Contents – Part I

Human Centred AI

Human-Robot Interaction and Intelligent Interfaces

User Centred Design

AI and Big Data

Automated Fashion Clothing Image Labeling System

Jun-Oh Lim[1], Woo-Jin Choi[2], and Bong-Jun Choi[3(✉)]

[1] Dongseo University, 47 Jurye-ro, Sasang-gu, Busan 47011, Korea
dh03219@naver.com
[2] Department of Fashion and Textiles, Seoul National University, Seoul, South Korea
aasseev@snu.ac.kr
[3] Department of Software, Dongseo University, 47 Jurye-ro, Sasang-gu, Busan 47011, Korea
bongjun.choi@dongseo.ac.kr

Abstract. As interest in fashion clothing has recently increased, the importance of fashion services using deep learning technology is increasing. These services are used in a variety of fields, including customized clothing recommendations, fashion style analysis, and improving online shopping experiences. However, the image labeling work, which is the core of these services, is still performed manually, which requires high costs and time, and relies on manpower. This study seeks to automate this process of fashion clothing labeling by using three steps image preprocessing technology and Yolov8 for dataset learning. During the research process, rembg and OpenCV were used to preprocess image data, and OpenPose was introduced to obtain object joint information for the accurate location of clothing items. Through this, we were able to effectively remove the background of the image, extract and classify the joints of the object, distinguish tops and bottoms, and then teach them to Yolo. As a result, this study presents an automated labeling approach that can maintain high accuracy while reducing the time and cost required for manual labeling of fashion clothing images.

Keywords: Training Data Labeling · Image Preprocessing · Item Classification

1 Introduction

Today, with the increasing interest in fashion, South Korea's fashion and clothing market sees annual growth [1]. As we enter the digital age, fashion consumption is increasingly shifting online. According to data from the National Statistical Office, the online fashion market grew from 43 trillion won in 2019 to 49.7 trillion won in 2021, showing a growth rate of 9.2%, equivalent to about 6.7 trillion won, compared to 2019 [2]. This indicates that many consumers are actively engaging in fashion activities online. Numerous online fashion companies label their products and outfit images, providing keywords to users for conveying clothing information [3]. However, up to this point, this work has been done manually by inputting labeling information for clothes one by one. Currently, there is a trend toward requiring more advanced performance due to accuracy issues with automated labeling systems but when a large number of images need labeling, it takes a considerable amount of time, accuracy issues still arise, and there is no professional

B. J. Choi et al. (Eds.): IHCI 2023, LNCS 14532, pp. 3–8, 2024.
https://doi.org/10.1007/978-3-031-53830-8_1

manpower to do this. Furthermore, considering the presence of agencies and services that offer image labeling work for a fee, image labeling incurs substantial costs. In this paper, we propose an automated system for labeling fashion clothing images. We leverage three libraries in conjunction with Yolo to address these challenges. To ensure precise extraction of labeling targets from images, our system employs rembg to remove backgrounds and employs OpenCV to generate virtual bounding boxes using object border coordinates. This significantly reduces recognition range errors in the images. Subsequently, we employ OpenPose to augment the image data with joints for the target objects, enabling Yolo to be trained on this data.

2   Related Research

Korea's Online Fashion Commoners Entrepreneur 'Musinsa' conducted research using Amazon Sagemaker's Auto Labeling system and Amazon Sagemaker Ground Truth to develop a model for automated image inspection. This system improves the accuracy of predictions by learning from large datasets of at least 1,250 and 5,000 objects that users have previously worked on. After manually classifying unlabeled images, training data and validation data are collected. Using these two data, a model for automatic classification is created. Afterwards, the probability value (probability that an image belongs to a specific label) calculated by the model is compared to manual work to determine how reliable it must be to set a threshold. Inference is conducted on unlabeled images using the model, and if the probability exceeds the threshold, an "auto-annotated" tag is assigned. No further intervention is performed on tagged images, and the label is gradually expanded by re-performing the operation on untagged images. As a result of research, the coverage increased by up to 8.7%, and the accuracy improved by up to 1.6% for the same model. Table. 1 shows the indicators for each iteration during a total of 7 iterations, and training and validation are the cumulative values up to the previous iteration. However, it took about 5 days to inspect about 20,000 images before work, and 22 workers were assigned to classify a total of 10,000 images for 5 days through double check. Additionally, the cost of labeling through this process was approximately $960, and a total of 28 h were spent manually training the model for auto labeling [4].

Table 1. Indicators for each iteration

Iteration	Training	Validation	Inference	Labeled	Labeled (auto)
1	0	0	0	1,000	0
2	0	0	0	40	0
3	1,000	1,040	0	1,000	0
4	2,570	470	1,040	1,000	3,839
5	3,717	323	470	1,000	142
6	4,898	142	323	1,000	444
7	0	0	142	868	73
Sum				5,908	4,498

3 Fashion Clothing Image Labeling System

This paper aims to improve the efficiency of fashion clothing image labeling tasks and contributes to ongoing research towards automation the labeling process. The system presented in this paper utilizes data preprocessing techniques and a custom model to enhance the accuracy of the Yolov8 model used for image labeling. As a result, user-provided images intended for labeling undergo preprocessing, and the labeled results are then provided to the user.

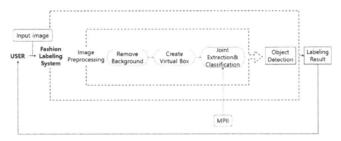

Fig. 1. System concept

3.1 Dataset Collection

This paper utilizes snapshot images referenced by users on an online fashion platform. However, to address the problem of generating labeling keywords for worn clothing and to reduce recognition range errors during the image preprocessing phase, this study employed model images featuring individuals wearing clothing rather than images of clothing products [5].

3.2 Image Preprocessing

Image preprocessing is performed to improve the accuracy of image labeling work [6]. The image preprocessing process proceeds in three steps as shown in Fig. 1

Remove Background. In the first step, the image's background is removed to selectively extract the target object, addressing the issue of mislabeling the wrong target in the image. For background removal, this study employed two libraries, OpenCV and rembg. Initially, differentiation of the target from the background was attempted using OpenCV, with the goal of eliminating everything except the target object. However, an issue arose where backgrounds of similar colors to clothing were incorrectly recognized along-side the object. Furthermore, accessories like hats were classified as separate objects and consequently removed, even though they were essential components of the object, leading to unsuitable results for this study. When rembg was employed, objects unrelated to the target and the background were effectively distinguished and removed, rendering it a more suitable choice for our study. In Fig. 2, the left side displays the OpenCV result, while the right side displays the rembg result.

Fig. 2. Remove background results

Virtual Box for Target in Image. In the second step, the recognition range error is reduced through OpenCV to clearly extract the joints of the target. While the results obtained from the first step successfully eliminated the background, excluding only the target, the size of the background within the image remains unchanged from its original size of image. To refine this process, the maximum and minimum values of the x-axis and y-axis, denoted as x_max, x_min, y_max, and y_min, are determined based on the target's border within the image, as illustrated in Fig. 3. Subsequently, a virtual bounding box is created with dimensions width=x_max-x_min and height=y_max-y_min, defining the range for joint extraction. This approach reduces recognition range errors when extracting joints using OpenPose without altering the original image size [7].

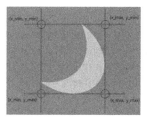

Fig. 3. Coordinate settings

Joint Extraction and Classification. In the third step, The extraction of the target's joints within the image proceeds, enabling a clear differentiation between the upper and lower body of the target. For performing joint extraction using OpenPose on the resulting image from the second phase, the Human Pose Dataset MPII Dataset is leveraged, a rich dataset comprising 25,000 images encompassing approximately 40,000 individuals [8]. This dataset facilitates the extraction of a total of 15 joints, encompassing the Head, Neck, Shoulders (both left and right), Elbows (both left and right), Wrists (both left and right), Hips (both left and right), Knees (both left and right), Ankles (both left and right), and Chest (or Background). Notably, the attire worn by the target within the image can be discerned by focusing on the hip joints (both left and right, denoted as numbers 8 and 11), as exemplified in Fig. 4.

Fig. 4. OpenPose key point extraction results

3.3 Yolo Training and Labeling Results

To detect targets and items within the images, Yolov8, the latest version known for its high performance in detecting multiple objects, has been adopted. Specifically, the Yolov8n model was selected, which is one of the smallest pre-trained models based on the extensive COCO val2017 image dataset and achieves a mean average precision (mAPval) of 37.3 [9], serving as an indicator for object detection model performance. Following image preprocessing steps, including background removal, joint extraction, and classification, the resulting image was divided into upper and lower garments based on the key points extracted through OpenPose, specifically points 3 (left hip) and 11 (right hip). This division enabled the creation of a new training dataset, which was then used for Yolov8 training. For precise results, the learning conditions were configured with epoch (number of learning iterations) = 200, batch size = 20, and image size = 640 [10]. Additionally, a threshold range of 0.5 to 0.95 was set to calculate the mAP50-95, providing a comprehensive measurement. The outcome of the training displayed an average mAP50-95 value of approximately 0.79, with the highest value recorded at 0.97795 during the 93rd iteration. Figure 5 showcases the results achieved through this process.

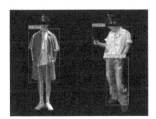

Fig. 5. Labeling test results

4 Conclusion

In this paper, we discussed the labeling approach and research content using three libraries for the efficiency and convenience of fashion clothing image labeling work. The problem of OpenPose's joint key points being out of sync, which was a problem that occurred during the experiment, was partially solved by removing the background through rembg and OpenCV. In addition, the recognition range error was reduced by creating a virtual box on the target in the image using OpenCV. A comparison between the final results obtained through this study's system and the existing manually labeled train data using Yolov8's mAP value showed a high agreement rate. However, this result only appeared in images where the top and bottom were clearly distinguished, and there were cases where onepiece clothing such as dresses in some images were not properly recognized. This problem can be solved by using the l, x model, which is more accurate than Yolov8n, and by subdividing the clothing classes, and creating a system and dataset with better performance.

Acknowledgement. This research was supported by the MIST(Ministry of Science, ICT, Korea, under the National Program for Excellence in SW), Supervised by the IITP(Institute of Information & communications Technology Planning & Evaluation) in 2023 (2019-0-01817)

References

1. Jung, J.S.: Domestic fashion market in 2022 records a 5.2% increase, surpassing 45.8 trillion won, marking two consecutive years of growth. ktnews. https://www.ktnews.com/news/articl eView.html?idxno=126297. Last accessed 1 Sep 2023
2. Jung, J.S.: Statistics Korea reports 9.2% growth in online fashion transaction volume, reaching 49.7192 trillion won. Ktnews. https://www.ktnews.com/news/articleView.html?idxno= 122454. Last accessed 1 Sep 2023
3. Han, S.Y., Cho, Y.J., Lee, Y.R.: The effect of the fashion product classification method in online shopping sites. J. Korean Soc. Cloth. Text. **40**(2), 287–304 (2016)
4. Amazon. https://aws.amazon.com/ko/blogs/korea/develop-an-automatic-review-image-ins pection-service-with-amazon-sagemaker/. Last accessed 31 Aug 2023
5. An, H.S., Kwon, S.H., Park, M.J.: A case study on the recommendation services for customized fashion styles based on artificial intelligence. J. Korean Soc. Cloth. Text. **43**(3), 349–360 (2019)
6. Seo, J.B., Jang, H.H., Cho, Y.B.: Analysis of image pre-processing algorithms for efficient deep learning. J. Korea Inst. Inf. Commun. Eng. **24**, 161–164 (2020)
7. Khan, R., Raisa, T.F., Debnath, R.: An efficient contour based fine-grained algorithm for multi category object detection. J. Image Graph. **6**(2), 127–136 (2018)
8. Cao, Z., et al.: OpenPose: realtime mulit-person 2D pose estimation using part affinity fields. arXiv:1812.08008v2 (2018)
9. Ultralytics. https://docs.ultralytics.com/tasks/detect/. Last accessed 17 Aug 2023
10. Kim, J.S., Kwon, J.H, Lee, J.H., Bae, J.H.: Data construction using generated images and performance comparison through YOLO object detection model. In: Proceedings of KIIT Conference, pp. 722–726(2023)

AI-Based Estimation from Images of Food Portion Size and Calories for Healthcare Systems

Akmalbek Abdusalomov[1]([✉]) [iD], Mukhriddin Mukhiddinov[2]([✉]) [iD],
Oybek Djuraev[2] [iD], Utkir Khamdamov[2] [iD], and Ulugbek Abdullaev[3] [iD]

[1] Department of Computer Engineering, Gachon University Sujeong-Gu, Gyeonggi-Do,
Seongnam-Si 461-701, South Korea
akmalbek@gachon.ac.kr

[2] Department of Communication and Digital Technologies, University of Management and
Future Technologies, Tashkent 100208, Uzbekistan
{mukhiddinov,djuraev}@umft.uz

[3] Department of Hardware and Software of Management Systems in Telecommunication,
Tashkent University of Information Technologies named after Muhammad al Khwarizmi,
Tashkent 100084, Uzbekistan

Abstract. In the realm of nutrition science, it is well-recognized that individuals' dietary needs vary based on factors such as age, gender, and health status. This divergence in nutritional requirements is particularly critical for vulnerable groups, including newborns, the elderly, and individuals with diabetes, as their dietary choices can have profound implications for their health. Moreover, the dearth of Uzbek recipes in mainstream culinary literature, which predominantly focuses on Western cuisine, exacerbates the issue. To address these challenges, this study undertakes the ambitious task of constructing a comprehensive AI system, comprising both backend and frontend components, tailored to the nuances of Uzbek cuisine. The primary objectives encompass recipe classification, ingredient identification and localization, and the estimation of nutritional content and calorie counts for each dish. Convolutional Neural Networks (CNNs) are employed as the cornerstone of this computational solution, proficiently handling image-based tasks, including the recognition of diverse food items and portion size determination within Uzbek recipes. Food classification is executed using MobileNet, while the You-Only-Look-Once (YOLO) network plays a pivotal role in the dual functions of ingredient classification and localization within dishes. Upon rigorous training, testing, and system deployment, users can effortlessly capture images of food items through the smartphone application, facilitating the estimation of nutritional data and calorie counts. Ultimately, this vital information is presented to users via the smartphone interface, bridging the accessibility gap and enhancing comprehension of nutritional aspects within Uzbek cuisine.

Keywords: Food calories prediction · Deep learning · Uzbek dishes · Food classification · MobileNet · CNN · YOLO

B. J. Choi et al. (Eds.): IHCI 2023, LNCS 14532, pp. 9–19, 2024.
https://doi.org/10.1007/978-3-031-53830-8_2

1 Introduction

Contemporary generations exhibit a heightened consciousness regarding dietary choices and calorie intake due to the recognition of the adverse implications of excessive caloric consumption, particularly on body weight, a pressing public health concern. The integration of Artificial Intelligence (AI), specifically deep learning, presents a promising avenue for comprehending the nutritional content of consumed food items. Food consumption is an integral facet of daily existence, with profound repercussions on body weight. Alarming data from the World Health Organization (WHO) underscore the surge in global rates of overweight and obesity over the past four decades. In 2016, the global count of overweight adults exceeded 1.9 billion individuals, with over 650 million among them classified as obese. Closer examination of national statistics, such as those from Norway in the year 2017, as reported by the Norwegian Institute of Public Health (NIPH) in their late 2017 update, reveals that 25% of men and 20% of women in the country grappled with obesity. This epidemiological trend is of grave concern, as it substantially elevates the risk profile for debilitating diseases, including diabetes, cardiovascular conditions, musculoskeletal disorders, and various types of malignancies [1, 2].

AI holds promise as a tool for promoting healthier lifestyles. Employing sophisticated neural networks like ResNet, it becomes feasible to identify food items accurately. The process of segmentation, involving the subdivision of images into smaller, meaningful units, is facilitated through the application of the YOLO object detection, enabling the comprehensive identification of pertinent components within the images [3–5]. Subsequently, these segmented constituents of food can be leveraged to compute the weight of the food depicted in the image, achieved through the utilization of an inception network. This proposed solution marks a pioneering endeavor in the domain of estimating food weight solely from a single image. The outcomes gleaned from this thesis investigation underscore the network's capacity to make reliable weight predictions, demonstrating a standard error of 8.95 across all categories and a notably lower 2.40 for the most accurately predicted category, namely, bread.

While food classification has been extensively explored in prior research [6–8], the challenge of determining the nutritional content of food items depicted in images remains a significant gap in the field. A prominent hurdle in assessing nutritional levels from images is the inherent two-dimensional nature of images, making it intricate to accurately gauge the portion size of the food items contained within. Consequently, it becomes imperative to devise methodologies for quantifying the volume of food represented in the images. Furthermore, variations in food preparation methods introduce another layer of complexity. Discerning the nutritional content of an apple, for instance, is more straightforward compared to a complex dish like "plow" (as illustrated in the top-left corner of Fig. 1), where numerous ingredients may be obscured within the image, rendering precise estimation challenging, if not unattainable. Nevertheless, it is plausible to derive approximations based on known recipes for such intricate dishes, albeit with some degree of discrepancy. Conversely, simpler food items present a more tractable scenario for assessing nutritional levels, as the image can encompass all constituents, facilitating a more accurate estimation process.

This paper presents a foundational demonstration in the realm of estimating food calories from images. In summary, our contributions encompass the development of a method for approximating the weight of food within a single image, achieved through the training of a neural network on established weight data for various food items. Consequently, the weight of the food can be predicted exclusively from a single image, thereby enabling the computation of calorie content by combining this weight information with data sourced from a comprehensive food database. Additionally, this thesis introduces a novel dataset comprising weight annotations for each individual food element depicted in the images.

2 Dataset of Uzbek Foods

To effectively identify food items and derive calorie information from images, the necessity for comprehensive training data is evident. In this paper, we have leveraged three distinct datasets, comprising two publicly accessible ones and a novel dataset created specifically for this study. The most prevalent dataset utilized for food classification purposes is the Food-101 dataset [9], featuring 1000 images for each of the 101 most commonly encountered food categories, thereby amassing a total of 101,000 images. Our initial experimentation with this dataset revealed the formidable challenges associated with the project. Accurate calorie estimation hinges on precise food classification across a broader spectrum of food classes, necessitating a more extensive dataset. It became apparent that the 101 food categories provided insufficient coverage, and the utilization of a significantly larger dataset would introduce complexities in terms of training and model iteration. Despite the prevailing reliance on conventional approaches in existing research, where both food classification and calorie estimation revolve around such datasets, we opted for an alternative path due to the intricate nature of food classification within this context.

A novel dataset has been curated specifically for the purpose of calorie estimation within this research, hereafter referred to as the "Uzbek dishes" dataset. This dataset encompasses a collection of food images, accompanied by a scale or ruler for reference, facilitating the determination of food proportions and weights. In addition to the visual content, the dataset includes a document detailing the nutritional content for each food item depicted in the images. The "Uzbek dishes" dataset encompasses a total of 10 distinct categories of Uzbek cuisine, encompassing approximately 12,000 images, with a select few exemplified in Fig. 1. Each category is further subdivided into six distinct groups for estimating the quantity of food, denoted as E1, E2, E3, E4, E5, and E6, as illustrated in Fig. 2.

Fig. 1. Sample images from custom Uzbek dish dataset.

E1   E2   E6

Fig. 2. Sample images for six amount estimation group such as E1, E2, and E6.

Our methodology comprises two primary components: Object and Food classification utilizing YOLO, and Food Quantity estimation employing Wide-Resnet50 [10]. To facilitate these tasks, we harnessed two distinct datasets tailored to each algorithm's specific requirements. Initially, a set of 12,000 images was employed solely for food item classification and detection, categorized into 10 distinct classes. These base images possessed dimensions of 4032×3024 pixels.

Subsequently, we adopted classical computer vision techniques with predefined parameters to segment and extract the ingredients within the images. These methods involve the computation of pixel intensities and their conversion into real-sized regions. Building upon our prior research efforts [11–13], we employed a range of image processing techniques to identify and crop objects from sequences of images. The pixel regions corresponding to non-uniform components were segmented, enabling the determination of nutritional information and calorie content through reference to a standardized information table [14].

In the context of this study, we developed a comprehensive, full-stack system that operates based on the RESTful protocol, utilizing JSON format for communication between a smartphone and an artificial intelligence server.

3 Proposed Method

The YOLOv3 model architecture incorporates the DarketNet53 backbone [15]. Our proposed methodology encompasses food detection, food quantity estimation, and object detection, which entails the following steps:

- Identification of regions corresponding to Uzbek dishes through YOLOv3.
- Subdivision of the detected regions into multiple segments for the purpose of object detection.
- Extraction of food images based on the bounding box boundaries.
- Classification of food quantity estimates using Wide-Resnet-50

The initial phase of our proposed approach involves object detection and object classification utilizing YOLOv3. This convolutional network concurrently predicts multiple bounding boxes and the associated class probabilities for these boxes, streamlining the detection process. YOLOv3 adopts a unique approach by training on entire images and directly optimizing overall performance. This integrated model boasts several advantages over conventional object detection techniques. Notably, YOLOv3 excels in terms of rapidity in object recognition. It conceptualizes detection as a regression problem, eliminating the need for intricate pipelines [16, 17]. During testing, object detection is achieved by running a neural network on the new image.

Following object detection, we encountered certain challenges, notably that the YOLOv3 object detection model struggled to identify all objects due to limitations in the dataset size [18]. To address this issue, we devised a solution by partitioning the input image into several segments.

A. Quantity Estimation

The network architecture we employed is based on the Wide-Resnet-50, which is a deep convolutional neural network (CNN). Wide-Resnet-50 is an extension of the ResNet architecture, known for its ability to handle very deep networks effectively. It consists of 50 layers and is wider than the original ResNet, making it suitable for the complexity of food quantity estimation tasks. For the training data, we collected a diverse and extensive dataset of food images with corresponding ground truth labels for food quantity. This dataset includes images of various food items, each annotated with the respective quantity or portion size. The dataset is essential for training the Wide-Resnet-50 model to learn how to estimate food quantities accurately (Fig. 3).

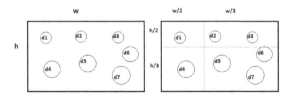

Fig. 3. Decreasing object with dividing the photo into several parts.

To enhance the effectiveness and resilience of our model, we employed data augmentation techniques. Specifically, we computed the mean and standard deviation for dataset normalization and incorporated methods such as Color-Jitter [19], RandomRotation, RandomAdjustSharpness, and RandomSharpness. Figure 4 provides visual representations of cropped images used for food quantity estimation, with E3, E6 representing 15%, 50%, and 100%, respectively. Additionally, Fig. 5 showcases the outcomes of data augmentation, illustrating the various augmentation types we previously mentioned.

Fig. 4. Cropped images for a food amount estimation.

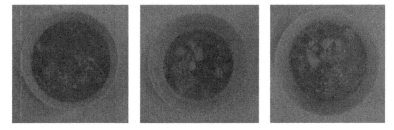

Fig. 5. The results of data augmentation.

As previously mentioned, we employed Convolutional Neural Networks (CNNs) for food quantity prediction. In our quest to select the most suitable CNN architecture for this purpose, we conducted training and testing across several CNN models, including MobileNetV2 [20], Resnet [21], EfficientNet [22], and Dilated CNN [23].

MobileNetV2, with a depth of 53 layers and comprising 3.5 million parameters, emerged as one of our candidate networks. It incorporates distinctive architectural features, including:

- Depth wise separable convolution
- Linear bottleneck
- Inverted residual

The concept of Depthwise Separable Convolution is visually depicted in Fig. 6.

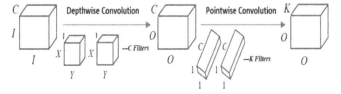

Fig. 6. The architecture of the Depthwise Separable Convolution.

MobileNetV2 employs a fusion of pointwise convolution and bottleneck structures, wherein pointwise convolution is employed to achieve the bottleneck effect. However, a challenge arises when applying ReLU activation after dimensionality reduction, resulting in the loss of valuable information. To mitigate this issue, a 1×1 convolutional layer, also known as a feature map pooling or projection layer, is introduced. This straightforward technique serves to reduce dimensionality by decreasing the number of feature maps while preserving their distinctive attributes. In our study, we utilized the max pooling technique for this purpose, and consequently, linear activation is implemented within the bottleneck to minimize information loss [25].

Furthermore, MobileNetV2 incorporates the concept of Inverted Residual blocks, following a pattern of narrow-wide-narrow configuration. This entails a sequence of 1×1 convolution, followed by 3×3 convolution, reducing parameter count, and concluding with another 1×1 convolution to further trim the number of channels. Importantly, shortcuts are established directly between these bottleneck stages [26].

4 Experimental Results and Analysis

In this section, we elucidate the experimental configuration and outcomes pertaining to the estimation of calories and quantity for Uzbek dishes using the YOLOv3 and MobileNetV2 models. The proposed method was implemented and carefully tested within Visual Studio 2019 C++, executed on a system equipped with an AMD Ryzen 9 5900X 12-Core Processor operating at 3.70 GHz, supported by 64 GB of RAM, and a NVIDIA RTX 3090 Graphics card. Our food dataset was employed for both training and testing purposes. Crucial parameters for our training experiments included a batch size of 32 pixels, an input image size of 416×416 pixels, a learning rate of 0.001, and a subdivision of 8. Our primary objective was to evaluate the classification performance, ensuring the reliable and accurate classification of Uzbek dishes. This study conducts a comprehensive analysis and comparison of various object detection and recognition techniques, encompassing YOLOv3 and MobileNetV2, in the training and evaluation of models for classifying Uzbek dishes. The findings underscore the precise detection capabilities of YOLOv3 in accurately identifying Uzbek dishes.

Our quantitative experiments entailed the application of object detection evaluation metrics, namely precision, recall, and accuracy, in line with our prior research endeavors [27]. These metrics serve as valuable tools for assessing and interpreting the outcomes. Precision gauges the classifier's capability to correctly identify relevant objects, quantifying the proportion of true positives detected. In contrast, recall measures the model's proficiency in identifying all pertinent instances, signifying the proportion of true positives among all ground truth cases. An effective model excels in recognizing the majority of ground-truth objects, showcasing high recall, while also limiting the inclusion of irrelevant objects, denoting high precision.

An ideal model would yield a false-negative value of 0 (recall = 1) and a false-positive value of 0 (precision = 1), signifying flawless performance. We computed precision and recall rates through the comparison of pixel-level ground-truth images with the outcomes generated by our proposed method. To calculate these metrics for food classification, we employed the following equations.:

$$Accuracy = TP + TN/TP + TN + FP + FN \tag{1}$$

$$Recall = TP/TP + FN \tag{2}$$

$$Precision = TP/TP + FP \tag{3}$$

In Fig. 7, it is depicted that the YOLOv3 model achieved an accuracy rate of 98.2% and exhibited a loss of 0.2 when detecting objects.

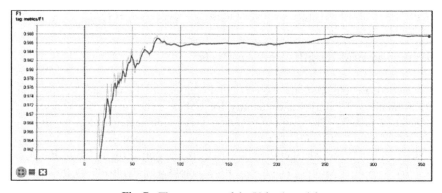

Fig. 7. The accuracy of the Yolov3 model.

We generated an additional dataset for quantity estimation purposes by resizing images to dimensions of 224 × 224 pixels for training with Wide-Resnet-50. Each of the food categories and quantity estimations comprises a set of 200 images. Consequently, this classification task incorporates a total of 5000 food images. The dataset was partitioned into training and validation subsets, with an 80% allocation to training and the remaining 20% allocated to validation. The results of the food quantity estimation using Wide ResNet-50 are shown in Table 1.

Table 1. Performance evaluation of the Wide ResNet-50 model for food quantity estimation.

Wide ResNet-50	Training	Testing
Accuracy	93%	90%
Recall	90%	88%
Precision	92%	91%

5   Limitation and Future Work

The proposed approach necessitates the incorporation of additional quantity estimation classes, as the utilization of just six such classes resulted in the omission of several categories by the MobileNetV2 model. In forthcoming research, we aspire to enhance our primary concept for quantity estimation. Additionally, we aim to optimize response times by deploying the service within a cloud computing environment, with the objective of reducing food identification duration [28].

Moreover, our future endeavors include the exploration of 3D modeling techniques for quantity estimation, which are expected to yield greater accuracy compared to the utilization of 2D images. We are currently in the process of acquiring expertise in the application of computer vision methodologies within the Unity environment. Our forward-looking plan entails the augmentation of algorithm complexity and the expansion of computer vision capabilities. For instance, our current approach involves the utilization of only ten quantity estimation positions per image; however, we aspire to incorporate a more extensive range of positions in the future.

6   Conclusions

This research introduces a Convolutional Neural Network (CNN) model designed to estimate the portion sizes of individual food items on a single plate based on single images. The envisioned outcome is the association of these portion sizes with a calorie database for each food unit. The primary objective of this paper is to leverage machine learning techniques for image classification, food segmentation, and the subsequent determination of weight and calorie content within the images. Additionally, we introduce a novel dataset termed "Uzbek dishes," encompassing food images along with their corresponding weights. The findings from our testing phase affirm the capability of deep neural networks to effectively classify food images.

Future investigations will focus on enhancing the accuracy of our methodology through the utilization of advanced deep learning techniques. Furthermore, we intend to optimize cache memory utilization in multi-core processors to enhance the efficiency of food calorie detection and estimation [29].

References

1. World Health Organization: Obesity and Overweight (2018)
2. NIPH: Overweight and obesity in Norway. Tech. rep. (2014)
3. Mukhiddinov, M., Djuraev, O., Akhmedov, F., Mukhamadiyev, A., Cho, J.: Masked face emotion recognition based on facial landmarks and deep learning approaches for visually impaired people. Sensors **23**(3), 1080 (2023)
4. Mukhiddinov, M., Jeong, R.G., Cho, J.: Saliency cuts: salient region extraction based on local adaptive thresholding for image information recognition of the visually impaired. Int. Arab J. Inf. Technol. **17**(5), 713–720 (2020)
5. Mukhiddinov, M., Akmuradov, B., Djuraev, O.: Robust text recognition for Uzbek language in natural scene images. In: 2019 International Conference on Information Science and Communications Technologies (ICISCT), pp. 1–5. IEEE. (2019)

6. Sathish, S., Ashwin, S., Quadir, M.A., Pavithra, L.K.: Analysis of convolutional neural networks on indian food detection and estimation of calories. In: Materials Today: Proceedings, 16 Mar (2022)
7. Li, S., Zhao, Y., Liu, S.: How food shape influences calorie content estimation: the biasing estimation of calories. J. Food Qual. 24 May (2022)
8. Kumar, R.D., Julie, E.G., Robinson, Y.H., Vimal, S., Seo, S.: Recognition of food type and calorie estimation using neural network. J. Supercomput. **77**(8), 8172–8193 (2021)
9. Bossard, L., Guillaumin, M., Gool, L.V.: Food-101–mining discriminative components with random forests. In: European Conference on Computer Vision, pp. 446–461. Springer, Cham (2014)
10. Keras: Resnet-50 [Online]. Available: https://www.kaggle.com/keras/resnet50 (2017).
11. Rakhmatillaevich, K.U., Ugli, M.M.N., Ugli, M.A.O., Nuruddinovich, D.O.: A novel method for extracting text from natural scene images and TTS. Eur. Sci. Rev. **1**(11–12), 30–33 (2018)
12. Mukhamadiyev, A., Mukhiddinov, M., Khujayarov, I., Ochilov, M., Cho, J.: Development of language models for continuous Uzbek speech recognition system. Sensors **23**(3), 1145 (2023)
13. Abdusalomov, A.B., Mukhiddinov, M., Whangbo, T.K.: Brain tumor detection based on deep learning approaches and magnetic resonance imaging. Cancers **15**(16), 4172 (2023)
14. Khamdamov, U., Abdullayev, A., Mukhiddinov, M., Xalilov, S.: Algorithms of multidimensional signals processing based on cubic basis splines for information systems and processes. J. Appl. Sci. Eng. **24**(2), 141–150 (2021)
15. Redmon, J., Farhadi, A.: Yolov3: an incremental improvement. arXiv preprint arXiv:1804. 02767 (2018)
16. Ege, T., Yanai, K.: Image-based food calorie estimation using knowledge on food categories, ingredients and cooking directions. In: Proceedings on Thematic Workshops of ACM Multimedia, pp. 367–375 (2017)
17. Mukhiddinov, M., Abdusalomov, A.B., Cho, J.: Automatic fire detection and notification system based on improved YOLOv4 for the blind and visually impaired. Sensors **22**(9), 3307 (2022)
18. Mukhiddinov, M., Cho, J.: Smart glass system using deep learning for the blind and visually impaired. Electronics **10**(22), 2756 (2021)
19. Jalal, M., Wang, K., Jefferson, S., Zheng, Y., Nsoesie, E.O., Betke, M.: Scraping social media photos posted in Kenya and elsewhere to detect and analyze food types. In: Proceedings of the 5th International Workshop on Multimedia Assisted Dietary Management, pp. 50–59 (2019)
20. Sandler, M., Howard, A., Zhu, M., Zhmoginov, A., Chen, LC.: Mobilenetv2: inverted residuals and linear bottlenecks. In: Proceedings of the IEEE conference on computer vision and pattern recognition, pp. 4510–4520 (2018)
21. Wu, Z., Shen, C., Van Den Hengel, A.: Wider or deeper: revisiting the resnet model for visual recognition. Pattern Recogn. **90**, 119–133 (2019)
22. Koonce, B.: EfficientNet. In: Convolutional Neural Networks with Swift for Tensorflow, pp. 109–123. Apress, Berkeley, CA (2021)
23. Yuldashev, Y., Mukhiddinov, M., Abdusalomov, A.B., Nasimov, R., Cho, J.: Parking lot occupancy detection with improved mobilenetv3. Sensors **23**(17), 7642 (2023)
24. Abdusalomov, A., Mukhiddinov, M., Djuraev, O., Khamdamov, U., Whangbo, T.K.: Automatic salient object extraction based on locally adaptive thresholding to generate tactile graphics. Appl. Sci. **10**(10), 3350 (2020)
25. Chen, G., et al.: Food/non-food classification of real-life egocentric images in low-and middle-income countries based on image tagging features. Front. Artif. Intell. **4**, 644712 (2021)
26. Mukhiddinov, M.: November. Scene text detection and localization using fully convolutional network. In: 2019 International Conference on Information Science and Communications Technologies, pp. 1–5 (2019)

27. Khamdamov, U.R., Mukhiddinov, M.N., Djuraev, O.N.: An overview of deep learning-based text spotting in natural scene images. Problems of Computational and Applied Mathematics. Tashkent, **2**(20), 126–134 (2020)
28. Muminov, A., Mukhiddinov, M., Cho, J.: Enhanced classification of dog activities with quaternion-based fusion approach on high-dimensional raw data from wearable sensors. Sensors **22**(23), 9471 (2022)
29. Farkhod, A., Abdusalomov, A.B., Mukhiddinov, M., Cho, Y.I.: Development of real-time landmark-based emotion recognition CNN for masked faces. Sensors **22**(22), 8704 (2022)

Blockchain Technology as a Defense Mechanism Against Data Tampering in Smart Vehicle Systems

Khasanboy Kodirov, HoonJae Lee, and Young Sil Lee[(✉)]

Dongseo University, Busan, Republic of Korea
hjlee@gdsu.dongseo.ac.kr, youngsil.lee0113@gmail.com

Abstract. Smart vehicle systems, integrated with cutting-edge technology, have evolved to become a crucial element of modern transportation. They are leading us into a fresh era of mobility, offering the potential for increased efficiency, upgraded safety, and enhanced user satisfaction. According to a study by Statista, in terms of IoT-related revenue, the automotive IoT market is expected to reach 23.6 billion US dollars in 2025 [1]. However, their reliance on data-driven technologies also exposes them to the ever-present threat of data tampering. In response to these emerging threats, the imperative to fortify the cybersecurity posture of smart vehicles has never been more pressing. Known for its decentralized, immutable, and transparent nature, blockchain is a promising solution for safeguarding critical vehicle data. As part of this research paper, we look at the fundamentals of blockchain technology and discuss how this technology can be used within smart vehicles to enhance the security and integrity of data generated, stored, and shared. By leveraging the blockchain's security features, stakeholders in the automotive industry can pave the way for safer and more reliable smart vehicle technologies. This research also invites further exploration and collaboration to harness the full potential of blockchain in fortifying the future of smart transportation.

Keywords: Smart vehicle systems · Automotive cybersecurity · Blockchain technology

1 Introduction

During the past few decades, a technological revolution has unfolded, transforming the world. As cloud computing transcended infrastructure, global connectivity was fostered, and the Internet of Things (IoT) connected everyday objects. As these technologies continue to evolve and intertwine, they have collectively woven a complex tapestry that covers the world in an intricate web of innovation and connectivity. It can be summed up by saying that all objects and things are being replaced by their smart substitutes. Smart vehicles are one of these substitutions. Vehicles embedded with cutting-edge technology are ushering in a new era of transportation that promises increased efficiency, enhanced safety, and improved user experiences. As of 2018, there were 330 million connected cars, according to Upstream. Those numbers are expected to rise to 775

B. J. Choi et al. (Eds.): IHCI 2023, LNCS 14532, pp. 20–25, 2024.
https://doi.org/10.1007/978-3-031-53830-8_3

million by 2023. In addition, approximately 25 GB of data will be produced per hour by connected cars by 2025. Those numbers jump to 500 GB an hour for fully autonomous vehicles. Upstream estimates that the automotive industry is projected to lose $505 billion by 2024 to cyberattacks [2]. In response to these emerging threats, the imperative to fortify the cybersecurity posture of smart vehicles has never been more pressing. In an era when smart vehicles are becoming an integral part of daily life, protecting them from cyber threats is not just an option but a critical necessity for their safety and integrity. A fundamental motivation for this research lies in the profound implications smart vehicles have on society, the economy, and public safety. The latest report (Table 1), released in early 2021 illustrates data from 2010 to 2020 and covers over 200 automotive cyber incidents across the world [3]. It is important to understand that smart vehicles are not just mechanical entities. They are intricate cyber-physical systems interacting with hardware, software, sensors, actuators, and external networks. The interplay of these intricate systems has created a range of cybersecurity threats that go beyond the conventional boundaries of automotive engineering.

Table 1. Automotive attack vectors: 2010–2020

#N	Hardware/software	Share of total
1	Cloud servers	32.9%
2	Keyless entry-Key fob	25.3%
3	Mobile app	9.9%
4	ODB port	8.4%
5	Infotainment system	7.0%
6	Sensors	4.8%
7	ECU-TCU-Gateway	4.3%
8	In-vehicle network	3.8%
9	Wi-Fi network	3.8%
10	Bluetooth	3.6%
11	USB or SD port	2.1%

2 Background and Literature Review

Data tampering in the context of smart vehicle systems refers to the unauthorized alteration, manipulation, or compromise of critical data that underpins the functioning and decision-making processes of these vehicles. Whether it involves falsifying sensor data, modifying route information, or tampering with vehicle-to-vehicle (V2V) communications, such attacks pose significant threats to the reliability, safety, and trustworthiness of smart vehicles. The consequences of data tampering incidents in this domain can range from minor inconveniences to life-threatening situations, making it imperative to address this issue comprehensively. Blockchain, the foundational technology behind cryptocurrencies like Bitcoin, has garnered significant attention across various industries due to its unique attributes. It is a decentralized and distributed ledger system that offers transparency, immutability, and trust through cryptographic mechanisms. These qualities position blockchain as a promising solution to secure the integrity of data generated, stored, and communicated within smart vehicle ecosystems. Smart vehicle systems represent a transformative paradigm in modern transportation. These systems integrate cutting-edge technologies, including Internet of Things (IoT) sensors, artificial intelligence (AI), machine learning (ML), and real-time data communication, to create vehicles that are not only intelligent but also interconnected. This convergence of technology promises to revolutionize road safety, traffic management, and overall driving experiences. However, this rapid evolution also introduces new challenges, particularly in terms of data security.

Data tampering, or the unauthorized alteration and manipulation of data, has emerged as a critical concern in the realm of smart vehicle systems. These systems rely heavily on data for various functions, including:

1. Sensors and telemetry: Smart vehicles use sensors to collect data on road conditions, weather, vehicle performance, and the environment. This data informs real-time decisions for navigation and safety.
2. Communication networks: Vehicle-to-vehicle (V2V) and vehicle-to-infrastructure (V2I) communication networks enable vehicles to share critical information. Tampered data in these networks can lead to accidents or disruptions in traffic management.
3. Automated decision-making: Many smart vehicles incorporate autonomous or semi-autonomous features that rely on data for decision-making. If this data is compromised, it can lead to incorrect or unsafe actions.
4. Fleet management: In commercial settings, fleet management relies on accurate data for route optimization, fuel efficiency, and maintenance scheduling. Tampered data can result in inefficiencies and increased costs.

In recent years, the convergence of blockchain technology and smart vehicle systems has gained substantial attention, with several noteworthy studies shedding light on the dynamic landscape of applications and innovative approaches. Notable among these are: (1) Smith et al. (2023), who introduce a blockchain-based communication framework designed to enhance secure vehicle-to-vehicle (V2V) communication in smart vehicles, thus ensuring data integrity and privacy[4]. (2) Brown and Lee (2022), whose comprehensive survey provides insights into diverse blockchain applications within the automotive industry, particularly emphasizing cybersecurity. This survey encompasses the latest trends in safeguarding smart vehicle systems against data tampering and cyber threats [5]. (3) Gupta et al. (2022), who explore scalable blockchain solutions for managing the burgeoning data volumes generated by smart vehicles [6]. These recent contributions illustrate the ongoing evolution of blockchain technology within smart vehicle systems, underscoring its potential in revolutionizing security, data management, and communication in the realm of connected and autonomous transportation.

3   Data Tampering in Smart Vehicle Systems

Data tampering in smart vehicle systems encompasses a range of malicious activities, including the unauthorized alteration, manipulation, or compromise of critical data. The consequences of such tampering are multifaceted and potentially severe:

1. **Safety risks:** One of the foremost concerns is the potential compromise of safety-critical data. For instance, tampering with sensor data, such as those responsible for collision avoidance or lane-keeping, can lead to incorrect decisions by autonomous driving systems, resulting in accidents and endangering lives.
2. **Traffic disruption:** Data tampering can disrupt the coordination and communication between smart vehicles. In V2V and V2I communication networks, altered data can mislead vehicles, leading to traffic congestion or collisions.
3. **Financial implications:** In commercial applications, data tampering can have financial repercussions. Fleet management relies on accurate data for route optimization, fuel efficiency, and maintenance scheduling. Tampered data can result in inefficiencies and increased operational costs.
4. **Privacy concerns:** Smart vehicles collect vast amounts of data, including location, behavior, and preferences. Unauthorized access or tampering with this data can lead to privacy breaches, identity theft, and misuse of personal information.

Before delving into further discussion, let's examine common data tampering incidents in smart vehicle systems in Table 2 [7].

Table 2. Common of data tampering incidents in smart vehicle systems

#N	Incident type	Description	Consequences
1	GPS spoofing	Falsification of GPS signals	Misleading navigation
2	Sensor manipulation	Altered LiDAR and radar data	Unsafe driving decisions
3	Communication disruption	Interference with V2V and V2I systems	Traffic congestion, accidents
4	Data theft	Unauthorized access and theft of data	Privacy breaches

Addressing data tampering in smart vehicle systems necessitates innovative and robust security measures. Blockchain technology emerges as a promising solution, offering inherent features that can help fortify the integrity of data and enhance trust among stakeholders. In the following sections, we will explore how blockchain technology can be leveraged to counter data tampering risks effectively within the context of smart vehicles.

4 Blockchain Technology as a Solution in Smart Vehicle Systems

Blockchain technology, when applied to smart vehicle systems, represents a robust and sophisticated approach to addressing data tampering vulnerabilities and ensuring the integrity of critical data and communications. In this section, we examine the technical aspects of how blockchains can be implemented within smart vehicle ecosystems, focusing on specific use cases, protocols, and underlying cryptography.

4.1 Technical Foundations of Blockchain in Smart Vehicle Systems

Blockchain technology has a lot to offer in the context of smart vehicles, but it's important to establish a technical foundation before looking at practical implementations. Key principles and components include:

- **Distributed ledger architecture**: Blockchain operates as a distributed ledger as shown in Fig. 1 [8], which consists of a network of nodes (computers) that collectively maintain a synchronized record of transactions. This architecture ensures redundancy and eliminates single points of failure.

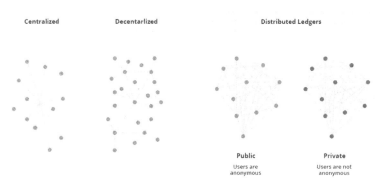

Fig. 1. Distributed Ledger Architecture of Blockchain

- **Consensus mechanisms**: Smart vehicle blockchain implementations often utilize consensus mechanisms like Proof of Work (PoW) or Proof of Stake (PoS) to validate and add new blocks to the chain. The choice of consensus mechanism impacts factors such as security, energy efficiency, and transaction speed.
- **Smart contracts**: Smart contracts are self-executing, programmable scripts deployed on the blockchain. They enable automation of predefined actions when specific conditions are met. In smart vehicle systems, smart contracts can govern agreements related to data access, sharing, and compensation.
- **Public vs. private blockchains**: The choice between public and private blockchains depends on the use case. Public blockchains (e.g., Ethereum) are open to anyone, offering transparency but with potential privacy concerns. Private blockchains provide controlled access, ensuring data privacy, making them suitable for business consortia and organizations.

4.2 Practical Implementations and Prototypes

Smart vehicle systems have been incorporated with blockchain technology in a number of practical implementations and prototypes, demonstrating its potential for enhanced data security and integrity. The initiatives include blockchain-powered data marketplaces for vehicles to share and monetize data securely, secure over-the-air software updates that verify update authenticity, and experimental decentralized autonomous vehicles (DAVs) that use blockchain for consensus on route planning and traffic coordination. Within smart vehicle ecosystems, these real-world endeavors demonstrate blockchain's feasibility and benefits in improving data tampering resistance and trust.

5 Conclusion

In conclusion, incorporating blockchain technology into smart vehicle systems is a pivotal step toward enhancing data integrity and security in the evolving landscape of connected transportation. Blockchain's technical foundations and practical implementations offer promising solutions to combat data tampering risks and establish trust within vehicular ecosystems. While challenges like scalability and interoperability persist, blockchain's potential to revolutionize smart transportation remains undeniable. It emerges as a transformative defense mechanism against data tampering, ushering in a future of safer and more reliable smart vehicles.

Acknowledgment. This research was supported by Korea Institute of Marine Science & Technology Promotion (KIMST) funded by the Ministry of Oceans and Fisheries (20210650).

References

1. Placek, M.: Automotive IoT Market Size 2021–2026. Statista, https://www.statista.com/statis tics/423083/iot-units-installed-base-within-automotive-segment/. Accessed 18 Sept 2023
2. Blum, B.: Cyberattacks on Cars Increased 225 Percent in Last 3 Years. Birmingham Times, http://www.birminghamtimes.com/2022/02/cyberattacks-on-cars-increased-225-percent-in-last-3-years/. Accessed 18 Sept 2023
3. Juliussen, E.: Now Your Car is a Cybersecurity Risk, Too. EETimes, https://www.eetimes.com/now-your-car-is-a-cybersecurity-risk-too/. Accessed 20 Sept 2023
4. Smith, J., Author, A.B., Author, C.D.: Blockchain-Based Secure Communication for Smart Vehicles. Journal of Advanced Transportation Technology **28**(3), 123–137 (2023)
5. Brown, A., Lee, S.: A Comprehensive survey of blockchain solutions for automotive cybersecurity. Int. J. Smart Vehicle Syst. **15**(4), 234–251 (2022)
6. Gupta, R.: Scalable blockchain solutions for smart vehicle data management. J. Connect. Vehicles Smart Transp. **10**(2), 87–101 (2022)
7. Trustonic: Top 10 Security Challenges in the Automotive Industry for Connected Cars. https://www.trustonic.com/opinion/top-10-security-challenges-for-connected-cars/. Accessed 26 Oct 2023
8. Lastovetska, A.: Blockchain Architecture Basics: Components, Structure, Benefits & Creation. https://mlsdev.com/blog/156-how-to-build-your-own-blockchain-architecture. Accessed 20 Sept 2023

Classification of Weeds Using Neural Network Algorithms and Image Classifiers

Rakesh Joshi[✉], Garima Sharma, Vikas Tripathi, and Ankita Nainwal

Department of Computer Science and Engineering, Graphic Era Deemed to Be University, Dehradun, Uttarakhand, India
rakeshjoshi21021369@gmail.com

Abstract. In agriculture it is very important to differentiate between crop seedling and weed from the farms, the traditional means to classify were mostly relies on manual methods which is not time and cost efficient. Automated Deep learning-based Weed classification plays crucial role to solve this in agriculture sector for effective crop yield and weed management. Also this study could help for precision agriculture, cost reduction and time efficiency in the field. In this research paper, we present several deep learning-based approach for automated classification between weed and crop seedlings. For the classification task we used popular image classification algorithms, which includes Convolution Neural Network (CNN), Artificial Neural Network (ANN), Support Vector Machine (SVM), K-Nearest Neighbors (KNN). The study utilizes a diverse dataset between crop seedling and weed image data. We also demonstrated the results of deep learning based algorithm and compare them to find the best for the classification task. The work highlights the capability of deep learning-based model for efficient weed management, aiding farmers in identifying and controlling weeds while growing crop seedlings, and hence leads to improved productivity.

Keywords: Weed classification · image classification algorithm · crop seedlings · Deep learning · CNN · ANN · SVM · KNN · Comparative analysis

1 Introduction

In traditional agriculture, weed classification which is fundamental task required manual labour [1] on the field, which is costly for the farm and lead to loss for crop yield. In this work we have presented the Weed Classification from the image dataset. We preprocessed the images by resizing, data augmentation and then used them to classify between crop seedlings and weeds.

In farms the early stage of a growing plant is called plant or crop seedlings which are seems similar as weeds. So, it became a tough task to differentiate between them and remove the weeds from the farm [2]. This activity requires focussed manual labour on field which is time consuming and costly. To automate the recognition of weed in the farm we can use technology like deep learning to differentiate the image [3] between weed and plant seedlings. Using technology like [4] Deep Learning will make automation recognition task easy and can deal with large image dataset in shorter amount of time. With improved model architecture the classification task between crop seedlings and weed will be easier and faster.

The Dataset [5] used in the classification work is divided into two classes weed and crop seedlings. But the manual dataset was unbalanced because images of crop seedling were very few. So, new images were added from a crop seedlings dataset [6], which helped to balance the dataset to fulfil the requirement for classification. The types of weed in weed class are like field pennycress (Thlaspiarvense), shepherd's purse (Capsella bursa-pastoris), field chamomile(Matricariaperforata), field pansy(viola arvensis). Similarly the crop seedlings in the dataset are like pumkin (cucurbita pepo), radish (Raphanus sativus var.sativus) black radish (Raphanus sativus var.niger).

With the help of basic pipeline for image classification deep learning concepts [7], various models were prepared. For achieving the required model goal, the procedure we followed started with Data collection, in simple terms it means collecting the required dataset for the image classification task. Dataset pre-processing which includes steps like images resizing, cropping or normalization techniques further if required data augmentation techniques are also employed to get a more diverse and variable dataset. After splitting the dataset into train, test and validation set, using train set model architecture is trained on image classifier algorithm and neural networks. To tune up the model validation set is used. Then prediction on test set is performed from which further model evaluation is done like accuracy, F1 score, Recall, Confusion matrix and Precision – recall graph. For image classification Neural Network algorithm like Convolutional Neural Network (CNN) [8, 9] and Artificial Neural Network (ANN) [10] are used to train the model. CNNs are specifically designed for processing images through grid by leveraging the concept of convolution. It consists of multiple layers, including convolutional layers, pooling layers and fully connected layers. ANNs are basic form of neural networks used for general-purpose machine learning tasks. It consists of multiple layers, including input, hidden, and output layers, with each layer interconnected nodes called neurons. Model based on image classification algorithm like K-Nearest Neighbors (KNN) [11] and Support Vector Machine (SVM) [12].

For image classification KNN is used as baseline algorithm by representing each image as a feature vector and comparing it to the feature vectors of labelled images in the training set. SVM is a powerful and versatile algorithm for image classification, it can be used to learn a decision boundary that separates different classes in a high-dimensional feature space. It aims to find a hyperplane that maximally separates the data points of different classes, with a margin that maximizes the distance between the hyperplane and the nearest data points.

We have derived the result after fitting every image classifier and neural network in the model and results have shown which classifier or neural network based model works best for the classification task between crop seedling and weed.

2 Literature Survey

Weed management is being shifted from conventional agricultural practices to technology-friendly practice [13] that employs Machine Learning, Deep Learning, big data and other modern technology in past decade. Many authors have used different approach to solve this problem.

Author in [14] worked on a dataset that provided weed plant of different species and also suggested a benchmark scale to readers for easy comparisons of classification outcomes. Classification using effective convolutional neural network [15] demonstrated the unsupervised feature representation of 44 different plant species with high metrics. Study over the recognition of plant diseases used image classification, the development of plant disease recognition model, based on leaf image classification [16] using deep convolutional networks. This developed model was able to recognize 13 different plant disease. Study on classification of plant seedling using CNN [17] where 12 different species are classified, and model achieved and accuracy of 99% which seems promising for the agriculture sector. Work on Feature extraction for disease leaf [18], showed the efficiency of algorithms like KNN and SVM for the image classification and furthered recognizing cop disease based on extracted features. Study on autonomous robotic weed control [19] showed how robotic technology may also provide a means to do the task like hand weed control. Not only deep learning, Machine Learning based technology [20] are also contributing to the automation of many manual task. Work on vegetable plant leaf classification through Machine Learning[21] based models where classifiers like Decision Tree, Linear Regression, Naïve Bayes, MLP are used for the image classification task and result showed that MLP acquired an accuracy of 90% for the task. It is also been seen that Machine Learning can also play role in environmental sustainability [22], a model using Regression Kriging is used to classify the radiative energy flux at the earth surface. Comparison of various deep learning techniques has also been done by many researchers keeping varied application into view[23]. Recent IOT-enabled model for weed seedling classification [24] have shown that the Weed-ConvNet model with color segmented weed images gives accuracy around 0.978 for the classification task.

Our work has conducted a comparative analysis between different neural networks like CNN, ANN and image classifiers models like KNN, SVM and did the metric study on the performance of these deep learning based models. From the metric result and precision-recall graph the study concluded that neural network like CNN work best for the weed classification task and image classifier like KNN and SVM also perform good whereas ANN based model performance for classification seems to be low.

3 Methodology

The procedure for classification task between weed and crop seedling is widely based on Deep Learning method, below Fig. 1 shows the pipeline flow of the work. Starting from Data collection and visualisation of dataset and getting some insights, then Dataset Pre-processing through which we resize the image and by Data augmentation we make the dataset more balanced, diverse and variable while preserving the original classes.

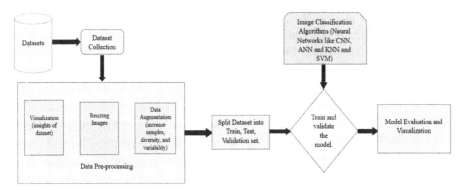

Fig. 1. Proposed methodology for weed classification system

Next step is to split the dataset into train, test and validation set and then applying the various image classification mode architectures like CNN (Convolutional Neural Network), ANN (Artificial Neural Network), KNN (K-Nearest Network) and SVM (Support Vector Machine). After training and validating the models we have evaluated the metrices of every model and tried to find out the comparative best model for the Classification.

3.1 Data Collection

The collected Weed image dataset used for classification have around 2,047 images of weed and crop seedlings. In the record 931 are under crop seedling class and 1116 are weed images. Types of weeds in the dataset are goosefoot (Chenopodium album), catchweed (Gallium aparine), field pennycress (Thlaspiarvense), shepherd's purse (Capsella bursa-pastoris), field chamomile (Matricariaperforata), field pansy (viola arvensis) and others.

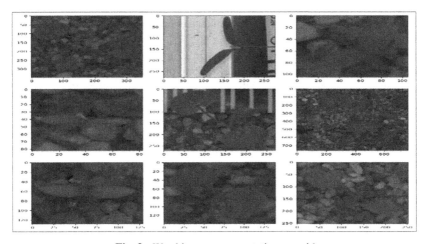

Fig. 2. Weed image representation on grid

Several types of plant seedlings are beetroot (Beta vulgari), carrot (Daucus carota var. Sativus), zucchini (Cucurbita pepo subsp. Pepo), pumkin (cucurbita pepo), radish (Raphanus sativus var.sativus) black radish (Raphanus sativus var.niger) and other seedling image data collected. Figures 2 and 3 show the grid plotting of the weed and plant seedling image respectively.

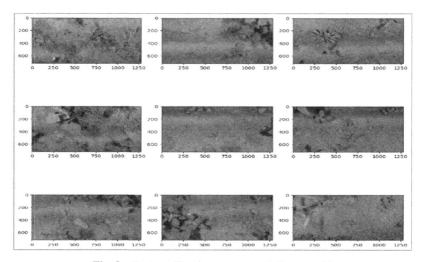

Fig. 3. Plant seedling image representation on grid

3.2 Data Pre-processing

This includes understanding the dataset through visualization and further pre-processing it, so that model can be applied on the dataset. Below Bar graph representation of Fig. 4 shows the image distribution of class weed and crop seedling in the dataset we acquired or collected.

As the raw image dataset have unequal size of images, and to perform any Deep Learning algorithm efficiently we perform Resizing of the images. Upon going through various trails for best resizing, we find the dimensions of 360 pixels width and 257 pixels height as the most fitting for the algorithms. Next, we performed Data Augmentation to increase the diversity and variability of dataset, techniques which we used here for data augmentation are random rotation of images, width shift, zoom range, horizontal and vertical flip which were applied randomly to images. After this new augmented dataset have around 6141 images, where Crop seedlings are 2793 and weed images were 3348. The very next step is to split the dataset into Train, Test and Validation set which will be further used in Model Architecture of different classification algorithms.

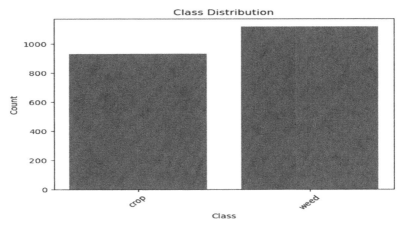

Fig. 4. Class representation in dataset

3.3 Classification Models Architectures and Algorithms

Image classification Deep Learning Algorithms are future applied through model in the dataset to achieve the classification task. First, we have used CNN (Convolutional Neural Network) model architecture using the Sequential model from keras. It consists multiple convolutional layers, a flatten layer, max-pooling layers and dense layers with dropout for regularization, more hyperparameter which were defined for CNN were Adam optimizer, 3 Convolutional layers of 32, 64, 128 filters, batch size of 32 and number of epochs were 7. After giving the architecture to the model, using fit method on Train set we trained the model, then we have complied the model by specifying the optimizer, loss function and accuracy for evaluation matric. Second classification algorithm we used is ANN (Artificial Neural Network), similar to CNN we defined the model architecture consists of flatten, dense, SoftMax layers and dropout further more hyperparameters were defined as Adam optimizer, dropout rate of 0.5, batch size of 32 and number of epochs were 6. Flatten is responsible for target image size, Dense layers are for learning patterns and to make predictions and dropout layers introduce regularization to reduce overfitting. Then we fit the model on Train set and after the compiled model we specify the evaluation matric.

For SVM (Support Vector Machine) we extracted the features from the training set and make a train features 2-D array, after features vectors are normalized SVM classifier is trained and now features are extracted from validation set called valid features and the SVM classifier is used to evaluate the performance through valid features. Similarly, for KNN (K-Nearest Neighbors) first the train images are converted into RGB format, then converted to NumPy array. Then while creating and training of KNN classifier we set the number of neighbor 'k' to 3, create an instance and fit the classifier on training images and labels. Validation set is used for prediction and for tuning up the model, then performance is evaluated. Every prediction of the above neural network algorithms or classifier algorithms are compared with the prediction on test set and the result performance of test is evaluated through which we get the metric of each image

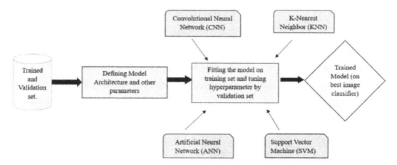

Fig. 5. Execution flow of the image classification models

classifier for weed Classification. This will help us to choose the best image classifier or neural network model to differentiate between weed and crop seedling.

4 Results

To evaluate the prediction result, we have calculated various metrics like Accuracy, F1 score and Recall. The evaluation displays the performance of pre trained model CNN, ANN, KNN and SVM on the classification task of test set between weed and crop seedling which is shown in Table 1.

Table 1. Comparative study of the accuracy of various algorithms

S. no.	Classifier	Accuracy (%)	F1 score	Recall
1	Convolution Neural Network (CNN)	96.1	96	90
2	Artificial Neural Network (ANN)	67.5	49.3	32.47
3	Support Vector Machine (SVM)	89.9	89	90
4	K-Nearest Neighbors (KNN)	88	87	87

To visualize the performance of the model's confusion matrix of the neural network algorithm and for KNN and SVM is displayed in Fig. 6.

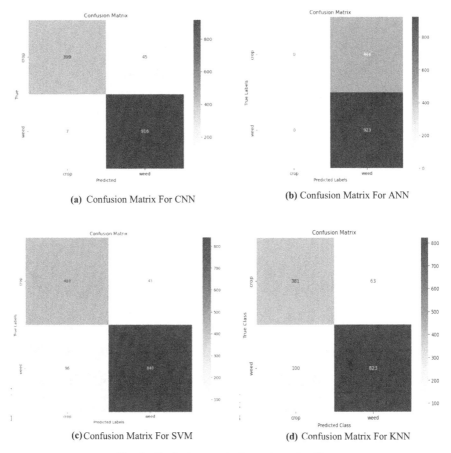

(a) Confusion Matrix For CNN

(b) Confusion Matrix For ANN

(c) Confusion Matrix For SVM

(d) Confusion Matrix For KNN

Fig. 6. Confusion matrix for various classifiers

Confusion matrix consists of four terms which are True Positive, True Negative, False Positives, False Negative which future determine the evaluation metrices of any model. From metric table and confusion matrix it is clear that CNN based model architecture perform the classification task most precisely whereas ANN based model accuracy is not quite good. SVM and KNN based classify model also perform well. For further evaluation, precision-recall curve is used as shown in Fig. 7.

Precision Recall is a graphical representation between precision and recall for different classification thresholds or decision boundaries. In general, a model with a higher precision-recall curve closer to the top-right corner indicates better performance as it achieves high precision and high recall, which we can see in CNN, SVM and KNN but not in ANN. Through all the evaluation of the models we can see the classification model like Convolutional Neural Network (CNN), Support Vector Machine (SVM) and K-Nearest Neighbors (KNN) perform very well for the classification task between crop seedling and weed.

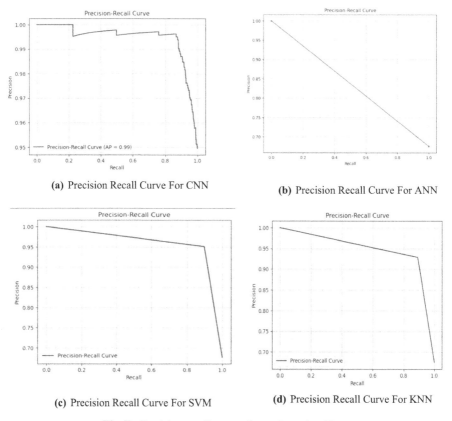

(a) Precision Recall Curve For CNN

(b) Precision Recall Curve For ANN

(c) Precision Recall Curve For SVM

(d) Precision Recall Curve For KNN

Fig. 7. Precision recall curves for various classifiers

5 Future Scope and Conclusion

The study shows that, Deep Learning based Neural networks and classifier used here helps to achieve the goal of classification between weed and crop on the image dataset. It provides us with the higher identification rate with more efficiency and less computation time. The strength of deep learning approach for Weed classification is it simplicity, accuracy, easy implementation for the required classification task. We have seen that using deep learning neural networks like CNN and classifier like KNN, SVM gives good results for the weed classification work in a farm. As in CNN based model under the defined hyperparameters gave accuracy around 96.1% and similarly in ANN based model under hyperparameters were gave accuracy around 67.5% which in comparison to CNN is very low, KNN and SVM have accuracy around 88% and 89.9% which is moderate. Although in this predefined dataset the computation time of all the models were very less but increase in dataset will increase the computation time and value of epochs can be increased to get the better result also. As the model will be applied more on real time data it would improve and eventually can be used for real time weed management in farm for agriculture which will be faster and more effective than the traditional ways. However, these models will also face limitation to new undiscovered data. For Future work, it is worth trying to apply more pre-processing in dataset to achieve more refined images suitable for new and improved neural networks or image classifiers. More experiments on bigger real time dataset will take these models to achieve more real-life approach to solve classification problems of weed in a agricultural farm. Future research and study in this field using deep learning and machine learning will provide better opportunities to farmers in the agriculture sector. Finally, based on the result we believed that the image classification algorithm like K-Nearest Neighbors (KNN), Support Vector Machine (SVM) and neural network like Convolutional Neural Network (CNN) works best for the classify weed and crop seedling. More work in them using technologies like deep learning and machine learning will help the farming sector both in time and cost in the following year.

References

1. Fennimore, S.A.: Weed control, management, ecology, and minutia. The role of hand weeding in weed management. UC Weed Science (2014)
2. Brown, R.B., Noble, S.D.: Site-specific weed management, sensing requirements—what do we need to see? Cambridge Core, Published online by Cambridge University Press,(2017)
3. Rajagopal, A., et al.: A deep learning model based on multi-objective particle swarm optimization for scene classification in unmanned aerial vehicles. IEEE Access, 8, pp. 135383–135393 (2020)
4. Kamilaris, A., Prenafeta-Boldú, F.X.: Deep learning in agriculture and survey. Comput. Electron. Agric. **147**, 70–90 (2018)
5. Vvatsalggupta: Weed detection image dataset contains food crop and plant seedling. Kaggle (n.d.)
6. Giselsson, T.M., Jørgensen, R.N., Jensen, P.K., Dyrmann, M., Midtiby, H.S: A Public Image Database for Benchmark of Plant Seedling Classification Algorithms. preprint arXiv:1711. 05458 (2017)
7. Affonso, C., Rossi, A.L.D., Vieira, F.H.A., Ponce de Leon Ferreira de Carvalho, A.C.: Deep learning for biological image classification. Expert Syst. Appl. **85**, 114–122 (2017)
8. Sharma, N., Jain, V., Mishra, A.: An analysis of convolutional neural networks for image classification. Procedia Comput. Sci. **132**, 377–384 (2018)
9. Khari, M., Garg, A.K., Crespo, R.G., Verdú, E.: Gesture recognition of RGB and RGB-D static images using convolutional neural networks. Int. J. Interact. Multim. Artif. Intell. **5**(7), 22–27 (2019)
10. Bala, R., Kumar, D.: Classification using ANN: a review. Int. J. Comput. Intell. Res. **13**, 1811–1820 (2017). ISSN 0973-1873
11. Amato, G., Falchi, F.: kNN based image classification relying on local feature similarity. In: Proceedings of the Third International Conference on Similarity Search and Applications, SISAP '10. pp. 101–108 (2010)
12. Chandra, M.A., Bedi, S.S.: Survey on SVM and their application in image classification. Int. J. Inf. Technol. **13**(5), 1–11 (2021)
13. Shaner, D.L., Beckie, H.J.: The future for weed control and technology. Pest Manag. Sci. **70**(9), 1329–1339 (2013)
14. T. Giselsson, R. Jørgensen, P. Jensen, M.Dyrmann, and H. Midtiby: A public image database for benchmark of plant seedling classification algorithms, CoRR, arXiv:1711.05458 (2017)
15. Sun, Y., Liu, Y., Wang, G., Zhang, H.: Deep learning for plant identification in natural environment. Comput. Intell. Neurosci. **2017**(4), 1–6 (2017)

16. Sladojevic, S., Arsenovic, M., Anderla, A., Culibrk, D., Stefanovic, D.: Deep neural networks based recognition of plant diseases by leaf image classification. **2016**, Article ID 3289801 (2016)
17. Ashqar, B.A.M., Abu-Nasser, B.S., Abu-Naser, S.S.: Plant seedlings classification using deep learning. Int. J. Acad. Inf. Syst. Res. (IJAISR) **3**(1), 7–14 (2019)
18. Nandhini, N., Bhavani, R.: Feature extraction for diseased leaf image classification using machine learning. In: 2020 International Conference on Computer Communication and Informatics (ICCCI), Coimbatore, India, pp. 1–4, ICCCI48352.2020.9104203 (2020)
19. Slaughter, D.C., Giles, D.K., Downey, D.: Autonomous robotic weed control systems: a review. Comput. Electron. Agric. **61**(1), 63–78 (2008)
20. Liakos, K.G., Busato, P., Moshou, D., Pearson, S., Bochtis, D.: Machine learning in agriculture: a review. Sensors **18**(8), 2674 (2018)
21. Kumar, C., Kumar, V.: Vegetable plant leaf image classification using machine learning models. LNNS, **612** (2023)
22. Asha, P., et al.: Role of machine learning in attaining environmental sustainability. Energy Rep. **8**, 863–871 (2022)
23. Verma, P., Tripathi, V., Pant, B.: Comparison of different optimizers implemented on the deep learning architectures for COVID-19 classification. Mater. Today **46**, 11098–11102 (2021)
24. Tiwari, S., et al.: IOT-enabled model for weed seedling classification - an application for smart agriculture. AgriEngineering **5**(1), 257–272 (2023)

Weed and Crop Detection in Rice Field Using R-CNN and Its Hybrid Models

Neha Shekhawat[1(✉)], Seema Verma[1], Manisha Agarwal[2], and Manisha Jailia[2]

[1] School of Physical Sciences, Banasthali Vidyapith, Jaipur, India
Shekhawatneha327@gmail.com, vseema@banasthali.in
[2] Department of Computer Sciences, Banasthali Vidyapith, Jaipur, India
jmanisha@banasthali.in

Abstract. Mostly, Weeds are the responsible for agricultural losses in recent years. Removing weeds is a challenging task as there are much similarity between weed and crop in terms of texture, color and shape. To deal with this challenge, a farmer needs to spray herbicides uniformly throughout the field. In addition to requiring a lot of pesticides, this method has an adverse effect on the environment and people's health. To overcome this, precision agriculture is used. Unmanned aerial vehicles (UAVs) have been shown great prospective for weed detection, as they can cover large areas of farmland quickly and efficiently. For this experiment, phantom p4 drone was used to take the images of rice field. Therefore, in this work, we propound a weed recognition system using UAVs and a combination of RCNN model and modified RCNN-LSTM and RCNN-GRU. The performance was compared using accuracy, precision, recall, and f1-score as evaluation criteria. Among all RCNN with GRU outperformed with 97.88%.

Keywords: Deep Learning · RCNN · Weed detection · Agriculture · Weeds · RCNN · RCNN-LSTM · RCNN-GRU

1 Introduction

India is the world' largest exporter and the 2nd largest producer of the rice. From FY 1980 to FY 2020–2021, From 53.6 million tonnes to 120 million tonnes, production has increased. However, weeds decrease the production of rice by competing with them for sunlight, soil, water and nutrients [1]. Sometimes weeds can be distinguished easily from crops by using some features like the color of leaves, shape, stems, and seeds. But at the same time, some weeds also have the same color and texture, so they are difficult to differentiate. To eradicate all weed types, numerous weed removal techniques have been developed over time. Manual weeding is one of the techniques to remove weeds by pulling them out. In this method, labourers are involved, which gives rise to many problems faced by farmers, like high costs, labour shortages, and crops damaged by the labourers. To steer clear these problems, the use of chemicals, i.e., herbicides, are also common.However, it causes other problems like environmental pollution and health issues for farmers spraying chemicals, and crops are also affected [2] because the

B. J. Choi et al. (Eds.): IHCI 2023, LNCS 14532, pp. 37–49, 2024.
https://doi.org/10.1007/978-3-031-53830-8_5

chemicals are sprayed on the entire field. To overcome this problem, precision agriculture has come into the picture. Precision farming has the potential to provide the proper chemical doses to the appropriate locations at an appropriate time [3].

For weed identification standard machine learning techniques including Support Vector Machines etc. depend on extracting foremost features including pattern, shape, and colour. However, the feature extraction method was time consuming. Deep learning gave the benefit of automatic feature extraction by using Convolutional Neural Networks (CNN). For weed identification, deep learning techniques that are extensively used include object detection and image classification.

By supplying the softmax outcome, image classification techniques are employed to identify one specie inside an image. The models for classifying images received training using transfer learning. When there is a small-scale dataset then transfer learning models are used to maintain their performance. In image classification, nothing is known about where the weeds are located and multiple species of weeds within an image. Therefore, to identify the precise locations of different weeds in an image, bounding boxes are utilised in object detection.

In literature, there are many acquisition methods among which unmanned aerial vehicle (UAV) is more advantageous than satellite or aircraft [4], as UAVs have great spatial resolution (up to 10 mm/pixel) and can scan a broad area quickly, enabling the detection of small things like weeds [5].

The main contribution of this paper are: This study proposes the hybrid networks of RCNN with LSTM and GRU. The CNN-LSTM and CNN-GRU algorithms were used for classification in traditional RCNN model. Annotations were used for object coordinates in an image. This experiment has been done on a dataset collected using UAVs over a rice field. Image resizing has been done. This experiment shows that the results of hybrid RCNN models are better than the traditional RCNN model. Accuracy, recall, precision, and f1-score were used to evaluate the performance of the recommended technique.

The rest of the article is as followed: Sect. 2 explores the previous work. Section 3 discusses the methodology. The propound network explained in Sect. 4. Section 5 discusses the outcome of this study. Section 6 summarizes this study.

2   Related Work

The combination of deep learning and unmanned aerial vehicle (UAV) techniques may provide effective solution in discriminating weeds from crop.

M Dan Bah et al., [6] propound a fully automatic learning algorithm using CNN with an unsupervised learning dataset. Firstly, there is detection of crop lines and for that purpose they use Hough transform to highlight alignments of the pixels. After that they use inter-row vegatation to constitute the training dataset. And then CNN was used to identify the weed and crop in images.

Arun Narenthiran et al., [7] used UAV based images of soybean field to detect the weeds. They has compared the two object detection methods i.e. Faster RCNN and Single Shot Detector (SSD). In terms of precision, recall, Intersection over Union and F1-score, both the models has good detection performances, but compared to the Faster RCNN, SSD's optimal confidence threshold was substantially lower. Abdur Rahman et al., [8] compared different weed detection models using one-stage and two-stage object detectors such as YOLOv5, RetinaNet, EfficientDet, Fast RCNN and Faster RCNN. They evaluate the speed of every model and came to conclusion that YOLOV5 showed good potential due to its speed.

Hu Zhang et al., [9] has developed a UAV-based weed field detection system. It uses the YOLOV4-tiny network structure for recognition of weeds. Firstly, they uses an Efficient Channel Attention (ECA) module is attatched to the Feature Pyramid Network (FPN) of YOLOv4-Tiny for better recognition of small target weeds. Orignal Intersection over Union (IOU) was replaced by Compelete Intersection over Union(IOU)loss so that the model can reach the convergence state faster. And also compared with YOLOv4-Tiny, YOLOv4, YOLOv5s, Swin-Transformer, and Faster-RCNN.

Nahina Islam et al., [10] collected RGB images of chilli farm to detect weeds. This study uses a Random forest based algorithm to generate weed maps in chilli field. Jian-Wen Chen et al., [11] proposed a smartphone based application for pest detection in which they compare Faster RCNN, Single shot multibox detectors(SSDs) and YOLOv4 models. The dataset in this study was annotated using labeling annotation tool, it is an open source software. Image augmentation was also applied to increase the dataset. YOLOv4 is the most promising among the three algorithms. Hafiza Sundus Fatima et al., [12] compares the YOLOv5 and SSD-Resnet network to detect weed. They collected weed images from six different farms in Pakistan. YOLOv5 shows the great potential.

From the above literature, it has been found that there are various deep learning techniques for weed detection performed by the researchers but almost none has carried out work using RCNN technique for detect weeds.

3   Methods and Materials

The method for weed detection outlined in this article consists of 4 phases. The first stage made up of capturing the images from the rice field, for which UAVs was used. The second stage consists of annotated images which were annotated manually. The third stage is exclusive to detection algorithms. The fourth stage made up of training of algorithms.

3.1   Experimental Site and Image Acquisition

The experimental site is located in Deoli in district Kota of India (25.0014 n, 76.1173 e). The UAV selected for data collection was phantom p4 drone, which is having built-in satellite position system to obtain UAV latitude and longitude information. The satellite positioning system in phantom p4 drone is GPS/GLONASS and lens is fov94 20 mm, f/2.8 focus and effective pixel is 12.4 million. The UAV was used to collect data in august 2021.

The constant height was maintained throughout the experiment. A video was shot and from which dataset has been created. Also, images have been separately captured Fig. 1 shows the images of rice field.

Fig. 1. Sample of images of Rice field captured by UAV

3.2 Data Pre-processing

3.2.1 Data Annotation

Image annotation is the process of labelling images to train the deep learning model. It often involves human annotators using an a tool for labelling images through annotation by assigning relevant classes to different entities in an image. Some of annotated images were shown in Fig. 2.

Fig. 2. Sample of annotated image of Rice field

3.2.2 Image Dataset

From the collected images and video of rice field, frames have been generated to create a large image dataset. Out of all the images were produced, of which 80% images were utilised for training the model and the other 20% images to test it.

It was also requested to resize the images, as the width and the height of the images were not same. After that images were manually annotated.

4   Weed Detection Method

In this study, RCNN [13] has been used for detection of weeds.

The RCNN model is made of 3 parts: Selective search algorithm, CNN for image classification and the last one is SVM. We have proposed that at the CNN section of RCNN model we can apply the hybrid models of CNN such as CNN-LSTM and CNN-GRU.

0.001 was used as the learning rate for implementing the SGD optimizer. The process is allowed to run for 50 epochs.

> **Algorithm:** Weed datasets using RCNN, RCNN-LSTM and RCNN-GRU
>
> Output: Detection of crop and weed
>
> Step1: annotated data taken in the form of .csv file
>
> Step2: applying selective search on an image
>
> Step3: Fed CNN, CNN-LSTM and CNN-GRU to extract the features of an image
>
> Step4: Divide annotated files into train and test set
>
> Step5: Model obtain after training data is feeded with the testing data
>
> Step6: Evaluate the results obtained on some parameters like accuracy

4.1   Regions with RCNN

4.1.1   Selective Search Algorithm

A selective search technique is employed in the standard RCNN to extract the regions of interest. Selective Search algorithm was originally proposed by Uijlings et al. [14] for object recognition.The superpixel algorithm is used in Selective Search to oversegment an image. Firstly, the image is segmented into numerous fragments using a graph-based segmentation algorithm. [15]. Subsequently, a larger zone is formed by merging similar regions based on fit, size, texture, and colour similarity. Eventually, these regions generate the final object locations (Region of Interest).

4.1.2   Convolutional Neural Network Layer

After the region has been chosen, the image with the selected regions is sent through a CNN, where the CNN model extracts the objects from the region with a 4096-dimensional size. In RCNN, AlexNet was used for feature extraction. However, in this work, pre-trained VGG16 was used as a feature extractor. The input size for VGG16 is $224 \times 224 \times 3$. So, this is necessary to resize the region proposal to the specified dimensions if the region proposals are small or large. From the VGG16 architecture, the final softmax layer was removed to get the feature vector. This feature vector passed to SVM and bounding box regressor.

4.1.3   Support Vector Machine

With the help of the feature vector generated by the previous CNN architecture, the SVM model generates a confidence score for the existence of an object in that region (Fig. 3).

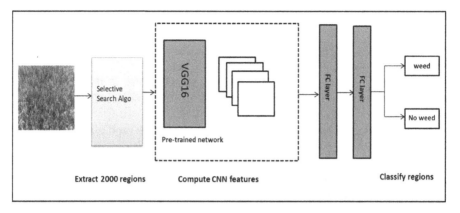

Fig. 3. R-CNN architecture for weed detection

4.2 Regions with RCNN-LSTM

4.2.1 Long-Short-Term-Memory

LSTM [16] deals with both the long-term and short-term memory and it uses the idea of gates. Three gates are used: an input gate, a forget gate, and an output gate. The current input is denoted by x_t, the new and previous cell states by c_t and c_{t-1}, and the current and prior outputs by h_t and h_{t-1}. The following illustrates how an LSTM input gate functions fundamentally.

$$i_t = \sigma\left(W_i \cdot \left[h_{t-1}, x_t\right] + b_i\right) \tag{1}$$

$$\tilde{C}_t = \tanh\left(W_i \cdot \left[h_{t-1}, x_t\right] + b_i\right) \tag{2}$$

$$C_t = f_t C_{t-1} + i_t \tilde{C}_t \tag{3}$$

The information that needs to be added is determined by transferring h_{t-1} and x_t through a sigmoid layer using Eq. (1) [17]. Equation (2) employed to obtain new information after h_{t-1} and x_t are passed through tahn layer [15]. In Eq. (3) [17], long-term memory information C_{t-1} and the current moment information $\tilde{C}_t$ are combined into C_t.

With the help of Eq. (4) [18], the decision is made whether to forget relevant information from the prior cell with a certain probability.

$$f_t = \sigma\left(W_f \cdot \left[h_{t-1}, x_t\right] + b_f\right) \tag{4}$$

Here, W_f refres to weight matrix, b_f is the offset and σ is the sigmoid function.

By applying Eqs. (5) and (6), the output gate of the LSTM ascertains the states necessary for continuation by the h_{t-1} and x_t inputs [18]. The final output is obtained and multiplied by the state decision vectors that pass new information, C_t, through the tanh layer.

$$O_t = \sigma\left(W_o \cdot \left[h_{t-1}, x_t\right] + b_o\right) \tag{5}$$

$$h_t = O_t \tanh(C_t) \tag{6}$$

The output gate's weighted matrices and LSTM bias are represented by the W_o and b_o, respectively.

4.2.2 Convolutional Layer with LSTM

In this study, the combined network of *Regions with CNN-LSTM* was developed to detect the weeds from the paddy field automatically. This architecture's framework was created by merging the CNN and LSTM networks at the classification part. First the convolutional layer is applied and after that LSTM layer. The input data is processed by the CNN to extract spatial features, and the data's temporal relationships are captured by the LSTM. The algorithm has 23 layers: 13 convolutional layers, 5 maxpooling layers, 1 LSTM layer and 1 output layer along with sigmoid layer. The feature extraction process, which is activated by the ReLU function, uses a convolutional layer with a 3×3 kernel size. An input image's dimensions are reduced using the max-pooling layer, which has a size of 2×2 kernels. In the final stage of the architecture, the function map is relocated to the LSTM layer in order to extract temporal information. Figure 4 shows the architecture of RCNN-LSTM.

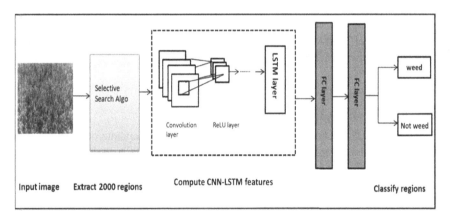

Fig. 4. RCNN-LSTM architecture for weed detection

The combined network of CNN-LSTM is described as follows:

1. *Convolutional layer:* The input to the convolutional layer is a 3D tensor of shape (batch_size, num_time_steps, num_features), denoted as X. The convolutional layer applies a set of filters withweights denoted as W_c, resulting in a 3D tensor of shape (batch_size, num_filters, convolved_time_steps), denoted as H_c.

   The convolutional operation can be expressed mathematically as:

$$H_c[:, f, t] = \text{activation\_func}(\text{sum}(X[:, :, t : t + \text{filter\_size}] * W_c[:, :, :, :, f]) + b_c[f]) \tag{7}$$

where, f is the filter index, t is time step index, filter_size is size of filter, activation_func is a nonlinear activation function, such as ReLU, b_c is bias term for the convolutional layer.

2. *Pooling layer:* In order to diminish the dimensionality of data, output of the convolutional layer is passed through a pooling layer. In this case maxpooling layer is used, and can be expressed mathematically as:

$$H_p[;, f, p] = \max(H_c[;, f, ppool\_size : ppool\_size + pool\_size]) \qquad (8)$$

Here, p is the pooling step index, pool_size is the size of the pooling window.

3. *LSTM layer:* The LSTM layer then accepts the output of pooling layer. The input to LSTM layer is 3D tensor shape (batch_size, num_convolved, time_steps, num_filters), denoted as H_p. The LSTM layer consists of number of LSTM cells each of which has a hidden state denoted as H_t and a memory cell denoted as C_t. The gates of the LSTM are expressed in Sect. 3.2.1.

4. Output layer: The output of LSTM layer is typically passed through a fully connected layer to produce the final output. The fully connected layer can be expressed mathematically as:

$$Y = \text{softmax}(W_y * H_t + b_y) \qquad (9)$$

Here, W_y is weight matrix, b_y is bias vector, softmax is activation function.

The entire combined CNN-LSTM network is trained end-to-end using backpropagation and stochastic gradient descent. During training, the weights of the convolutional layer, pooling layer, LSTM layer and output layer are updated to mininmize a loss function between the predicted output and the ground truth.

4.3 Regions with RCNN-GRU

4.3.1 Gated Recurrent Unit

GRU [19] is like LSTM but unlike LSTM. GRU don'thave separate cells. It has only one hidden state that is why it is simple architecture and easy to train. It has only 2 gates reset gate and updated gate. The reset gate regulates how much of the new input gets incorporated, while the updated gate decides how much of the previous concealed state is kept. The fundamental operation of the GRU reset gate is shown in the following Eq. (10) [20].

$$r_t = \sigma(W_r[h_{t-1}, x_t] + b_r) \qquad (10)$$

Here, W_r is weight matrix for reset gate, the input at time step t denoted by x_t, $[h_{t-1}, x_t]$ is the concatenation of the hidden state from the previous time step and the current input and b_r is the bias vector for the reset gate.

Equation (11) [20] tells that how much the past knowledge should bepassed into the future.

$$z_t = \sigma(W_z[h_{t-1}, x_t] + b_z) \qquad (11)$$

Here, W_z is the weight matrix for the update gate, x_t is the input at tme step t, $[h_{t-1}, x_t]$ is the concatenation of the hidden state from the previous time step and the current input and b_z is the bias vector for the update gate.

4.3.2 Convolutional Layer with GRU

In RCNN-GRU network, the hybrid of CNN-GRU is used for classificationof image which are fed into them after the process of section search algorithm.In CNN-GRU model, first layer is the convolutional layer is applied and after that GRU layer. The CNN is used to extract spatial features from the input data,. The CNN made up of an input layer,maxpooling layer and Batchnormalization layer. And after that the GRU layerwas applied with the neurons of 64. Figure 5 shows the architecture of RCNN-GRU.

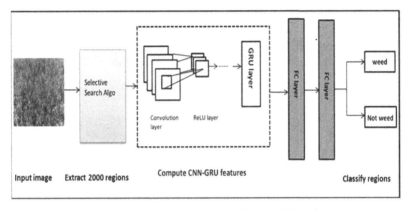

Fig. 5. RCNN-GRU architecture for weed detection

The mathematic expression for convolution and pooling layer of RCNN-GRU is same as the convolution and pooling layer of RCNN-LSTM the only difference is in the final layer i.e., the output layer.

The output layer can be written as:

$$Y = \text{softmax}\left(W_y * H_{final} + b_y\right) \tag{12}$$

Here, W_y is the weight matrix, b_y is bias vector and H_{final} is final hidden state of GRU network.

5 Results and Discussions

This section discusses the results based on RCNN and its hybrid models. The results of RCNN, RCNN-LSTM and RCNN-GRU were compared. The accuracy, precision, recall and f1-score were evaluated to know which model detects the weed properly.

Accuracy is the total observation rate of the correctly detect weeds observation and measured by Eq. (13)

$$Accuracy = \frac{TP + TN}{TP + TN + FP + FN} \tag{13}$$

Precision compute the weed that is classified perfectly. Precision refers to theweeds which are classified perfectly (TP, True Positive) over to the total numberof weeds classified actually.

$$Precision = \frac{TP}{TP + FP} \tag{14}$$

Recall calculates the weeds that are predicted precisely. Recall refers to theweeds classified precisely (TP, True Positive) over to the total number of weeds classified actually.

$$Recall = \frac{TP}{TP + FN} \tag{15}$$

F1-score considers both Recall and precision. F1-score is the ratio of twice the product of Recall and Precision over to the sum of Recall and Precision.

$$f1 - score = \frac{2 * Recall * Precision}{Recall + Precision} \tag{16}$$

Table 1. Parameter performance of different hybrid of RCNN architectures (in %)

Model	Accuracy	Precision	Recall	F1score
RCNN	96.7	95	95	95
RCNN-LSTM	97.9	96	95	95
RCNN-GRU	97.88	97	96	96

Table 1 presents the comparative analysis of models for crop and weed identication. In table models are arranged form lower to higher scores. In comparison to others RCNN-GRU model gave better results in terms of accuracy:97.88%, precision: 97% Recall: 96% and f1-score: 96%.

In other work, the regional convolutional neural network shows the better results in comparison to other deep learning algorithms for e.g., CNN and long short term memory. M. Vaidhehi et al. [21] compares the results of RCNN with the CNN in which RCNN performed better with the accuracy of 83.33% and precison of 83.56%. In comparison to their work the proposed algorithm shows the improvment of 13% and 12% in accuracy and precision.

Figure 6 presents the accuracy vs epoch curve of the RCNN models. As the accuracy curve goes higher, learning of model gets better. First the accuracy increases with the number of epochs, but after sometime the model stops learning. The graph of RCNN-GRU is more stable than the other models. As shown in Fig. 6, at first, the validation accuracy of RCNN-GRU is much less than the other two, but as the number of epochs increases validation accuracy increases and became equalized with the other two.

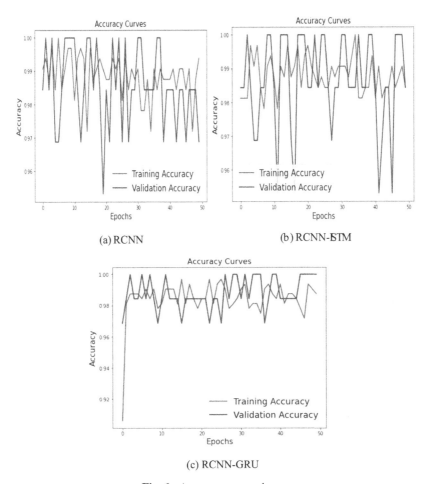

(a) RCNN

(b) RCNN-LSTM

(c) RCNN-GRU

Fig. 6. Accuracy vs epoch curves

6 Conclusion

One of the most vital crops in the world is rice, providing sustenance for millions of people. It is predominant food for over half of the world's popular, particularly in Asia, where it is consumed on a daily basis. Rice crops are grown in many countries around the world, with the top producers being China, India, Indonesia and Bangladesh. To detect the weeds in rice fields, need to effective methods. Nowadays, UAV based acquisition systems were popular.

In this study, the dataset of rice field was collected using unmanned aerial vehicle (UAV). After that images were resized and then they were annotated. Data augmentation techniques such as rotation-range, horizontal-flip, width and height-shift range etc., were applied to the training set. For weed detection hybrid of RCNN, RCNN-LSTM and RCNN-GRU were proposed. Other than accuracy, precision, recall and f1-score were also considered. Among which RCNN-GRU outperformed. For future work, object detection using segmentation techniques will be considered.

Acknowledgements. We are thankful to DST ASEAN (SERB) for providing us financial support under DST ASEAN India Collaborative R&D scheme.

Authors Contribution. Our research group has found very interesting results on the RCNN models used for weed detection. In this article we proposed the hybrids of RCNN model and achieved good results. Therefore, we are very excited to submit this article in your journal.

References

1. Asif, M., Iqbal, M., Randhawa, H., Spaner, D.: Managing and Breeding Wheat for Organic Systems: Enhancing Competitiveness Against Weeds. Springer Science & Business Media, Cham (2014)
2. Holt, J.S.: Principles of weed management in agroecosystems and wildlands1. Weed Technol. **18**(sp1), 1559–1562 (2004)
3. David, J.M.: Twenty five years of remote sensing in precision agriculture: key advances and remaining knowledge gaps. Biosyst. Eng. 114(4), 358–371. ISSN: 1537-5110 (2013)
4. Rango, A., et al.: Unmanned aerial vehicle-based remote sensing for rangeland assessment, monitoring, and management. J. Appl. Rem. Sens. **3**(1), 033542 (2009). https://doi.org/10.1117/1.3216822
5. Verlinden, M., De Boeck, H.J., Nijs, I.: Climate warming alters competition between two highly invasive alien plant species and dominant native competitors. Weed Res. **54**(3), 234–244 (2014)
6. Bah, M.D., Hafiane, A., Canals, R.: Deep learning with unsupervised data labeling for weed detection in line crops in UAV images. Remote Sens. **10**(11), 1690 (2018)
7. Veeranampalayam Sivakumar, A.N., et al.: Comparison of object detection and patch-based classification deep learning models on mid-to late-season weed detection in UAV imagery. Remote Sens. **12**(13), 2136 (2020)
8. Rahman, A., Lu, Y., Wang, H.: Performance evaluation of deep learning object detectors for weed detection for cotton. Smart Agri. Technol. **3**, 100126 (2023)
9. Zhang, H., et al.: Weed Detection in Peanut Fields Based on Machine Vision. Agriculture **12**(10), 1541 (2022)
10. Islam, N., et al.: Machine learning based approach for Weed Detection in Chilli field using RGB images. In: Advances in Natural Computation, Fuzzy Systems and Knowledge Discovery, pp. 1097–1105. Springer International Publishing, Cham (2021)
11. Chen, J.W., Lin, W.J., Cheng, H.J., Hung, C.L., Lin, C.Y., Chen, S.P.: A smartphone-based application for scale pest detection using multiple-object detection methods. Electronics **10**(4), 372 (2021)
12. Fatima, H.S., ul Hassan, I., Hasan, S., Khurram, M., Stricker, D., Afzal, M.Z.: Formation of a lightweight, deep learning-based weed detection system for a commercial autonomous laser weeding robot. Appl. Sci. **13**(6), 3997 (2023)
13. Girshick, R., Donahue, J., Darrell, T., Malik, J.: Rich feature hierarchies for accurate object detection and semantic segmentation. In: Proceedings of the IEEE Conference on Computer Vision and Pattern Recognition, pp. 580–587 (2014)
14. Uijlings, J.R., Van De Sande, K.E., Gevers, T., Smeulders, A.W.: Selective search for object recognition. Int. J. Comput. Vision **104**, 154–171 (2013)

15. Felzenszwalb, P.F., Huttenlocher, D.P.: Efficient graph-based image segmentation. Int. J. Comput. Vision **59**, 167–181 (2004)
16. Hochreiter, S., Schmidhuber, J.: Long short-term memory. Neural Comput. IEEE **9**(8), 1735–1780 (1997)
17. Pradhan, A.K., Das, K., Mishra, D., Chithaluru, P.: Optimizing CNN-LSTM hybrid classifier using HCA for biomedical image classification. Exp. Syst. e13235 (2023)
18. Zhou, F., Hang, R., Liu, Q., Yuan, X.: Hyperspectral image classification using spectral-spatial LSTMs. Neurocomputing **328**, 39–47 (2019)
19. Chung, J., Gulcehre, C., Cho, K., Bengio, Y.: Empirical evaluation of gated recurrent neural networks on sequence modeling (2014). arXiv preprint
20. Baghezza, R., Bouchard, K., Bouzouane, A., Gouin-Vallerand, C.: Gait-based age, gender and mobility recognition using cnn-Bgru on thermal silhouettes with an embedded implementation. Available at SSRN 4111792
21. Vaidhehi, M., Malathy, C.: An unique model for weed and paddy detection using regional convolutional neural networks. Acta Agri. Scandinavica, Sect. B—Soil & Plant Sci. **72**(1), 463–475 (2022)

Deep Learning

Spatial Attention Transformer Based Framework for Anomaly Classification in Image Sequences

Aishvarya Garg[1,3], Swati Nigam[2,3], Rajiv Singh[2,3(✉)], Anshuman Shastri[3], and Madhusudan Singh[4]

[1] Department of Physical Science, Banasthali Vidyapith, Rajasthan 304022, India
[2] Department of Computer Science, Banasthali Vidyapith, Rajasthan 304022, India
jkrajivsingh@gmail.com
[3] Centre for Artificial Intelligence, Banasthali Vidyapith, Rajasthan 304022, India
anshumanshastri@banasthali.in
[4] School of Engineering, Science and Management, Oregon Institute of Technology, Oregon 97601, USA
madhusudan.singh@oit.edu

Abstract. With the increasing number of crimes in crowded and remote areas, there is a necessity to recognize any abnormal or violent event with the help of video surveillance systems. Anomaly detection is still a challenging task in the domain of computer vision because of its changing color, backgrounds, and illuminations. In recent years, vision transformers, along with the introduction of attention modules in deep learning algorithms showed promising results. This paper presents an attention-based anomaly detection framework that focuses on the extraction of spatial features. The proposed framework is implemented in two steps. The first step involves the extraction of spatial features with the Spatial Attention Module (SAM) and Shifted Window (SWIN) transformer. In the second step, a binary classification of abnormal or violent activities is done with extracted features via fully connected layers. A performance analysis of pretrained variants of SWIN transformers is also presented in this paper for the choice of the model. Four public benchmark datasets, namely, CUHK Avenue, University of Minnesota (UMN), AIRTLab, and Industrial Surveillance (IS) are employed for analysis and implementations. The proposed framework outperformed existing state of the art methods by 18% and 2–20% with accuracy of 98.58% (IS) and 100% (Avenue) respectively.

Keywords: Anomaly classification · deep learning · Spatial Attention Module · shifted window transformer · convolutional neural networks

1 Introduction

In the current scenario, as the rate of crimes like robbing, vandalism, fights is surging, there is a lack of sense of security in humans that results in the requirement of surveillance systems. Several video surveillance cameras are functioning in public as well as remote

B. J. Choi et al. (Eds.): IHCI 2023, LNCS 14532, pp. 53–64, 2024.
https://doi.org/10.1007/978-3-031-53830-8_6

areas and crowded streets, for example, markets, stadiums, airports, metros and trains stations and many more. The vast majority of harmful incidents take place in remote and crowded locations. Anomaly detection systems have drawn a lot of attention in the domain of computer vision as the concerns for security have grown. This application focuses on the detection of any rare, irregular, odd, unexpected events or patterns that are very much dissimilar to the daily life's routines [1, 2].

With the introduction of deep learning techniques, there is huge help in detecting abnormalities with surveillance systems [4–7, 9]. Many approaches are proposed at frame level that comprises the use of deep learning algorithms, for example 2D and 3D convolutional neural networks (CNN), pretrained variants of CNN, long short term memory (LSTM) and autoencoders (AE) for extraction of spatial and temporal features were utilized for the abnormal activities detection [27, 28].

Attention modules and vision transformers (ViTs) are gaining trust for attaining good accuracy for classification of images [8, 10]. Many deep learning algorithms using attention modules, for examples: self-attention, channel and spatial attention were employed for the extraction of spatial and temporal features in video anomaly detection in automotives and surveillance videos [11–14]. As far as our study is concerned, we found two shortcomings:

- There are various methodologies proposed for anomaly detection employing ViT transformers. In [15–17], ViT was utilized for making future predictions with the reconstruction of spatio-temporal features, but models were not able to give a certain accuracy for the utilized dataset, namely, avenue dataset.
- Most works based on spatio-temporal features were proposed for the classification of anomalies. But as we are talking about the surveillance videos, the content is not very optimum, such that there is a requirement for paying attention to the locations of informative part of the video for better accuracy.

Keeping these points in mind, we proposed a deep learning-attention based framework that only focuses on the informative part of video for the classification of anomalies with the help of SWIN transformer and attention module. We also provide the performance analysis for the proposed method with various pretrained models, trained with weights of ImageNet V1 for four public available datasets, namely, UMN and CUHK Avenue, AIRTLab, and Industrial Surveillance datasets on the ground of their performance metrics. The key contributions of our work are summarized as below:

1. A hybrid attention based deep learning model is proposed that focusses on the binary classification of abnormal and violent activities with extracted spatial features on the frame level. A SAM is employed that depicts the details of informative part followed by SWIN transformer for the extraction of spatial features.
2. As we know CNN is infamous for image classification, and hence fully connected layers are utilized for the classification of anomalous activities.
3. The proposed model is tested for two types of anomalies (abnormal and violent), that outperformed the exiting classification models with the accuracy of 100% (Avenue) and 98.58% (Industrial Surveillance).

In this paper, Sect. 2 provides works related to anomaly detection with machine and deep learning algorithms along with attention modules. Section 3 provides a short description of SAM and SWIN transformer and a detailed version of the proposed methodology. In Sect. 4, experimental results from the dataset along with discussions for performance analysis are provided. Section 5 presents conclusion and future scope for proposed framework drawn from the analysis and results.

2 Related Study

This section includes the related works for anomaly detection using deep, machine, and transfer learning algorithms along with attention modules. Anomaly detection is still a challenging research area in computer vision because of various factors like backgrounds, light, colors and different viewpoints. Some works proposed for abnormal and violent activities detection utilizing the mentioned datasets are discussed here.

An industrial internet of thing (IIoT) based violence detection network (VD-Net) was proposed by Ullah et. al (2021) [20]. In this framework, lightweight CNN was utilized for the detection of any anomalous activity that is sent to an IIoT network. In this network, the features are extracted with ConvLSTM and then fed to gated recurrent unit (GRU) for classification of violent activities.

Ghadi et al. (2022) [25] proposed a framework that involves three types of features, viz, deep flow, force interaction matrix and force flow along with silhouettes for spotting of humans in formulated frames. The extracted features were then fed to maximum entropy Markov model for prediction of abnormal activities.

An anomaly detection framework is proposed by Alarfaj et al. (2023) in which first silhouettes of human in crowded locations were extracted followed by fuzzy c-mean clustering technique for identification of human and non-human silhouettes. From verified silhouettes, motion-related features are extracted which were given to XG-boost classifier for classification of abnormal activities [26].

Abdullah et al. (2023) [29] invented an idea for semantic segmentation technique for foreground extraction for identifying suspicious activities in crowded areas. For anomaly detection, spatial-temporal descriptors were utilized in features extraction. The resulting fused features were fed to multilayer neuro-fuzzy classifier for classification.

Sharif et al. (2023) designed a framework for decreasing error gap in the classification of anomaly with the help of different types of reconstruction networks and prediction networks. These rNet and pNet were fused to form a generalized network called rpNet. For formulation of this network, the fusion of convolutional autoencoders (ConvAE) and three types of U-Net frameworks (traditional, non-local and attention) were done [31].

Deepak et al. (2021) proposed an approach for anomaly detection with the help of end-to-end residual spatial-temporal autoencoders (STAE) model. This model comprised three residual blocks and convolution LSTM (ConvLSTM) layers for feature extraction as well classification of anomality at frame level [32].

A multi-modal semi supervised deep learning-based anomaly detection methodology is presented in [33]. Here features were extracted for two modalities namely, RGB and depth with MobileNet. The extracted features were then fed to bidirectional LSTM (BiLSTM) autoencoders for the classification of anomalies.

Ehsan et al. (2023) [34] proposed an unsupervised violence detection network called Spatial-Temporal Action Translation (STAT). This network was trained on non-violent behaviors and translating normal motion in spatial frame. The classification task was carried out with the means of similarity between actual and reconstructed frames.

Kumar et al. (2023) [42] proposed an unsupervised anomaly detection named residual variational autoencoder (RVAE). In this framework, ConvLSTM and residual network were employed for extraction of spatio-temporal features and vanishing gradients in encoder and decoder architectures.

An end-to-end hybrid deep learning-based model was proposed for surveillance video anomaly detection (SVAD) in [38]. Here the first spatial features are extracted with the help of EfficientNet V2 and these features were then fed to ViT transformer, where temporal features were extracted with temporal attention module. These spatio-temporal features were then given to one fully connected layer for classification.

Pillai et al. proposed a self-context aware based model that employed a one-class few-shot learning of ViT transformer for anomaly detection. In this framework, pretrained ResNet152 and FlowNet2 models were utilized for the extraction of spatial and temporal features from a few non-anomalies' frames. These computed features were then fused and fed to encoder that gives out results to decoder for prediction [39].

Sivalingam et al. (2023) proposed a SWIN transformer-based model for anomaly detection along with a new technique called anomaly scoring network. In this proposed framework, first the anomaly is predicted from the video frames using SWIN transformer. Then extracted features of the abnormal frame are then fed to the semi-supervised scoring network, where the value of the score is formulated by the means of a deep anomaly detector [40].

Many approaches have come into light since the introduction of vision transformers for the classification of human activities. Some works comprise combination of transformers with deep learning algorithms like CNN are presented for anomaly detection [35–37].

3   Proposed Framework

This section comprises brief details of components of the proposed framework namely, SAM and SWIN transformer. The proposed approach for the binary classification of anomalous activities is discussed here in a detailed manner.

3.1 Spatial Attention Module (SAM)

SAM is an attention module in CNN that concentrates on the location of informative part of the features of the given input. It helps to enrich significant information in the given feature by means of formulating spatial attention maps with the help of inter spatial relationships between features.

For the computation of spatial attention map, first step to restrict channel dimension of the given feature, by the means of fused results of formulated 2D maps from average pooling and max pooling operations (F_{mean} and F_{max}) that compute feature descriptor as shown in Fig. 1. This feature descriptor is helpful in enlightening informative part. A convolutional layer is employed to the extracted feature descriptor in order to get spatial attention map (M_s) that make decision which features has to be frozen and enhanced [22].

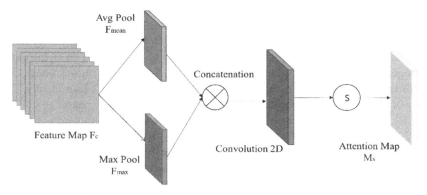

Fig. 1. Architecture for Spatial Attention Module (SAM)

3.2 Shifted Window Transformer (SWIN)

SWIN transformer is a type of ViT transformer that introduces formulation of hierarchical features maps with concept of shifted windows as in CNN algorithms. It works the same as ViT but in place of global features, self-attention module produces features maps within local windows that results in linear computation and good accuracy. This specialty helps the framework to classify more complex recognition and general-purpose tasks [23].

The patch merging block produces hierarchical feature maps. The main function of this block is to downscale the number of patches by concatenating features of neighboring patches and then applying a linear layer for dimensionality reduction of features.

The SWIN block consists of window multi-head self-attention (W-MSA) and shifted window multi-head self-attention (SW-MSA) modules in place of multi-head self-attention module in ViT architecture. Like multi head self-attention (MSA) module, W-MSA computes the local feature within window and hence lessen computational cost and complexity. In W-MSA, there is one limitation that is there is no link between these windows and hence limits modelling power, in order to overcome this problem, SW-MSA is introduced. This module enables cross window linkage by shifting windows and cyclic shift for getting meaningful results. The concept of shifted window enables model to establish the connection between its next window without collapsing each other and hence helpful in gaining good results in image and object detection.

3.3 SST for Anomaly Detection

In our paper, an attention-transformer deep learning-based framework for the binary classification of abnormal and violent activities is proposed. For this work, we mainly focus on the locations of informative part of images by the means of SAM in the process of feature extraction for the binary classification of abnormal activities. The framework is divided into two steps: feature extraction using SWIN transformer along with SAM and classification using fully connected layers from CNN architectures.

In the proposed methodology, videos are first segmented into frames with pre-processing techniques: normalization and resizing. For the extraction of robust features, the base version of SWIN transformer and SAM are employed. For base SWIN trans-former, the number of channels is taken as 128. There are four stages in architecture; each stage has patch merging block and SWIN block as shown in Figs. 2 and 5. The depth of the SWIN block in each stage is taken as 2, 2, 18, and 2 respectively. The number of attention head having dimension 32 for MSA operation is taken as 4, 8, 16, and 32 for four stages respectively. As the input dimensions of input image is taken as $224 \times 224 \times 3$, so the size of window is taken as 7×7. For the hierarchical structure, we used default downscaling factor of 4, 2, 2, and 2 for the patch merging block.

The input images are passed through SAM as we want to focus on the informative part of input for more robust features. The extracted features were then fed to the patch merging block followed by SWIN block in each stage of architecture. The input and output dimensions from the SAM are same such that there is no hurdle in computing features as accordance to the architecture of transformer.

The resultant tensor from the last stage is then passed on to an adaptive average pooling layer with size of 3 for dimensionality reduction and then a flatten layer is applied for the task of classification of anomalous activities. Five linear layers are utilized in which four layers have 1024, 256, 64, and 16 units with leaky ReLU activation function and last layer are with 2 units as the number of classes is 2. As the binary classification is performed here, we used sigmoid as activation function for last linear layer. At the end, we utilized batch normalization and dropout layers between the linear layers to decrease the complexity and training time and minimize the risk of overfitting of model. The architecture of the proposed framework is shown in Fig. 2.

4   Experimental Results and Discussions

This section provides details of used public benchmark datasets for the validation of proposed framework on the basis of performance metrics, confusion matrix and training accuracy and loss per epoch graph. A study of performance analysis of three pretrained models of SWIN transformer and results attained with proposed framework are presented in Tables 3 and 4 along with discussions. The implementations have been done on four datasets comprising clips of violent/non-violent and normal/abnormal behaviors.

The implementations on the datasets were performed on Jupyter notebook of Google Colaboratory Pro and Anaconda using PyTorch library. For the experiments, different split ratios and batch size were utilized as per data provided in the dataset. The loss and optimizer for datasets were taken as cross entropy loss and SGD respectively. The statistics for the implementation for datasets and formulated confusion matrices are shown in Tables 1 and 2 respectively.

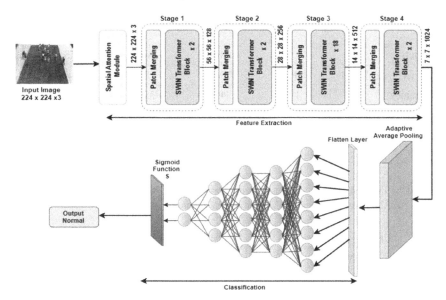

Fig. 2. Proposed Framework for Anomaly Detection

4.1 Datasets

The UMN dataset consists of three scenes: park, mall and bank taken in outdoor and indoor environments. There is a total of 11 videos having normal crowd in starting and sudden panic attacks at the ending of video stating abnormal activity. Each video has dimensions of 320×240 and frames per second (fps) as 30 [18]. The formulated graph is shown in Fig. 3(a).

The CUHK Avenue dataset consists of three abnormal events: strange actions (throwing of bag, dancing, running, standing, and jumping), wrong direction and human with objects (human with bicycle) and normal events taken place outside metro station. It consists of 16 training videos that have normal activities and 21 testing videos that have normal and abnormal clips [19]. Each video has dimensions of 320×640 and fps as 25. For the implementation, we first segmented the testing videos into normal and abnormal categories. The formulated graph is shown in Fig. 3(b).

The AIRTLab dataset comprises violent (kicks, punches, slapping, clubbing, stabbing and gunshots) and non-violent (hugging, high fives, clapping, exulting, and gesticulating) activities [20]. There is a total of 350 clips in which 230 depict violent and the rest 120 clips represent non-violent behaviors. Each video is of dimensions of 1920×1080 with 30 fps. The formulated graph is shown in Fig. 3(c).

The Industrial Surveillance dataset comprises the violent and non-violent activities occurring in different industries, stores, offices, and petrol pumps that are linked to the industries [21]. It consists of 150 videos of both categories having variable dimensions and fps as 20–30. The formulated graph is shown in Fig. 3(d).

Table 1. Different Hyperparameters for Implementation

Parameters	UMN	Avenue	AirTLab	IS
Frames taken per Video	90	75	10	40
Total Frames	1980	3450	3500	12000
Split Ratio	9:1	8:2	8:2	8:2
Batch Size	6	6	14	20
Epochs	30	40	30	20

Table 2. Confusion Matrices for proposed framework on UMN, Avenue, AirTLab and IS Datasets (Normal (N), Abnormal (A), Violent (V), and Non-Violent (NV))

Dataset	UMN		Avenue		Dataset	AIRTLab		IS	
Activity	N	A	N	A	Activity	V	NV	V	NV
N	95	0	311	0	V	456	5	1176	26
A	8	95	0	379	NV	7	232	8	1190

Table 3. Performance metrics for various models and proposed framework for UMN and Avenue Datasets

	UMN Dataset				Avenue Dataset			
Model / Metrics (%)	SWIN (Tiny)	SWIN (Small)	SWIN (Base)	SWIN (Base)+SAM	SWIN (Tiny)	SWIN (Small)	SWIN (Base)	SWIN (Base)+SAM
Accuracy	95.95	92.92	92.42	**95.95**	99.85	99.85	99.85	**100**
Precision	100	91.04	97.68	**100**	100	100	100	**100**
Recall	92.11	94.81	86.66	**92.23**	99.71	99.71	99.71	**100**
F1 Score	95.89	92.89	91.84	**95.95**	99.85	99.85	99.85	**100**

Table 4. Performance metrics for various models and proposed framework for AIRTLab and Industrial Surveillance Datasets

	AIRTLab Dataset				Industrial Surveillance Dataset			
Model	SWIN (Tiny)	SWIN (Small)	SWIN (Base)	SWIN (Base)+SAM	SWIN (Tiny)	SWIN (Small)	SWIN (Base)	SWIN (Base)+SAM
Accuracy	98.28	98.85	98.57	**98.28**	97.83	98.16	98.5	**98.58**
Precision	96.29	97.51	97.48	**97.89**	96.29	97.59	98.61	**98.59**
Recall	98.73	99.15	98.31	**97.07**	99.30	98.61	98.61	**98.58**
F1 Score	97.50	98.32	97.89	**97.47**	97.77	98.10	98.44	**98.58**

4.2 Results and Discussions

The performance analysis for the datasets is done on the grounds of performance metrics namely, accuracy, precision, recall, f1-score, and confusion matrix. On the implementation of proposed framework, an accuracy of 95.95%, 100%, 98.28%, and 98.58% is attained with UMN, Avenue, AIRTLab and Industrial Surveillance datasets respectively with different parameters.

In case of normal/abnormal datasets, during implementation process, different hyperparameters have been varied for example, batch size 2, 3, and 6, ratio split from 70% to 90% for which we get best results with batch size 6 and 9:1 ratio split in UMN whereas batch size 6 and 8:2 split ratio for avenue datasets. The implementations were done with range of epochs from 20 to 50, in which if number of epochs greater than or equal to 50, the model resulted in overfitting. In the case of UMN dataset, initially we attained an average accuracy resulting in problem of having imbalance dataset. For the solution of this problem, augmentation technique is employed. For further analysis frames were taken between 90 to 120, out of which, good results with 90 frames were attained as compared to previous results and as shown in Table 3, it outperforms all models. In the terms of accuracy, some of the works in literature section outperforms our proposed methodology with 2–3% in case of UMN dataset while in case of avenue dataset, our proposed methodology outperforms by huge percentages as compared to works mentioned in literature as well as Table 3. For validation of the model, the avenue dataset was split into 70% training, 20% validation and 10% for testing. From the graph plotted for validation/training accuracy, both accuracies are almost same with the increasing number of epochs.

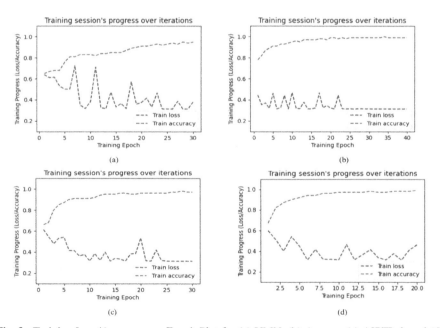

Fig. 3. Training Loss/Accuracy per Epoch Plot for (a) UMN, (b) Avenue, (c) AIRTLab and (d) Industrial Surveillance datasets

In case of violent/non-violent datasets, during implementation process, variations of different hyper-parameters are done for example, batch size in range of multiple of 7 and 10, for which best results were attained with batch size 14 in AIRTLab whereas batch size 20 for industrial surveillance datasets. The implementations were performed with a range of epochs between 20 to 50, in which if number of epochs greater than 30, the model resulted in overfitting. In the case of AIRTLab, the attained accuracy is very much similar to the pretrained models shown in Table 4. In the case of the second dataset, our proposed model outperformed the original work by almost 18% and pretrained models in Table 4.

For the optimizer, the model is trained with Adam, RMS Prop and SDG, out of which SDG performed well. For loss, the proposed model is implemented with cross entropy loss and BCE loss for binary classification, in which cross entropy loss gave out good results. As per the second gap mentioned in Introduction section, the content quality of avenue and industrial surveillance datasets is adequate, the proposed model can classify the anomalous activities with the highest accuracy.

5   Conclusions

In this paper, we proposed a deep learning attention-transformer based anomaly detection framework that focuses on the locations of informative part of images by means of SAM. This framework comprises two steps. The first step is to extract features with the help of base version of SWIN transformer and SAM. Then the extracted features are then fed to the fully connected layers of CNN for binary classification of abnormal activities in the second step. We have also provided a tabular performance analysis for different pretrained variants of SWIN transformers for better understanding of choice for model. For the validation of our proposed framework, four public benchmark datasets, namely UMN and Avenue, AIRTLab and Industrial Surveillance were utilized that gained accuracy of 95.95%, 100%, 98.28%, and 98.58% respectively. The attained results show that the proposed framework is successful in classifying the anomalous activity irrespective of quality of content. In future, the work on the imbalance dataset can be done and this framework can be utilized for more complex video anomaly detection datasets with different image processing techniques.

References

1. Franklin, R.J., Dabbagol, V.: Anomaly detection in videos for video surveillance applications using neural networks. In: Fourth International Conference on Inventive Systems and Controls (ICISC), p. 632. IEEE (2020)
2. Anomaly Detection in Videos using LSTM Convolutional Autoencoder, https://towardsdatas cience.com/prototyping-an-anomaly-detection-system-for-videos-step-by-step-using-lstm-convolutional-4e06b7dcdd29. Last accessed 25 Apr 2023
3. Garg, A., Nigam, S., Singh, R.: Vision based human activity recognition using hybrid deep learning. In: 2022 International Conference on Connected Systems & Intelligence (CSI), pp. 1–6. IEEE (2022)
4. Berroukham, A., Housni, K., Lahraichi, M., Boulfrifi, I.: Deep learning-based method for anomaly detection in video surveillance: a review. Bull. Electr. Eng. Inf. 2(1), 314–327 (2023)

5. Suarez, J.J.P., Naval Jr, P.C.: A survey on deep learning techniques for video anomaly detection. arXiv preprint arXiv: 2009.14146 (2020)
6. Ramzan, M., et al.: A review on state-of-the-art violence detection techniques. IEEE Access **7**, 107560–107575 (2019)
7. Chandrakala, S., Deepak, K., Revathy, G.: Anomaly detection in surveillance videos: a thematic taxonomy of deep models, review and performance analysis. Artif. Intell. Rev. 1–50 (2022)
8. Jamil, S., Jalil Piran, M., Kwon, O.J.: A comprehensive survey of transformers for computer vision. Drones **7**(5), 287 (2022)
9. Nigam, S., Singh, R., Misra, A.K.: A review of computational approaches for human behavior detection. Arch. Comput. Meth. Eng. **26**, 831–863 (2019)
10. Guo, M.H., et al.: Attention mechanisms in computer vision: a survey. Comp. Visual Media **8**(3), 331–368 (2022)
11. Kukkala, V.K., Thiruloga, S.V., Pasricha, S.: Latte: LSTM self-attention based anomaly detection in embedded automotive platforms. ACM Trans. Embedded Comput. Syst. **20**(5s), 1–23 (2021)
12. Ma, H., Zhang, L.: Attention-based framework for weakly supervised video anomaly detection. J. Supercomput. **78**(6), 8409–8429 (2022)
13. Nasaruddin, N., Muchtar, K., Afdhal, A., Dwiyantoro, A.P.J.: Deep anomaly detection through visual attention in surveillance videos. J. Big Data **7**(1), 1–17 (2020)
14. Li, Q., Yang, R., Xiao, F., Bhanu, B., Zhang, F.: Attention-based anomaly detection in multiview surveillance videos. Knowl.-Based Syst. **252**, 109348 (2022)
15. Du, Z., Zhang, G., Gan, J., Wu, C., Liu, X.: VadTR: video anomaly detection with transformer. In: 2022 5th International Conference on Advanced Electronic Materials, Computers and Software Engineering (AEMCSE), pp. 711–714. IEEE (2022)
16. Aslam, N.K., Narayanan, S., Kolekar, M.H.: Bidirectional motion learning using transformer based Siamese network for video anomaly detection (2023)
17. Pang, W., He, Q., Li, Y.: Predicting skeleton trajectories using a skeleton-transformer for video anomaly detection. Multimedia Syst. **28**(4), 1481–1494 (2022)
18. Monitoring Human Activity – Detection of Events. http://mha.cs.umn.edu/proj_events.shtml#crowd. Last accessed 15 Aug 2023
19. Avenue Dataset for Abnormal Event Detection. http://www.cse.cuhk.edu.hk/leojia/projects/detectabnormal/dataset.html last accessed 2023/08/14
20. Bianculli, M., et al.: A dataset for automatic violence detection in videos. Data in Brief 33, 106587 (2020)
21. Ullah, F.U.M., et al.: AI-assisted edge vision for violence detection in IoT-based industrial surveillance networks. IEEE Trans. Industr. Inf. **18**(8), 5359–5370 (2021)
22. Woo, S., Park, J., Lee, J.Y., Kweon, I.S.: CBAM: convolutional block attention module. In: Proceedings of European Conference on Computer Vision (ECCV), pp. 3–19 (2018)
23. Liu, Z., et al.: Swin transformer: hierarchical vision transformer using shifted windows. In: Proceedings of the IEEE/CVF International Conference on Computer Vision, pp. 10012–10022 (2021)
24. Nigam, S., Singh, R., Singh, M.K., Singh, V.K.: Multiview human activity recognition using uniform rotation invariant local binary patterns. J. Ambient. Intell. Humaniz. Comput. **14**(5), 4705–4725 (2022)
25. Ghadi, Y.Y., et al.: Extrinsic behaviour prediction of pedestrian via maximum entropy Markov model and graph-based features mining. Appl. Sci. **12**(12), 5985 (2022)
26. Alarfaj, M., et al.: Automatic anomaly monitoring in public surveillance areas. Intell. Autom. Soft Comput. **35**(3), 2655–2671 (2023)

27. Ullah, W., Ullah, A., Hussain, T., Khan, Z.A., Baik, S.W.: An efficient anomaly recognition framework using an attention residual LSTM in surveillance videos. Sensors **21**(8), 2811 (2021)
28. Ilyas, Z., Aziz, Z., Qasim, T., Bhatti, N., Hayat, M.F.: A hybrid deep network based approach for crowd anomaly detection. Multimed. Tools Appl. **80**, 24053–24067 (2021)
29. Abdullah, F., Jalal, A.: Semantic segmentation based crowd tracking and anomaly detection via neuro-fuzzy classifier in smart surveillance system. Arab. J. Sci. Eng. **48**(2), 2173–2190 (2023)
30. Aziz, Z., Bhatti, N., Mahmood, H., Zia, M.: Video anomaly detection and localization based on appearance and motion models. Multimed. Tools Appl. **80**(17), 25875–25895 (2021)
31. Sharif, M.H., Jiao, L., Omlin, C.W.: Deep crowd anomaly detection by fusing reconstruction and prediction networks. Electronics **12**(7), 1517 (2023)
32. Deepak, K., Chandrakala, S., Mohan, C.K.: Residual spatiotemporal autoencoder for unsupervised video anomaly detection. SIViP **15**(1), 215–222 (2021)
33. Khaire, P., Kumar, P.: A semi-supervised deep learning based video anomaly detection framework using RGB-D for surveillance of real-world critical environments. Forensic Sci. Int. Digit. Investig. **40**, 301346 (2022)
34. Ehsan, T.Z., Nahvi, M., Mohtavipour, S.M.: An accurate violence detection framework using unsupervised spatial-temporal action translation network. Vis. Comput. 1–21 (2023). https://doi.org/10.1007/s00371-023-02865-3
35. Yuan, H., Cai, Z., Zhou, H., Wang, Y., Chen, X.: Transanomaly: video anomaly detection using video vision transformer. IEEE Access **9**, 123977–123986 (2021)
36. Yang, M., et al.: Transformer-based deep learning model and video dataset for unsafe action identification in construction projects. Autom. Constr. **146**, 104703 (2023)
37. Lee, Y., Kang, P.: AnoViT: unsupervised anomaly detection and localization with vision transformer-based encoder-decoder. IEEE Access **10**, 46717–46724 (2022)
38. Ullah, W., Hussain, T., Ullah, F.U.M., Lee, M.Y., Baik, S.W.: TransCNN: hybrid CNN and transformer mechanism for surveillance anomaly detection. Eng. Appl. Artif. Intell. **123**, 106173 (2023)
39. Pillai, A., Verma, G.V., Sen, D.: Transformer based self-context aware prediction for few-shot anomaly detection in videos. In: 2022 IEEE International Conference on Image Processing (ICIP), pp. 3485–3489. IEEE (2022)
40. Sivalingan, H., Anandakrishnan, N.: Crowd localization and anomaly detection using video anomaly scoring network. Math. Stat. Eng. Appl. **72**(1), 825–837 (2023)
41. Sernani, P., Falcionelli, N., Tomassini, S., Contardo, P., Dragoni, A.F.: Deep learning for automatic violence detection: Tests on the AIRTLab dataset. IEEE Access **9**, 160580–160595 (2021)
42. Kumar, A., Khari, M.: Efficient video anomaly detection using variational autoencoder. In: 2023 International Conference on Communication System, Computing and IT Applications (CSCITA), pp. 50–55. IEEE (2023)

Development of LSTM-Based Sentence Generation Model to Improve Recognition Performance of OCR System

Jae-Jung Kim[1], Ji-Yun Seo[1], Yun-Hong Noh[2], Sang-Joong Jung[1],
and Do-Un Jeong[1(✉)]

[1] Dongseo University, 47 Jurye-ro, Sasang-gu, Busan 47011, Korea
{sjjung,dujeong}@dongseo.ac.kr
[2] Busan Digital University, 57 Jurye-ro, Sasang-gu, Busan 47011, Korea
yhnoh@bdu.ac.kr

Abstract. In recent development of deep learning algorithms, recurrent neural net-work models that can effectively reflect dependencies between input entities and LSTM models developed from them are being used in language models. In this study, a next sentence prediction model based on LSTM was implemented with the goal of improving the sentence recognition accuracy of OCR systems. The implemented model can ensure accurate recognition by generating the next sentence using a sentence generation model and comparing its similarity with the sentence entered the OCR system in case the OCR system does not recognize it accurately due to misrecognition or loss.

Keywords: LSTM · Sentence generator · OCR system

1 Introduction

In recent times, there has been a rapid surge in the digitization of analog documents across various industries, aimed at enhancing operational efficiency[1]. Analog documents are susceptible to loss and damage, making them challenging to manage. Consequently, research endeavors are actively underway in the field of Optical Character Recognition (OCR) systems to extract diverse information from images or scanner. OCR systems utilize optical recognition devices such as cameras and scanners to convert the text present in traditional analog documents into a computer-readable text format. During the process of digitizing analog documents using OCR technology, issues related to document damage and misinterpretation due to specific fonts are prevalent. Particularly, the failure to accurately distinguish paragraph or sentence boundaries often leads to incomplete recognition. When the extracted text fails to faithfully replicate the intended flow of the original document, it can convey incorrect meanings, necessitating substantial post-processing efforts, including time, manpower, and costs, for correction. In this research paper, we have implemented an LSTM-based neural network model to edit OCR results into a coherent and pristine form. The implemented model excels at identifying the flow

© The Author(s), under exclusive license to Springer Nature Switzerland AG 2024
B. J. Choi et al. (Eds.): IHCI 2023, LNCS 14532, pp. 65–69, 2024.
https://doi.org/10.1007/978-3-031-53830-8_7

and boundaries of sentences within the document, enabling the accurate reproduction of the intended content from the original document. Furthermore, it offers automated editing capabilities, thereby minimizing unnecessary expenses.

2   Related Works

LSTM models were developed to address the issue of long-term dependencies in recurrent neural network (RNN) models. LSTM models can predict future data by considering not only the immediate past data but also a wider temporal range of past data. The LSTM model takes sequential information one by one, stores it in internal nodes, and incorporates it into the model. This processing method is applied to learning data with strong correlations between previously inputted information and subsequently inputted information, such as language models. In a study by Anusha Garlapati et al. [2], natural language processing and bidirectional LSTM models were used to classify harmful ratings of comments on online social platforms with an accuracy of over 90%. Shivalila Hangaragi et al. [3], an LSTM model was employed for spelling correction in handwritten text recognized by an OCR system. LSTM models are being utilized not only for prediction but also as language models in various fields[4–7]. In this paper, we implemented an LSTM-based sentence generation model that utilizes the characteristics of LSTM to predict subsequent words when incomplete sentences are inputted, enabling the generation of complete sentences. Fig. 1 shows the structure of LSTM.

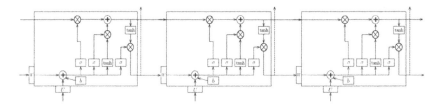

Fig. 1.   Structure of LSTM model.

3   LSTM-Based Sentence Generation Model

3.1   Training Dataset

In this paper, we aim to accurately predict sentences that have been incompletely recognized by OCR and similar systems using an LSTM-based sentence generation model. We used Kaggle's public dataset (New York Times Comments) as the training data for our sentence generation model. This dataset contains information about comments on articles published in the New York Times from January to May 2017 and January to April 2018, encompassing approximately 2 million comments and 9,000 news articles. Prior to use, we removed missing values and outliers labeled as 'Unknown' from the dataset. Furthermore, we performed text preprocessing on the English text found in articles and comments, which included converting all text to lowercase letters, removing special symbols, and eliminating unnecessary spaces. Subsequently, we tokenized sentences into individual words using Word Tokenizer to create a vocabulary set. Each word in this set was assigned an index value; based on these indexed pairs of words, we incrementally added new phrases to generate a list type. Finally, zero-padding and one-hot encoding were applied to ensure uniform length across all samples in the generated list. Fig. 2 shows an example of dataset creation.

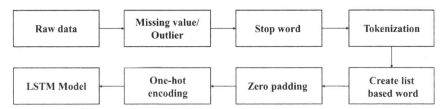

Fig. 2. Example of dataset creation.

3.2 Implemented LSTM Model

The LSTM model operates by sequentially inputting the input values into the Input Layer, allowing it to learn the context from the preceding word sequence in order to construct complete sentences. Subsequently, at the Output Layer, the model outputs probability values for each word in the vocabulary as the next word to follow an incomplete sentence. The architectural diagram of the implemented LSTM model is illustrated in Fig. 3. A new sentence is received at the Input Layer and passed on to the Embedding Layer. The Embedding Layer maps the received values into an embedding space of a specific dimensionality, producing word vectors for each word. These word vectors are then propagated through hidden layers for sequential data processing, enabling sentence prediction. At the Output Layer, probability values are generated for all words present in the dictionary. Ultimately, among these candidate words, it becomes possible to generate a new sentence by selecting one with highest probability.

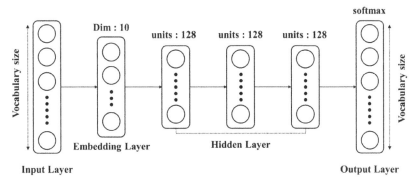

Fig. 3. Structure of implemented LSTM model.

4 Experiments and Results

The sentence generation model was trained using the cross-entropy loss function, and the choice of optimization function was determined through experiments. Three models were created using Adam, RMSProp, and SGD as optimization functions, and the loss values and perplexity were evaluated during training. Perplexity is a metric that evaluates how accurately a language model can generate sentences, with lower values indicating better performance in predicting and generating sentences. The evaluation results are shown in Table 1, confirming that the Adam function performed the best.

Table 1. Performance results according to Optimizer.

Optimizer	Loss	Perplexity
Adam	1.065	1.932
RMSProp	1.078	2.133
SGD	1.876	8.165

The second experiment was a qualitative evaluation of the sentence generation model. The evaluation focused on identifying errors in the sentences generated by the model using the Adam optimizer. The generated sentences are presented in Table 2. Upon review, it was confirmed that the generated sentences did not contain any grammatical errors, such as improper noun combinations, verb usage, or sentence endings.

Table 2. Result of sentence creation using LSTM model.

Keyword	Generated sentence
I completely	I completely agree with the author's perspective on this issue. It is crucial for us to address climate change and take immediate action to protect our planet.
The article	The article raises important questions about income inequality and its impact on society. We need to find sustainable solutions that promote fairness and equal opportunities for all.
I appreciate	I appreciate the author's in-depth analysis of the current political landscape. It is essential for us as citizens to stay informed and actively participate in shaping our democracy.
The piece	The piece offers a fresh perspective on the challenges faced by the education system today. We must prioritize investment in quality education to empower future generations.

5   Conclusion

In this paper, an LSTM-based sentence generation model was implemented to accurately predict incomplete sentences recognized by OCR and similar systems. An open dataset was used as the training data for the implemented system, and the choice of optimization function was determined through performance evaluation. Three functions, Adam, RMSProp, and SGD, were compared based on their loss values and perplexity scores. Among them, Adam achieved a loss value of 1.065 and a perplexity score of 1.932, making it the optimal optimization function for the model. Subsequently, a qualitative evaluation was conducted on the generated sentences. The evaluation results confirmed that the generated sentences did not contain any grammatical errors such as improper noun combinations, verb usage, or sentence endings. As a direction for future research, we plan to conduct studies to enhance the language generation capabilities of sentence generation models, utilizing machine reading comprehension and OpenAI, in order to improve the performance of OCR systems.

Acknowledgement. This work was supported by the Technology development Program(RS-2023-00224316) funded by the Ministry of SMEs and Startups(MSS, Korea)

References

1. Moudgil, A., Singh, S., Gautam, V.: An Overview of Recent Trends in OCR Systems for Manuscripts. Cyber Intelligence and Information Retrieval: Proceedings of CIIR **2021**, 525–533 (2022)
2. Garlapati, A., Malisetty, N., Narayanan, G.: Classification of Toxicity in Comments using NLP and LSTM. In 2022 8th International Conference on Advanced Computing and Communication Systems (ICACCS), 1, 16–21, 2022
3. Hangaragi, S., Pati, P. B., Neelima, N.: Accuracy Comparison of Neural Models for Spelling Correction in Handwriting OCR Data. In Proceedings of Fourth International Conference on Communication, Computing and Electronics Systems: ICCCES 2022(Springer Natue Singapore), 229–239, (2023)
4. Cui, Y., Liu, L.: Investor sentiment-aware prediction model for P2P lending indicators based on LSTM. Plos one, 17(1), e0262539(2022)
5. Tuli, P., Patra, J.P.: Symbol question conversion in structured query language using fuzzy with deep attention based rain LSTM. Multimedia Tools and Applications **81**(22), 32323–32349 (2022)
6. Jian, L., Xiang, H., Le, G.: Lstm-based attentional embedding for English machine translation, Scientific Programming. (2022)
7. Wang, H., Li, F.: A text classification method based on LSTM and graph attention network. Connect. Sci. **34**(1), 2466–2480 (2022)

Satellite Imagery Super Resolution Using Classical and Deep Learning Algorithms

T. A. Kuchkorov$^{(\boxtimes)}$ [ID], J. X. Djumanov[ID], T. D. Ochilov[ID], and N. Q. Sabitova

Tashkent University of Information Technologies Named After Muhammad Al-Khwarizmi, Tashkent 100084, Uzbekistan
t.kuchkorov@tuit.uz
http://www.tuit.uz

Abstract. The single image super resolution is very essential and needed in applications like image analysis, recognition, classification, better analysis and diagnosis of complex structured images. The applications in the different tasks of satellite imagery, medical image processing, CCTV image analysis, and video surveillance where a zoom is required, the super resolution becomes crucial in this case for a particular region of interest. In this paper, we have analyzed both classical and deep learning algorithms of satellite image enhancement using super resolution approach. The main focus of this paper lies in the comparison results of existing image enhancement algorithm such us bicubic interpolation, discrete wavelet transforms (DWT) based algorithms and deep learning based EDSR and WDSR architectures. These Deep learning (DL)- based improvement technique is presented to increase the resolution of the low-resolution satellite images. When compared to the existing classical methodologies, the DL-based algorithms significantly improve the PSNR while appropriately enhancing the satellite image resolution.

Keywords: Super resolution · image enhancement · deep learning · low-resolution · high-resolution · satellite imagery

1 Introduction

The Copernicus program and its Sentinel missions have made Earth observation data more widely available and more reasonably priced. With the use of Sentinel- 2 multispectral images, every location of earth may be freely monitored around every 5–10 days. According to the Sentinel missions, spatial resolution of satellite images for the RGBN (RGB + Near-infrared) bands will be different like 60, 20 or 10 m, which is enough for some tasks but inadequate for others [1]. Single image super resolution (SISR) is the process of estimating a high-resolution (HR) version of a low-resolution (LR) input image. This is a well-studied problem, which comes up in practice in many applications, such as zoom-in of still and text images, conversion of LR images/videos to high-definition screens, and more [2]. The desired results cannot be achieved through low resolution images, therefore low, medium and high-resolution images have different

B. J. Choi et al. (Eds.): IHCI 2023, LNCS 14532, pp. 70–80, 2024.
https://doi.org/10.1007/978-3-031-53830-8_8

possibilities and image resolution is important in performing various tasks regarding accuracy. In this case, step-by-step image quality improvement increases the accuracy of satellite image analysis tasks and decision-making processes without any additional costs. Since, high resolution images are very costly and difficult to use and implement for comprehensive land monitoring applications. So, in this article, existing classical methodologies and deep learning based algorithms for satellite image enhancement and super resolution are reviewed and conclusions are presented based on comparative analysis.

2 Background

2.1 Image Resolution

There are several resolution options available from satellite imagery companies including Planet, Sentinel, DigitalGlobe, and others. Most crucially, how can we tell if low-resolution images will do or if high-resolution ones are required? There are three different types of resolution namely low, medium and high (see Fig. 1). Imagery with low and medium resolutions has a pixel density of 60 m and 10–30 m, respectively [3]. While satellite images covering larger land areas serve their purpose, they fall short when it comes to detecting small features within those images. Platforms like Landsat and Sentinel offer low and medium- resolution imagery for land and water monitoring. For example, Landsat data has a 30-m resolution, meaning each pixel represents a 30 m × 30 m ground area. While it can cover entire cities, it lacks the detail needed to distinguish specific buildings or vehicles.

Conversely, higher resolution means smaller pixel sizes and more detailed images. The Planet satellites, like SkySat with a resolution of 30 cm–50 cm/pixel and PlanetScope with a resolution of 3–5 m/pixel, offer the highest resolutions. This means each pixel can capture objects that are as small as 30 cm to 50 cm in height and 3 to 5 m in length. However, obtaining these high-resolution images can be challenging, as they require satellites with high to very high-resolution capabilities. Consequently, improving the quality of low- and medium- resolution satellite images and upscaling them to high resolution emerges as a critical concern in this context [4].

2.2 Interpolation for Image Enhancement

Finding the values of a continuous function from discrete samples is the process of interpolation. Interpolation is used in image processing for a wide range of purposes, such as image enlargement or reduction, subpixel image registration, the correction of spatial distortions, and image decompression, among others [5]. Despite the fact that regularly used linear methods like pixel duplication, bilinear interpolation, and bicubic convolution interpolation have advantages in terms of simplicity and speed of implementation. The edge preservation condition makes it difficult for the conventional linear interpolation algorithms to perform adequately. To keep edges crisp, some nonlinear interpolation methods were suggested. In general, nonlinear interpolation techniques outperform linear techniques at maintaining edges [6].

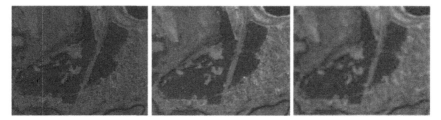

Fig. 1. High, medium and low resolution image samples.

2.3 DWT and Noise Removal Techniques

In the realm of image processing, wavelets have found extensive application. They enable the representation of images in terms of their local spatial and frequency characteristics. While global frequency characteristics can be obtained using the Fourier transform and DCT, they do not provide insights into local frequency characteristics. Wavelet transforms address this issue effectively. The discrete wavelet transform (DWT) employs discrete sampling of wavelets for both numerical and functional analysis, allowing it to capture both frequency and time information.

The discrete wavelet transform (DWT) works by decomposing signals into sub-bands characterized by smaller bandwidths and slower sample rates. These sub-bands include Low-Low (LL), Low-High (LH), High-Low (HL), and High- High (HH). At one level of transformation, four sub-bands are obtained: the first low-pass sub-band, denoted as LL, represents a coarse approximation of the source image, while the remaining three high-pass sub-bands, namely HL (horizontal details), LH (vertical details), and HH (diagonal details), capture image details in various orientations [8].

A novel image enhancement approach based on DWT is introduced by G. Saravanan et al. [8]. This method effectively combines smoothing and sharpening via a piecewise non-linear filter technique. It begins by converting the RGB (Red, Green, and Blue) values of each pixel in the original image into the HSV (Hue, Saturation, and Value) color space. The image enhancement architecture, as shown in the following figure, applies wavelet transform to the luminance value of the V component, decomposing the input image into four sub-bands using Discrete Wavelet Transform (DWT) [8].

Image enhancement result well be better after preprocessing and denosing process. In [11] paper deblurring techniques are analyzed well as a domain of DWT and Fourier transform algorithms. The blur in image may seem as irreparable operation, as each pixel in image is turned to spot and every-thing is mixed up. If blur cover whole are of photo, we will get flat color all over the image. However, we do not lose information, but we get redistributed information in accordance with some rules. Indeed, in the context of blurred images, a common characteristic is the smoothness, which implies that the edges within the image are not clearly defined. This smoothing effect is typically achieved using a filter known as a low-pass filter. The term "low-pass" is used because this filter permits low-frequency components to pass through while suppressing high-frequency components. This is especially relevant around edges, where pixel values change rapidly. In the case of a blurred image, the smoothness of the image necessitates the removal or filtering out of high-frequency details (Fig. 2).

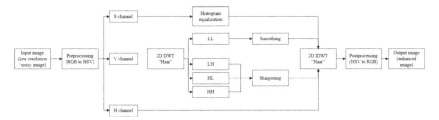

Fig. 2. DWT based image enhancement architecture.

Gaussian blur is produced using the Gaussian function and is primarily employed as a noise removal filter in image processing.

Motion blur occurs in images when there is camera movement or object motion during the exposure time. It is defined by its point spread function (PSF), and the PSF's parameters are intricately connected to the nature of the motion itself.

However, in real-world scenarios, motion can be considerably more complex than this straightforward linear motion blur. Related works shows that for image degraded by motion blur with one dimensional kernel at an angle is defined as following way:

$$g(x, y) = f^a * m \cong \sum_{k=0}^{K-1} m_k * f(x + k \cos(\alpha) * y + k \sin(\alpha)) \tag{1}$$

Wiener filter is a widely used and one of the earliest methods that takes into consideration the presence of noise in an image [9]. It treats both the image and the noise as random processes and seeks a value for f', a distortion-free version of the image f, which minimizes the mean square deviation between these values.

3   Deep Learning Based Architectures for Image Super Resolution

3.1   High Level Architecture

The predominant approach in contemporary super-resolution models involves the acquisition of the majority of the mapping function within the low-resolution space. Subsequently, one or more upsampling layers are appended towards the conclusion of the network. This particular upsampling method is commonly referred to as "post-upsampling super resolution," as illustrated in Fig. 5. The up-sampling layers themselves are learnable and undergo training alongside the preceding convolution layers in a holistic end-to-end fashion.

In earlier methodologies, a different strategy was employed, where the initial step involved upsampling the low-resolution image through an operation previously defined. Subsequently, the mapping was learned within the high-resolution (HR) space. This approach is referred to as "pre-upsampling." However, a limitation of this method is the resultant increase in the number of learnable parameters, necessitating more substantial computational resources for training deeper neural networks.

3.2   Enhanced Deep Residual Networks for Single Image Super-Resolution(EDSR)

EDSR [14] is a model of the winner in the NTIRE2017 Super-Resolution Challenge. The architecture of the EDSR network is derived from the SRResNet by making specific alterations. One notable adjustment is the removal of batch normalization within the ResNet layers. Furthermore, another departure from the original residual networks is the elimination of the ReLU activation layer that typically follows the residual networks (Fig. 3).

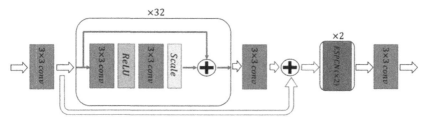

Fig. 3. Residual block of EDSR.

Additionally, the model does not incorporate residual scaling layers as it relies solely on 64-dimensional features for each convolutional layer. The overall structure of EDSR residual network is illustrated in Fig. 5.

The inclusion of the batch normalization (BN) layer in SRResNet was originally adopted from the original ResNet architecture. However, given that the original ResNet was primarily engineered to tackle intricate computer vision challenges such as classification and detection, its direct application to low-level tasks like super-resolution often resulted in suboptimal outcomes. The decision to eliminate batch normalization layers in this context has yielded significant benefits, notably a notable reduction in GPU memory consumption. Batch normalization layers consume a comparable amount of memory to convolutional layers, and this adjustment has resulted in a noteworthy 40 percent reduction in GPU memory usage during the training process, as compared to the SRResNet.

Simultaneously, given that the BN layer consumes an equivalent amount of memory as the preceding convolutional layer, its removal serves as a means to conserve memory resources.

3.3   Wide Activation for Efficient and Accurate Image Super-Resolution(WDSR)

WDSR [16] is a super-resolution framework proposed by JiaHui Yu et al. in 2018. Simultaneously, it's worth noting that the image super-resolution approach built upon the principles of WDSR achieved top-ranking positions in all three real tracks of the NTIRE 2018 challenge, securing the first place. WDSR stands as an enhanced algorithm, grounded in the CNN optimization model. Within the realm of CNN-based super-resolution algorithms, optimization can be pursued in the following four key directions (Fig. 4).

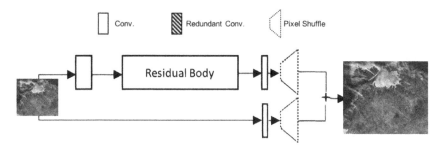

Fig. 4. Residual block of WDSR.

The primary modification introduced in WDRS pertains to the upsampling algorithm. Conventional upsampling methods like bilinear interpolation and deconvolution prove inadequate for restoring the fine texture of low-resolution images, often leading to noticeable artifacts. In 2016, a novel image super-resolution technique, the pixel shuffle convolution algorithm, was introduced. By leveraging this innovative pixel shuffling method, which periodically inserts low-resolution features into specific locations within low-resolution images, the loss of details due to artifacts is significantly mitigated. Secondly, WDSR demonstrates that weight normalization alone, without relying on data-dependent initialization, contributes to enhanced accuracy in deeper WDSR models. Lastly, WDSR further refines EDSR's normalization technique while partially modifying the EDSR structure. In this context, it involves the removal of redundant convolutional layers, which serves to reduce memory consumption and accelerate computational speed.

3.4 Deep Alternating Network(DAN)

In this paper [23], an end-to-end algorithm for blind SR is presented. This algorithm relies on alternating optimization, wherein both parts are realized through convolutional modules, denoted as Restorer and Estimator. The alternating process is expanded to establish a network that is trainable end-to-end. Through this approach, information from both LR and SR images can be harnessed by the Estimator, thereby simplifying the task of blur kernel estimation. DAN algorithm has several improvements. Firstly, an alternating optimization algorithm is employed for the estimation of the blur kernel and the restoration of the SR image in the context of blind SR within a single network (DAN). This promotes compatibility between the two modules and is expected to yield superior final results compared to previous two-step solutions. Next, two convolutional neural modules are devised, capable of being alternated iteratively and subsequently unfolded to construct an end-to-end trainable network, free from the need for any pre/post-processing. This network is easier to train and offers higher speed in comparison to previous two-step solutions. Lastly, through extensive experiments conducted on synthetic datasets and real-world images, it is demonstrated that their model excels in comparison to state-of-the-art methods and is able to generate more visually appealing results at significantly higher speeds.

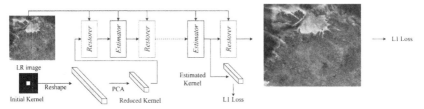

Fig. 5. Architecture of DAN

4 Evaluation Metrics

PNSR. The Peak Signal-to-Noise Ratio (PSNR) is an engineering metric that quantifies the ratio between the highest attainable power of a signal and the power of disruptive noise that impacts the faithfulness of its representation. Given the extensive dynamic range of many signals, PSNR is typically expressed logarithmically, using the decibel scale. It finds widespread application in evaluating the quality of reconstruction in images and videos subjected to lossy compression, and its noisy approximation K, MSE is defined as:

$$MSE = \frac{1}{mn} \sum_{i=0}^{m-1} \sum_{j=0}^{n-1} [I(i,j) - K(i,j)]^2 \tag{2}$$

The PSNR (in dB) is defined as:

$$PSRN = 10 * \log_{10}\left(\frac{MAX_I^2}{MSE}\right) = 20 * \log_{10}\left(\frac{MAX_I}{\sqrt{MSE}}\right)$$
$$= 20 * \log_{10}(MAX_I) - 10 * \log_{10}(MSE) \tag{3}$$

IFC. Information Fidelity Criterion (IFC) represents a full-reference metric employed for the evaluation of image quality [17]. This metric primarily relies on the analysis of natural scene statistics. Empirical research has substantiated that these statistics, characterizing the space occupied by natural images, can be effectively modeled using diverse approaches, including Gaussian scale mixtures. Importantly, any form of image distortion tends to disrupt these inherent statistics of natural scenes, consequently leading to the perception of unnatural images.

LPIPS (Learned Perceptual Image Patch Similarity), as detailed in reference [18], constitutes a learned metric designed for the assessment of image quality with reference to a known standard. To elaborate, LPIPS [18] is derived by quantifying the difference between the reference image and the test image within a deep feature space. This deep feature space has been shown to align closely with human perceptual judgments, enhancing its applicability in evaluating image quality.

5   Comparisons and Challenges

In this section, we conduct a comparative analysis between both traditional image upscaling methods and a selection of representative Single Image Super Resolution (SISR) techniques. Specifically, we consider several algorithms, namely, Bilinear interpolation [94], Bicubic interpolation, a DWT-based algorithm, EDSR [14], WDSR[16], and DAN [96], encompassing a variety of methodologies. We use the official models provided by the authors for these super-resolution methods. The experiments are carried out on a computer running the Ubuntu operating system, equipped with an Intel i9 11900 K, 64 GB RAM, and a GeForce RTX 3070 Ti. The performance of these methods is evaluated using image quality assessment metrics such as PSNR, IFC [11], and LPIPS [12]. Traditional methods tend to create some artefacts in the upscaling but their inference speed are quite short compared to dL-based algorithms. Besides accuracy, the size of the model and its execution speed also hold significance for super-resolution algorithms. Notably, due to its larger number of parameters, EDSR lags behind WDSR and DAN in terms of inference speed. The results presented in this section may slightly be different from official repositories and papers owing to variations in test environment settings, hyperparameters, and other factors. It's widely acknowledged that achieving a completely equitable comparison among these competing methods is challenging, since there a lot of factors to consider (Table 1).

Table 1. Comparisons PSNR value of analyzed image enhancement algorithms.

Type of Image enhancement	PSNR	IFC	LPIPS
Bilinear interpolation	21.23	1.942	0.3127
Bicubic interpolation	22.08	2.034	0.2952
DWT based algorithm	22.92	2.128	0.2601
DL based EDSR	28.87	3.881	0.1949
DL based WDSR	29.13	4.109	0.1961
DAN	30.19	4.381	0.1713

The Fig. 6 shows the results of bicubic interpolation, DL-based EDSR, WDSR and DAN for real world satellite image. DAN has more accurate result regarding deeper feature upsampling compared to the other algorithms.

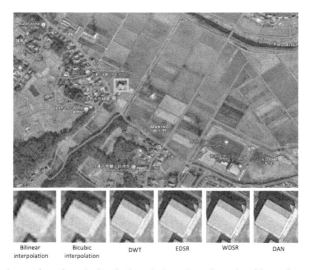

Bilinear   Bicubic   DWT   EDSR   WDSR   DAN
interpolation   interpolation

Fig. 6. Comparison of analyzed classical and deep learning algorithms for super-resolution satellite images.

6   Comparisons and Challenges

This article provides a concise overview of both classical and deep learning-based approaches to image enhancement and super-resolution. Satellite images, with their intricate features and complex structures, have been the focus of analysis and comparison. The findings reveal that classical techniques exhibit limitations when it comes to achieving superior image enhancement and quality.

To illustrate this, a sample satellite image sourced from Google Earth Engine, featuring a centrally located building, was selected as the target for up-sampling. Traditional methods such as bilinear and bicubic interpolation tend to introduce artifacts when upscaling low-resolution images. In contrast, deep learning-based algorithms significantly reduce these artifacts. Notably, the DAN algorithm yields satisfactory results in the upscaling of complex-structured satellite images than ESDR and WDSR.

References

1. Galar, M., Sesma, R., Ayala, C., Albizua, L., Aranda, C.: Super-resolution of Sentinel-2 images using convolutional neural networks and real ground truth data. Remote Sensing **12**(18), 2941 (2020). https://doi.org/10.3390/rs12182941
2. Romano, Y., Isidoro, J., Milanfar, P.: RAISR: rapid and accurate image super resolution. IEEE Trans. Comput. Imag. **3**(1), 110–125 (2017). https://doi.org/10.1109/TCI.2016.2629284
3. Sentinel-2 MSI User Guide, https://sentinel.esa.int/web/sentinel/user-guides/sentinel-2-msi. Copernicus, Last accessed July 10, 2022
4. High-Res vs Mid-Res vs Low-Res: What Type of Satellite Imagery Fits the Bill, https://med ium.datadriveninvestor.com/high-res-vs-mid-res-vs-low-res-whattype-of-satellite-imagery-fits-the-bill-78cbd45f5d79. Published in Data Driven Investor, 11 Mar 2021

5. Parth, B., Ankit, S.: Image Enhancement Using Various Interpolation Methods **2**(4), 799–803 (2012). ISSN 2249-9555
6. Liyakathunisa, Ravi Kumar, C.N.: Advances in super resolution reconstruction of noisy, blurred and aliased low resolution images. Int. J. Comput. Intell. Res. **4**(2), 215–222 (2010)
7. Khaire, G.M., Shelkikar, R.P.: Resolution Enhancement of images with Interpolation and DWT-SWT Wavelet Domain Components. Int. J. Appl. Innov. Eng. Manage. (IJAIEM) **2**(9), 242–248 (2013). ISSN 2319-4847
8. Saravanan, G., Yamuna, G., Vivek, R.: A color image enhancement based on discrete wavelet transform. International Journal of Computer Applications® (IJCA). Proceedings on National Conference on Emerging Trends in Information and Communication Technology (NCETICT), pp. 1–8 (2013). ISSN 0975-8887
9. Sharma, A., Khunteta, A.: Satellite Image Enhancement using Discrete Wavelet Transform, Singular Value Decomposition and its Noise Performance Analysis. International Conference on Micro-Electronics and Telecommunication Engineering 2016, ICMETE, pp. 594–599. IEEE (2016). https://doi.org/10.1109/ICMETE.2016.32
10. Kuchkorov, T.A., Urmanov, Sh.N., Nosirov, Kh.Kh., Kyamakya, K.: Perspectives of deep learning based satellite imagery analysis and efficient training of the U-Net architecture for land-use classification. Developments of Artificial Intelligence Technologies in Computation and Robotics: Proceedings of the 14th International FLINS Conference 2020, (FLINS), pp. 1041–1048 (2020). https://doi.org/10.1142/9789811223334_0125
11. Kuchkorov, T.A., Sabitova, N.Q.: Deblurring techniques and fuzzy logic methods in image processing. Descend. Muhammad al-Khwarizmi Sci. Pract. Info. Analy. J. **1**(19), 12–16 (2019). ISSN 2181-9211
12. Poulose, M.: Literature survey on image deblurring techniques. Int. J. Comp. Appl. Technol. Res. **3**, 286–298 (2013)
13. Yang, W., Zhang, X., Tian, Y., Wang, W., Xue, J.-H., Liao, Q.: Deep learning for single image super-resolution: a brief review. IEEE Trans. Multimedia **21**(12), 3106–3121 (2019). https://doi.org/10.1109/TMM.2019.2919431
14. Lim, B., Son, S., Kim, H., et al.: Enhanced deep residual networks for single image super-resolution. IEEE Conference Computer Vision and Pattern Recognition Workshops, pp. 136–144. USA (2017)
15. Kuchkorov, T., Ochilov, T., Gaybulloev, E., Sabitova, N., Ruzibaev, O.: Agro-field Boundary Detection using Mask R-CNN from Satellite and Aerial Images. In: 2021 International Conference on Information Science and Communications Technologies (ICISCT), pp. 1–3. Tashkent, Uzbekistan (2021). https://doi.org/10.1109/ICISCT52966.2021.9670114
16. Yu, J., et al.: Wide activation for efficient and accurate image super-resolution. arXiv preprint arXiv:1808.08718 (2018)
17. Sheikh, H.R., Bovik, A.C., De Veciana, G.: An information fidelity criterion for image quality assessment using natural scene statistics. IEEE Trans. Image Process. **14**(12), 2117–2128 (2005)
18. Zhang, R., Isola, P., Efros, A.A., Shechtman, E., Wang, O.: The unreasonable effectiveness of deep features as a perceptual metric. In: Proceedings of the IEEE Conference on Computer Vision and Pattern Recognition, pp. 586–595. CVPR (2018)
19. Safarov, F., et al.: Improved agricultural field segmentation in satellite imagery Using TL-ResUNet architecture. Sensors **22**, 9784 (2022). https://doi.org/10.3390/s22249784
20. Djumanov, J., Abdurashidova, K., Rajabov, F., Akbarova, Sh.: Determination of Characteristic Points Based on Wavelet Change of Electrocardiogram Signal. ICISCT, 1–5 (2021)
21. Kutlimuratov, A., Khamzaev, J., Kuchkorov, T., Anwar, M.S., Choi, A.: Applying enhanced real-time monitoring and counting method for effective traffic management in tashkent. Sensors **23**, 5007 (2023). https://doi.org/10.3390/s23115007

22. Nasimova, N., Muminov, B., Nasimov, R., Abdurashidova, K., Abdullaev, M.: Comparative Analysis of the Results of Algorithms for Dilated Cardiomyopathy and Hypertrophic Cardiomyopathy Using Deep Learning. ICISCT, 1–5 (2021)
23. Luo, Z., Huang, Y., Li, S., Wang, L., Tan, T.: Unfolding the alternating optimization for blind super resolution. In: Conference on Neural Information Processing Systems (NeurIPS) (2020)

Traffic Sign Recognition by Image Preprocessing and Deep Learning

U. R. Khamdamov[1]([✉]) [ID], M. A. Umarov[1] [ID], S. P. Khalilov[1] [ID], A. A. Kayumov[2], and F. Sh. Abidova[3]

[1] University of Management and Future Technologies, Tashkent, Uzbekistan
utkir.hamdamov@mail.ru
[2] Samarkand Branch of Tashkent University of Information Technologies named after Muhammad al Khwarizmi, Tashkent, Uzbekistan
[3] Tashkent University of Information Technologies named after Muhammad al Khwarizmi, Tashkent, Uzbekistan

Abstract. Due to the improvement in the car manifacture, the rate of road traffic accidents is increasing. To solve these problems, there is loads of attention in research on the development of driver assistance systems, where the main innovation is traffic sign recognition (TSR). In this article, a special convolutional neural network model with high accuracy compared to traditional models is used for TSR. The Uzbek Traffic Sign Dataset (UTSD) applied in the zone of Uzbekistan was created, consisting of 21.923 images belonging to 56 classes. We proposed a parallel computing method for real-time processing of video haze removal. Our utilization can process the 1920 × 1080 video series with 176 frames per second for the dark channel prior (DCP) algorithm. 8.94 times reduction of calculation time compared to the Central Processing Unit (CPU) was achieved by performing the TSR process on the Graphics Processing Unit (GPU). The algorithms used to detect traffic signs are improved YOLOv5. The results showed a 3.9% increase in accuracy.

Keywords: Traffic sign recognition · deep learning · image dehazing · UTSD dataset · parallel processing · data augmentation

1 Introduction

In the world, special importance is attached to the preprocessing of images, and the development of algorithms for identifying, extracting, and classifying important objects from images. Traffic sign recognition acts as a visual guide to help drivers navigate the road infrastructure [1]. The goal is to detect a road sign and process images in real-time through a surveillance camera installed in a car. It is necessary to ensure road safety for human life, which is considered the most important factor [2]. During the last 10 months of 2022, a total of 7.681 accidents occurred in Uzbekistan. 1.964 people died, 6.886 people were injured, and 54% were caused by subjective factors (in 23% of cases, speed increase). From the point of view of the issue of mathematical and software development of intelligent systems for rapid monitoring of situations, the United States, Germany, Russia, India, China, Japan, and other developed countries are addressing to solve the theoretical and practical problems of developing traffic sign identification and classification algorithms [3]. Increasing both economic and social efficiency by solving such issues with the help of artificial intelligence technologies is one of the urgent issues of today.

Today, Google, Uber, Ford, Tesla, Mercedes-Benz, Toyota, and many other multinational companies are conducting scientific research using modern technologies [4]. To present the state in a process of scientific research, a simple search was conducted to find published and cited articles in the journals of the Scopus database on the term "road sign detection and recognition". The results showed a relatively rapid increase in the number of publications and citations for this term. TSR from images consists of two main stage, the first is image preprocessing for video haze removal and the second is traffic sign detection from video [5]. Noise appears in the image as a result of insufficient light transmission (fog, cloud) and the occurrence of various natural phenomena [6]. Such images make it difficult to recognize objects in computer vision problems and require preprocessing of the images (Fig. 1). He et. All proposed the DCP algorithm which durable the haze concentration is high [7]. However, during processing, it is difficult to remove noise from the image by DCP algorithm for high-definition videos. In the research, the task of accelerating the fog removal algorithm using parallel processing on GPU was set. We use widely used deep learning algorithms to recognize traffic signs from videos. In 2015, R. Joseph proposed the YOLO algorithm for one-step object detection [8]. We improve the YOLOv5 algorithm, which can detect objects in actual time.

Fig. 1. Traffic signs in a foggy weather.

The remaining of this article is structure as follows: Sect. 2 devolopes associated works about the traditional approach and deep learning-based method. Section 3 discusses the details of the proposed methodology to recognize the traffic signs efficiently. In Sect. 4, the results and analysis are extends. The conclusion is described in Sect. 5.

2 Related Works

2.1 Traditional Approach

Currently, due to the rapid development of the automobile industry, most of the automobile manufacturing countries such as Germany,the United States, China, and Japan have relevant research institutes that research automated control systems [9]. Among them, traffic sign recognition systems from images are an important part, and the main purpose of the system is to help drivers prevent road accidents and perform related tasks. Such research is crucial for intelligent vehicles [10, 23]. Object recognition approaches on traditional Hue Saturation Value color models based on image segmentation and contour analysis methods were studied.

In 2009, Xiamen University of China proposed a Histogram of Oriented Gradients for visual feature detection and a Support Vector Machine for classification in TSR research [11]. The detection rate reached 88.3% and still, it is not enough. In 2010, the Massachusetts Institute of Technology developed a traffic sign recognition system that uses color boundary segmentation and principal component analysis algorithms to recognize objects. The detection rate of this system reached 91.2% [12]. However, the level of detection of traffic signs in difficult weather conditions of the system is low and cannot face the requirements of real-time. In addition, object detection algorithms based on shape features can be used or combined. In such methods, the noise in the images makes the level of object detection difficult and does not allow improvement.

2.2 Deep Learning Based Methods

Object detection algorithms based on deep learning CNN (Convolutional Neural Network) are currently Single Shot MultiBox Detector (SSD), Region-based Fully Convolutional Networks (R-FCN), Fast R-CNN, Faster R-CNN, and YOLO (You Only Look Once) is used in recent advances in neural network architectures and showed high results [13]. There are many types of algorithms based on CNN for detecting traffic signs, but they can be divided into two main types: one-stage and two-stage models. Two phase object detection models are R-CNN, Faster R-CNN, SPP Net, and FPN-based methods. These work by moving a window across the image and predicting object connectivity and objectivity estimates at each position [14]. One-stage object detection models are regression or classification-based methods such as SSD, Retina Net, and YOLO. One-shot detectors like this do not use a window, instead, they predict the bounding box and object directly from the feature map in one pass, but in some cases at multiple scales [15, 22].

SSD type detectors are usually slightly more accurate, but much faster than RCNN networks. YOLO is one of the most popular models in this category of object detection, it is fast, reliable, and accurate [16]. R. Zhang proposed an improved YOLOv5 model for detecting traffic signs and the mAP increased from 90.2% to 92.5% [17]. Xiang Gao proposes an improved traffic sign detection algorithm based on the Faster R-CNN algorithm [18]. Then, climb the recognition precision of distant road signs (90.6% on a sunny and 86.9% on a foggy day). The YOLO model shows good results in terms of accuracy and speed of object detection from images.

3 Our Methodology

3.1 Dark Channel Prior Based Image Dehazing

In computer vision problems, the image in different weather is represented by the following formula (1)-(3):

$$I(x, y) = J(x, y)t(x, y) + A(1 - t(x, y)) \tag{1}$$

where $I(x, y)$ is the input image, $J(x, y)$ is the scenery image, A is the global atmospheric lights, $t(x, y)$ is the initial transmission describing the portion of the ligh [19].

Generating a quality image is done by reconstructing using the $I(x, y)$ and A.

Creating a dark channel of input image $I(x, y)$ is done using the following formula:

$$I^d(x, y) = min_{c \in \{r,g,b\}} \left(min_{(x,y) \in B_i} (I_c(x, y)) \right) \tag{2}$$

where $I_c(x, y)$ is a color channel of $I(x, y)$, $B_i (i = \overline{1, k})$ is a local patch centered, $I^d(x, y)$ is the dark channel of input image.

The execution sequence of the image quality improvement algorithm based on the processing of the dark channel for the input image is as follows:

Step 1. The gray image of an input image is generated using the following formula:

$$I(x, y) = 0.299*I_r + 0.587*I_g + 0.114*I_b \tag{3}$$

where $I(x, y)$ is the input image, I_r is the red channel of the image, I_g is the green channel, I_b is the blue channel.

Step 2. $I^d(x, y)$ - formation of the dark channel in the following formula (2).

Step 3. Based on the dark channel and the input image, highest 0.1% pixel value indexed of the dark channel to determine the atmospheric light intensity. Then, A is taken as the average value of the indexed pixels in the input image.

Step 4. Based on the dark channel and initial transmission $\hat{t}(x, y)$ is determined using the following formula:

$$\hat{t}(x, y) = 1 - \omega * min_{c \in \{r,g,b\}} \left(min_{(x,y) \in B_i} \frac{I_c(x, y)}{A_c} \right) \tag{4}$$

where the size of the image piece was set as $\Omega = 15*15$, $\omega = 0.95$.

Step 5. Creating a quality image based on the Guided filter and (4) is given in the following formula:

$$t(x, y) = \overline{a}(x, y) * \hat{t}(x, y) + \overline{b}(x, y) \tag{5}$$

Step 6. Restore the scenery image is performed by the following formula:

$$J(x, y) = \frac{I(x, y) - A}{max(t(x, y), t_0)} + A \tag{6}$$

where $t_0 = 0.1$.

Table 1. An indicator of the time spent to improve image quality on the CPU.

Steps (ms)	800*600	1280*720	1600*720	1920*1080
Create dark channel of input image	1.25	2.14	2.60	6.56
Approximate atmospheric lights A	0.011	0.018	0.020	0.050
Estimation the initial transmission	2.19	5.84	7.56	17.71
Purify the transmission with guided filter	4.59	9.07	11.42	26.27
Repair the scenery image	0.03	0.05	0.06	0.15
Total	8.071	17.118	21.66	50.74

Table 1 showed the results of the time consumption analysis for the steps of the above algorithms on the central processor.

As can be seen in the Table 1 above, the most time spent to computation the initial transmission and cleance the transmission with guided filter [21].

The minimum filter is used to create a dark channel of the input image and to calculate the initial transmission. The main process of the guided filter used to improve image quality is the mean filter. The implementation of the parallel minimum filtering algorithm for calculating the initial transmission is carried out using the following formulas based on the capabilities of the graphics processors:

$$g_x = \begin{cases} f_x \\ min[g_{x-1}, f_x] \end{cases}$$

$$h_x = \begin{cases} f_x \\ min[h_{x+1}, f_x] \end{cases} \tag{7}$$

$$n_x = min[g_{x+(r-1)/2}, h_{x-(r-1)/2}]$$

where r is the measurement of the filter kernel. f_x - the incoming matrix is separate into small arrays g_x and h_x of size r. n_x is result.

The larger the stream k, the more unnecessary calculations are installed, but if k is too small, GPU devices are wasted. Therefore, stream configuration is essential to improve the efficiency of the mean filtering algorithm (Fig. 2), the number of streams per block is set to 128.

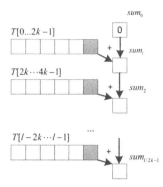

Fig. 2. Parallel mean filtering process in each block.

3.2 Improved Detection Model

We evaluated three YOLOv5, YOLOv6, and YOLOv7 models (Table 2). We include the total of parameters in the model to provide a good comparison point.

Table 2. Specifications of YOLO detectors.

	YOLOv5	YOLOv6	YOLOv7
AP (%)[b]	55.8	52.5	56.8
Model parameters (million)	20.9 M	34.3 M	70.8 M

The structure of YOLOv5 consists feature map, detection and classification (Fig. 3).

Neural network structure consists of a Focus module, Convolutional module (Conv), C3 module, and Spatial Pyramid Pool modules [19]. The size of the input image is 416*416. The structure of the Feature Pyramid Network and Pixel Aggregation Network was applied to object detection. Nonmaximum suppression and IoU as a loss function were used in object classification. The improvement of the C3 module, which is considered the main TSR, has been changed as follows (Fig. 4).

The LeakyReLU activation function is another version of the ReLU activation function, which is good at extracting and recognizing the features of objects.

$$LeakyReLU(x) = \begin{cases} a*(x), & if\ x \le 0 \\ x, & if\ x > 0 \end{cases} \tag{8}$$

$$ReLU(x) = \begin{cases} 0,\ if\ x \le 0 \\ x,\ if\ x > 0 \end{cases} \tag{9}$$

where a = 0.01.

YOLO neural network architecture is distinguished from other algorithms by its high recognition speed [20]. Algorithm combined with regression problem compared to CNN. The development of a neural network model of traffic sign recognition and

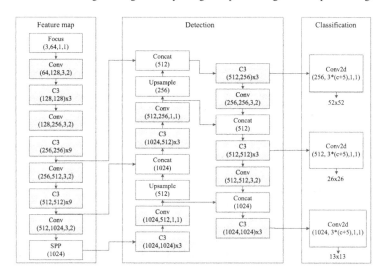

Fig. 3. YOLOv5 neural network structure.

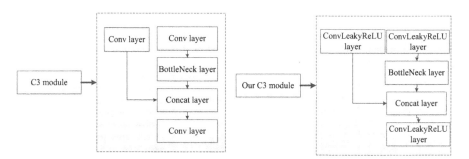

Fig. 4. C3 module.

deep learning of the dataset was carried out based on the following hyper-parameters (Table 3).

These hyperparameters were selected based on deep learning experiments on UTSD.

Table 3. Hyperparameters.

№	Name	Parametrs
1	Batch size	16
2	Learnig rate	0.001
3	Loss function	IoU_Loss
4	Epochs	300
5	IoU	0.45
6	Activation function	leakyReLU
7	Classification	Softmax
8	Optimization function	Adam

3.3 Data Augmentation

Traffic signs are organized to provide drivers with important statistics about the road infrastructure, and are designed with specific shapes and colors that drivers like. In the formation of a UTSD in the territory of Uzbekistan, in different seasons of the year (spring, summer, autumn, winter) and at different times of the day and in different weather conditions images were used. Figure 5 shows different views of rock fall danger traffic signs in different countries.

Countries	USA	Australia	Sweden	Poland	Uzbekistan
Traffic sign					

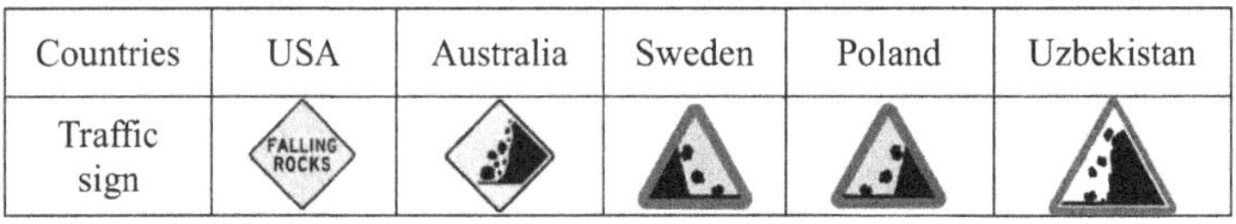

Fig. 5. Rock fall danger traffic signs in different countries.

In this dataset, the input image size is 1280 × 720 (HD) and was taken by a surveillance camera installed in cars. In this research work, annotation files of each image were created using the labeling software tool. The formation of the dataset requires a lot of work and a lot of time, so data augmentation methods (scaling, rotation, cropping, resizing) have been proposed. These methods serve to improve the accuracy of object recognition from images based on deep learning. In this case, it is important to create a high-quality dataset in a short-time period and increase the accuracy of the system of object recognition from images. This dataset was created in three steps: extracting frames from video, labeled, and data augmentation of the dataset [20].

4 Experiments and Analysis

4.1 Results

YOLO's neural network architecture is modifiable, allowing customizations such as adding layers, removing blocks, changing image processing, optimization, and activation functions. As a result of the training, the accuracy of TSR increased by 2.6% than other models (Table 4).

Table 4. Results obtained through road sign detection models.

№	Models	mAP(%)
1	YOLOv5 + ShuffleNet-v2 + BIFPN + CA + EIOU (Haohao Zou)	92.5
2	Yolov5l + GSconv + PDCM + SCAM (Jie Hu)	89.2
3	Faster R-CNN + FPN + ROI Align + DCN (Xiang Gao)	86.5
4	YOLOv5 + DA + IP + C3 (our)	**95.1**

YOLOv5 is distributed in four different versions namely YOLOv5s, YOLOv5m, YOLOv5l, and YOLOv5x. We have implemented our methodology using the YOLOv5s algorithm because it is fast as well as accurate enough to perform real-time classification, easy install to on mobile GPU (Fig. 6).

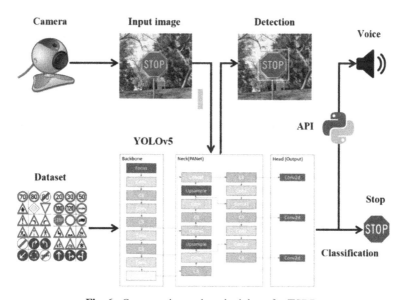

Fig. 6. Our experimental methodology for TSDR.

4.2 Experimental Analysis and Discussion

We used the GTSDB dataset in this study to compare the performance of faster RCNN and SSD models, with the proposed YOLOv5 algorithm. YOLOv5s network model is optimized to develop to the embedded intelligent mobile platform [21]. The above enhance methods improved the detection efficiency of our model by 3.9% overall (Table 5).

Table 5. Results of traffic sign recognition using the improved YOLOv5s.

Models	Data augmentation	Image preprocessing	C3 module	mAP(%)
YOLOv5s	–	–	–	91.2
Improvement I	+	–	–	92.9
Improvement II	+	+	–	93.2
Improvement III	+	+	+	**95.1**

The following software and hardware tools was used for the experiment (Table 6).

Table 6. Training environment.

CPU: Core i5 10400F	OS: Ubuntu 20.04
GPU: NVIDIA RTX 3050Ti	PyTorch version: 1.9
Memory RAM: 16G	CUDA version: 10.1
Python version: 3.9	YOLO version: YOLOv5s
Torch-vision: 0.9	Numpy version: 1.18
Compiler environments: Anaconda	

In this world, two minimum filters and six mean filters were used in the most time-consuming refining of the transmission with the guided filter step Algorithms for parallel calculation of these minimum and mean filters were developed on graphics processors (Table 7). When programming with Compute Unified Device Architecture technology, a stream block is allocated for each line of the image, and 128 streams are set for each stream block [21]. To improve image quality, a parallel mean filtering algorithm was developed based on the Prefix sum algorithm.

Table 7. Evaluation of image processing algorithm execution time on CPU and GPU.

Image size	800*600	1280*720	1600*720	1920*1080
CPU	8.071	17.118	21.66	50.74
GPU	1.52	2.60	3.24	5.67
Efficiency	5.30	6.58	6.68	8.94

5   Conclusion

In this article, we developed YOLOv5 to solve the disadvantages existing in road sign detection and recognition. Compared to the aforementioned latest methods, our methods are more accurate and faster. A parallel processing model and algorithms for removing noise and increasing the brightness of dynamic images in graphic processors were developed. The real-time image processing operation was reduced to 5.67 ms by parallel implementation on GPU. A neural network model of traffic sign recognition was developed and the accuracy of traffic sign recognition increased by 3.9%. These algorithms are an effective tool for recognizing traffic signs from video taken in different weather conditions. A 5-fold increase in dataset size was achieved by using data augmentation techniques to the traffic sign dataset. Based on the neural network model and software tools, the functional structure of the software complex that creates the textual and sound representation of the traffic signs determined from the images and the inter-program communication interface was improved. In our further research, we will focus on the development of object detection algorithms adapted to self-driving cars.

References

1. Fu, M.-Y., Huang, Y.-S.: A survey of traffic sign recognition, 119–124 (2010). https://doi.org/10.1109/ICWAPR.2010.5576425
2. Mukhiddinov, M., Cho, J.: Smart glass system using deep learning for the blind and visually impaired. Electronics **10**, 2756 (2021)
3. Umarov, M., Muradov, F., Azamov, T.: Traffic sign recognition method based on simplified gabor wavelets and CNNs. In: 2021 International Conference on Information Science and Communications Technologies (ICISCT), pp. 1–5. Tashkent, Uzbekistan (2021)
4. Wang, X., et al.: Real-time and efficient multi-scale traffic sign detection method for driverless cars. Sensors **22**, 6930 (2022). https://doi.org/10.3390/s22186930
5. Laguna, R., Barrientos, R., Blazquez, L.F., Miguel, L.J.: Traffic sign recognition application based on image processing techniques. In: 19th IFAC World Congress Cape Town, August 24–29. South Africa (2014)
6. Tan, R.: Visibility in bad weather from a single image. In: Proceedings of the 2008 IEEE Conference on Computer Vision and Pattern Recognition, 23–28, pp. 1–8. Anchorage, AK, USA (2008)
7. He, K., Sun, J., Tang, X.: Single image haze removal using dark channel prior. IEEE Trans. Pattern Anal. Mach. Intell. **33**, 1956–1963 (2009)
8. Redmon, J., Divvala, S., Girshick, R., Farhadi, A.: You only look once: unified, real-time object detection. In: IEEE Conference on Computer Vision and Pattern Recognition, pp. 779–788. Las Vegas (2016). https://doi.org/10.1109/CVPR.2016.91

9. Gudigar, A., Chokkadi, S., Raghavendra, U.: A review on automatic detection and recognition of traffic sign. Multimed Tools Appl **75**, 333–364 (2016). https://doi.org/10.1007/s11042-014-2293-7

10. Shustanov, A., Yakimov, P.: CNN design for real-time traffic sign recognition. Procedia engineering **201**, 718–725 (2017)

11. Yao, C., Wu, F., Chen, H.-J., Hao, X., Shen, Y.: Traffic sign recognition using HOG-SVM and grid search. Int. Conf. Sig. Proc. Proce. ICSP. **2015**, 962–965 (2015). https://doi.org/10.1109/ICOSP.2014.7015147

12. Le, T.T., Tran, S.T., Mita, S., Nguyen, T.D.: Real time traffic sign detection using color and shape-based features. ACIIDS 2010. Lecture Notes in Computer Science(), vol 5991. Springer, Berlin, Heidelberg (2010). https://doi.org/10.1007/978-3-642-12101-2_28

13. Wu, W., et al.: Application of local fully Convolutional Neural Network combined with YOLOv5 algorithm in small target detection of remote sensing image. PLoS ONE **16**(10), e0259283 (2021). https://doi.org/10.1371/journal.pone.0259283.Retraction.In:PLoSOne.202 3Sep7;18(9):e0291288.PMID:34714878;PMCID:PMC8555847

14. Girshick, R.: Fast R-CNN, arXiv preprint arXiv: 1504.08083 (2015)

15. Mukhiddinov, M., Abdusalomov, A.B., Cho, J.: Automatic fire detection and notification system based on improved YOLOv4 for the blind and visually impaired. Sensors **22**, 3307 (2022). https://doi.org/10.3390/s22093307

16. Jin, Y., Fu, Y., Wang, W., Guo, J., Ren, C., Xiang, X.: Multi-feature fusion and enhancement single shot detector for traffic sign recognition. IEEE Access **8**, 38931–38940 (2020). https://doi.org/10.1109/ACCESS.2020.2975828

17. Zhang, R., Zheng, K., Shi, P., Mei, Y., Li, H., Qiu, T.: Traffic sign detection based on the improved YOLOv5. Appl. Sci. **13**, 9748 (2023). https://doi.org/10.3390/app13179748

18. Gao, X., Chen, L., Wang, K., Xiong, X., Wang, H., Li, Y.: Improved traffic sign detection algorithm based on faster R-CNN. Appl. Sci. **12**, 8948 (2022). https://doi.org/10.3390/app12188948

19. Wu, X., Wang, K., Li, Y., Liu, K., Huang, B.: Accelerating haze removal algorithm using CUDA. Remote Sens. **13**, 85 (2021). https://doi.org/10.3390/rs13010085

20. Khamdamov, U., Umarov, M., Elov, J., Khalilov, S., Narzullayev, I.: Uzbek traffic sign dataset for traffic sign detection and recognition systems. In: 2022 International Conference on Information Science and Communications Technologies (ICISCT), pp. 1–5. Tashkent, Uzbekistan (2022). https://doi.org/10.1109/ICISCT55600.2022.10146832

21. Umarov, M., Elov, J., Khalilov, S., Narzullayev, I., Karimov, M.: An algorithm for parallel processing of traffic signs video on a graphics processor. In: 2022 International Conference on Information Science and Communications Technologies (ICISCT), pp. 1–5. Tashkent, Uzbekistan (2022). https://doi.org/10.1109/ICISCT55600.2022.10146809

22. Rakhimov, M., Elov, J., Khamdamov, U., Aminov, S., Javliev, S.: Parallel Implementation of Real-Time Object Detection using OpenMP. In: 2021 International Conference on Information Science and Communications Technologies (ICISCT), pp. 1–4. Tashkent, Uzbekistan (2021). https://doi.org/10.1109/ICISCT52966.2021.9670146

23. Khamdamov, U., Zaynidinov, H.: Parallel algorithms for bitmap image processing based on daubechies wavelets. In: 2018 10th International Conference on Communication Software and Networks (ICCSN), pp. 537–541. Chengdu, China (2018). https://doi.org/10.1109/ICCSN.2018.8488270

Convolutional Autoencoder for Vision-Based Human Activity Recognition

Surbhi Jain[1,3], Aishvarya Garg[2,3], Swati Nigam[1,3], Rajiv Singh[1,3(✉)],
Anshuman Shastri[3], and Irish Singh[4]

[1] Department of Computer Science, Banasthali Vidyapith, Rajasthan 304022, India
jkrajivsingh@gmail.com
[2] Department of Physical Science, Banasthali Vidyapith, Rajasthan 304022, India
[3] Centre for Artificial Intelligence, Banasthali Vidyapith, Rajasthan 304022, India
anshumanshastri@banasthali.in
[4] Departmenet of Computer Science & Engineering, Oregon Institute of Technology, Oregon,
USA
irish.singh@oit.edu

Abstract. Human activity recognition (HAR) is a crucial component for many
current applications, including those in the healthcare, security, and entertainment
sectors. At the current state of the art, deep learning outperforms machine learning
with its ability to automatically extract features. Autoencoders (AE) and convolutional neural networks (CNN) are the types of neural networks that are known
for their good performance in dimensionality reduction and image classification,
respectively. As most of the methods introduced for classification purposes are
limited to sensor based methods. This paper mainly focuses on vision based HAR
where we present a combination of AE and CNN for the classification of labeled
data, in which convolutional AE (conv-AE) is utilized for two functions: dimensionality reduction and feature extraction and CNN is employed for classifying the
activities. For the proposed model's implementation, public benchmark datasets
KTH and Weizmann are considered, on which we have attained a recognition rate
of 96.3%, 94.89% for both, respectively. Comparative analysis is provided for the
proposed model for the above-mentioned datasets.

Keywords: Human Activity Recognition · Deep Learning · Convolutional
Neural Networks · Autoencoders · Convolutional Autoencoders

1 Introduction

Over the decades, there has been an increase in the demand for a system that can perform human activity recognition has attracted a significant amount of attention towards
HAR. Human activity refers to the movement of one or several parts of the person's
body and it is basically determined by kinetic states of an object at different times [1].
HAR is the system that can automatically identify and analyze different human activities
from the different body movements by using machine learning and deep learning techniques. HAR provides applications not only in the field of healthcare and surveillance

but also for human computer interaction. Recent advancements in the field of neural networks allow for automatic feature extraction instead of using hand-crafted features to train HAR models. These models are primarily based on convolutional neural network (CNN), autoencoders (AE), and long short-term memory (LSTM). These methods extract important features from an image without any human supervision and provide a better feature representation of images to recognize objects, classes, and categories.

Generally, most of them follow the steps as shown in Fig. 1. Firstly, there is a requirement of raw data can be collected using sensors and vision-based tools. In sensor based, data is gathered from variety of sensors such as accelerometer, gyroscope, barometer, compass sensor, and other wearable sensors. For vision-based datasets, it uses visual sensing facilities, for example single camera or stereo and infrared to capture activities, and the collected data can be in the form of either videos or images [6]. Then the collected data goes through a pre-processing stage where processes such as segmentation, cropping, resizing is applied, unwanted distortions are suppressed so that the quality of the image is improved. Pre-processed data is then fed to the feature extractor; the process of feature extraction is attained by using either machine learning or deep learning techniques. In machine learning, features are extracted manually which are known as hand crafted features with the help of descriptors such as local binary pattern (LBP), histogram of oriented gradient (HOG), Scale-invariant feature transform (SIFT). In deep learning, features are extracted by using neural networks such as in CNN it is done at convolution layer and pooling layer. And lastly, activities are classified with the help of support vector machines in ML and in DL by utilizing a fully connected layer of CNN.

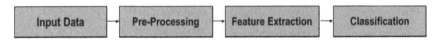

Fig. 1. Framework of Human Activity Recognition.

Many approaches have been implemented for recognizing human activities using deep learning model. From the surveys provided by [2–5], we have realized that most work in the field of activity recognition is done on sensor-based models. Hence, one of our objectives is to generate a model for vision based HAR. Going through different types of deep learning algorithms in depth, we came to know that AE can extract unique features along with dimensionality reduction property. So, we employed AE for the classification of activities. We have implemented a small framework in which AE was unable to predict the activities such that we can say that the efficiency of this algorithm degraded when the labeled data as input was given and the output obtained from it had some loss but, on the contradictory, the extracted features were far better than other techniques resulted from the above-mentioned property.

As we know, CNN is known for the best results in classification. Keeping this point in mind, to utilize AE's functionality for labeled data, we proposed a hybrid model for the classification of activities in which we employed two models i.e., Conv-AE and CNN models. In this proposed model, Conv-AE is utilized for feature extraction and dimensional reduction purposes and CNN for classification of activities from two standard public vision-based datasets: KTH and Weizmann. For the performance analysis on both datasets, we have used a confusion matrix and accuracy for our proposed model.

In this paper, Sect. 2 reviews the existing literature including different deep learning approaches used for vision based HAR. Section 3 provides a brief introduction to the methods used and describes the methodology and tools and techniques that are utilized in the proposed system. Section 4 provides the complete overview of the result and comparative analysis of the system. It also includes the work plan of the proposed model. And lastly Sect. 5 represents the conclusion and the future scope based on the study done for the paper.

2 Literature Review

This section reviews the existing literature including different deep learning approaches utilized for vision based HAR for KTH and Weizmann datasets. Several vision-based activity recognition approaches were published in the last two decades. A considerable amount of research has been done on HAR and different techniques of deep learning have been published by the researchers over the decades to determine human activities. It is reviewed on the basis of extracted features which include spatial features, temporal features, and both spatial-temporal features.

Kiruba et al. [7] proposed a work that is based on three different stages. To reduce the computational time and cost, discriminative frames were considered, Volumetric Local Binary Pattern (VLBP) was utilized in the second stage. They discovered that the hexagonal volume local binary pattern (H-VLBP) descriptor outperforms all other novel geometric shape-based neighborhood selection techniques for the identification of human action. To achieve multi-class recognition and to reduce the dimensions, the decoder layer is replaced with the softmax layer of the deep stacked autoencoder (SAE). They were able to obtain an accuracy of 97.6% 98.6% for the KTH and Weizmann datasets, respectively. Nigam et al. [9] proposed a method utilizing background subtraction for human detection and employed it with LBP (local binary pattern) and multiclass-SVM for classification of human activities. Their method was able to achieve an accuracy of 100% for Weizmann dataset.

Ramya et al. [11] proposed a method for HAR utilizing the distance transformation and entropy features of human silhouettes. They tested their method for KTH and Weizmann dataset and were able to obtain an accuracy of 91.4% for KTH and 92.5% for Weizmann dataset. Karuppannan et al. [13] proposed the HAR method in three orthogonal planes (TOP) pyramidal histogram of orientation gradient-TOP (PHOG-TOP) and dense optical flow-TOP (DOF-TOP) were also utilized with a SAE for reducing the dimensions and lastly the output from SAE is fed to LSTM for classification. They were able to achieve an accuracy of 96.50% and 98.70% for the KTH dataset and Weizmann datasets, respectively.

For characteristics representation Gnouma et al. [8] took human actions as a series of silhouettes. Furthermore, a stacked sparse autoencoder (SSA) system is provided for automated human action detection. The updated history of binary motion image (HBMI) is utilized as the first input to softmax classifier (SMC). They evaluated their method on the KTH and Weizmann dataset and obtained an accuracy of 97.83% and 97.66%

respectively. Song et al. [10] proposed the method of feature extraction by using the recurrent neural network (RNN) and the AE. The features of RNN effectively express the behavior characteristics in a complex environment, and it is merged with AE via feature similarity to generate a new feature with greater feature description. The experiments using KTH and Weizmann datasets demonstrate that the technique suggested in the research has a high recognition rate as they achieved an accuracy of 96% and 92.2% for the KTH and Weizmann datasets, respectively.

Mahmoud et al. [12] proposed an end-to-end deep gesture recognition process (E2E-DGRP) which is used for large scale continuous hand gesture recognition using RGB, depth and grey-scale video sequences. They achieved 97.5% and 98.7% accuracy for the KTH and Weizmann dataset, respectively. Garg et al. [14] proposed a model where they used a hybrid model of CNN-LSTM for feature extraction and classification of human actions. On testing their model for KTH and Weizmann datasets they were able to achieve an accuracy of 96.24% and 93.39% on KTH and Weizmann datasets, respectively.

Singh et al. [15] analyzed the real world HAR problem. They proposed a model utilizing discrete wavelet transform and multi-class support vector machine classifier to recognize activities. They used KTH and Weizmann datasets to test their model and obtained a recognition rate of 97% for both the datasets. A deep NN-based approach is suggested by Dwivedi et al. [16] for identifying suspicious human activity. The deep Inception V3 model in this instance extracts the key features for discrimination from images. They feed features into LSTM. The proposed system is tested against eleven benchmark databases, including the KTH action database and the Weizmann datasets. The suggested method had a 98.87% recognition rate.

To extract features, Badhagouni et al. [17] suggested that a CNN classifier be combined with the Honey Badger Algorithm (HBA) and CNN. The projected classifier is used to recognize human behaviors like bending, strolling, and other similar ones. HBA can be used to improve CNN performance by optimizing the weighting parameters. The proposed method was tested using the Weizmann and KTH databases. Saif et al. [18] proposed a convolutional long short-term deep network (CLSTDN) consisting of CNN and RNN. In this method CNN uses Inception-ResNet-v2 to classify by extracting spatial features and laştly RNN uses Long Short-Term Memory (LSTM) for prediction based on temporal features. This method achieved an accuracy of 90.1% on KTH dataset.

3   Proposed Methodology

We have proposed a hybrid model using convolutional type of autoencoder (Conv-AE) and CNN in which Conv-AE is utilized for two objectives- one is for dimensionality reduction and the other is for extracting features, and CNN model is utilized for the classification of activities by using vision-based datasets i.e., KTH and Weizmann.

3.1   Convolutional Autoencoder (Conv-AE)

An autoencoder is a feed-forward neural network that has the same output as input and is based on unsupervised learning technique. As shown in Fig. 2 input layer's size is the same as the size of output layer. They consist of three main layers- an encoder, a code, and lastly a decoder. Latent space representation is the output obtained from the encoder layer after compression. It encodes the input image in a compressed form into a reduced dimension. The compressed image obtained from the encoder layer will be a distorted form of the original input image. The second layer is known as the code layer which is the compressed input going to be fed to the decoder.

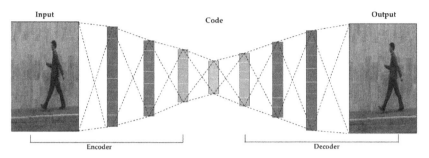

Fig. 2. Basic Architecture for the Autoencoder (AE).

And lastly, the decoder layer decodes the input compressed image back to its original dimensions. The decoded form of image is regenerated from latent space representation, and it's the reconstruction of the original image consisting of some loss. There are different types of AE and for our proposed method we are going to use Conv-AE which is same as the simple AE, as it is used for dimensionality reduction and it also has same number of input and output layers, the only difference is that the network of encoder layers converts into a network of convolutional layers and the decoder layer changes into a transpose of convolutional layers.

3.2 Convolutional Neural Network (CNN)

Convolutional neural network is a type of feed-forward neural network commonly utilized for extracting features and classifying images. As you can see in Fig. 3 it is made up of three layers namely convolutional layer, pooling layer, and fully connected layer. At the convolutional layer, convolution operations are performed on the passed image. The output of this layer gives us information about corners and edges of an image. This operation reduces the size of the numerical presentation so that only the important features are sent further for image classification which helps in improving accuracy. Then they obtained feature map is provided to a pooling layer as input which further helps in reducing the size of the feature map which forces the network to identify key features in the image. The extracted key features from the pooling layer are then passed through a flattening layer which converts the pooled feature map into a single column known as vectors. The column of vectors is then passed to the fully connected layer which consists of weights, biases, and neurons. Here, the process of classification takes place as the vectors go through a few more fully connected layers. And to resolve the problem of over-fitting batch normalization is used. And, lastly the output from fully connected layers passes through a softmax activation function which computes the probability distribution and gives the most probabilistic output as to which input image belongs to which class [19].

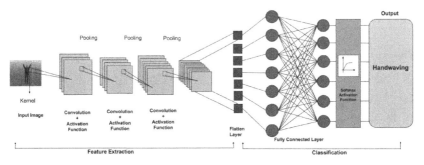

Fig. 3. General Architecture of the Convolutional Neural Network (CNN).

As we know that, Conv-AE works best for classifying the images for unlabeled data which makes it difficult to predict the data for labeled classes but because of its ability of dimensionality reduction, we are going to employ it with CNN where Conv-AE is utilized for feature extraction and dimensionality reduction purposes and 2D-CNN is used for classification.

3.3 Pre-processing

For the input to the model, pre-processing is done on the datasets. Operations such as segmentation is performed to convert into frames from videos at the given rate of frames per second, normalization is implemented to transform the dataset's value into a common scale, resizing converts the input shape of the image from $120 \times 260 \times 3$ to $100 \times 140 \times 1$ for the KTH dataset and $144 \times 180 \times 1$ for the Weizmann dataset.

3.4 Feature Representation

To perform the process of feature extraction, we have taken 3 convolutional layers for the encoder and the number of filters for the same are 64, 32, 16. The kernel's size is the same for all the layers that is (3, 3) with strides of size (1, 1) and the padding used is kept valid for the model. We have also added a pooling layer, more specifically a max-pooling layer of size (2, 2) in all Conv2D layers, as you can see from Fig. 4. Increasing the depth by adding number of layers not only helps with better representation of features but also causes the problem of over-fitting, to overcome that a layer of batch normalization is applied to the first and third layer of Conv2D of the encoder and the loss continues decreasing.

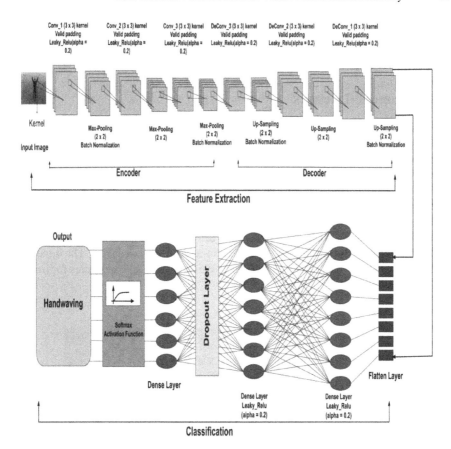

Fig. 4. Framework of the proposed method for Human Activity Recognition.

For the decoder, the same number of layers with exact number of filters are applied, but in transposed order of encoder, as the decoder is just the transpose of the encoder, the only difference is that instead of using a max-pooling layer we have used an up-sampling layer of size (2,2) between the three Conv2D layers. Lastly the output from the decoder is passed through a Conv2D layer of filter size 1 which represents present channels in the image. For all the layers an activation function is used known as leaky relu with the slope value of 0.2.

3.5 Classification

The 2-dimensional output from the decoder can't be a direct input to the fully connected layer, it is then transformed into a column of vectors by passing through a flatten layer so that it can be fed to the CNN model. For fully connected layer we have employed 3 dense layers with filter size of 32, 16 and last one has the number of filters same as the number of classes of dataset, in between we have also utilized a dropout layer to reduce over-fitting, dropping 20% of the neurons. And lastly there's a softmax activation function which provides us with the output based on probability and classifies the activity based on their classes. As you can see from diagram 4 an input image of KTH dataset is given to the model, and it provides hand-waving as the highest probability.

4 Experimental Results

4.1 Datasets

For data processing, initially the frames are extracted from the videos to provide to the model as input. Our proposed model is implemented on the vision based standard public benchmark datasets i.e., KTH and Weizmann. KTH dataset is a recorded video database consists of videos which are 600 in number of 6 activities named as boxing, handclapping, hand-waving, jogging, running and lastly walking that is performed by 25 people in four different environments with different angle, illumination conditions, clothing and gender and having frame rate of 25 frames per second (fps). The dimension of each video in the database is 120 x 160 x 3. We have used all 6 activities for training and classification [20].

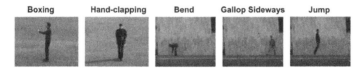

Fig. 5. Still images consisting of different actions and viewpoints from the KTH and Weizmann dataset.

Weizmann dataset consists of 90 videos containing 10 different activities such as bend, gallop sideways, jump, jump in place, jumping jack, one hand wave, two hands wave, run, skip, walk performed by 9 different subjects with 30 frames per second frame rate and having dimension of each video is 144 x 180 x 3, all 10 activities are considered for classification purposes [21] as one can see in Fig. 5.

For the implementation of the model, we have utilized the Jupyter notebook environment of Google Colaboratory with Keras Deep Learning API. The model was compiled with the help of categorical cross-entropy and for optimizer RMS-prop was used. Dividing the data in the ratio of 2:8 for testing and training for both the datasets where the random state is kept zero and shuffle is set to true. Table 1 gives the idea about different hyper-parameters used for the KTH and Weizmann dataset.

Table 1. Hyper-parameters used in the model for KTH and Weizmann dataset.

Hyper-parameters	KTH dataset	Weizmann Dataset
Videos	600	90
Epochs	30	35
Batch size	15	5
Train-test ratio	8:2	8:2
Frames per class	50	50
Total number of frames	30000	4500
Dimension of video	120 x 160 x 3	144 x 180 x 3

4.2 Results and Discussions

During implementation on Google Colaboratory, we faced the problem of crashing of model during the execution time when 60 frames were taken from each video in case of KTH dataset. The accuracy varied when the model was trained on different compute engine backend such as TPU and GPU of Google Colaboratory and more time was taken for the implementation. But TPU obtained the best results with foremost training accuracy and prediction rate. We have implemented this model with varying hyper-parameters such as train and test ratio, batch-size, epochs, resizing the frames, number of layers, optimizers, and filter size to achieve the best prediction accuracy rate. When the data was divided into 9:1 or 7:3 train test ratio it had a visible effect on the accuracy as the model was unable to predict correctly. We also trained our model between a range of 5 to 40 epochs for both datasets and with different batch sizes. But the most accurate result was obtained when we set the epochs at 30 and 35 with batch size 15 and 5 for the KTH and Weizmann datasets, respectively.

Relu activation function was considered to employ in the model as the time taken to learn by relu is less and is computationally less expensive than other activation functions, but it kept causing a problem of dying relu that is when many relu neurons only output values of 0. That's why instead of utilizing relu we implemented leaky relu activation function.

Initially we implemented the model containing only two Conv-2D layers, but the results were better with three layers and only two batch normalization layers in between to avoid over-fitting, as we know the increasing number of layers is directly proportional to the better classification of activities if the problem of over-fitting is kept in consideration. RMS-prop was decided to use as an optimizer when implementing the model with Adam and Stochastic Gradient Descent (SGD) optimizers didn't have any effect on the model's accuracy. As the amount of data given as input to the model also has an effect on the prediction accuracy, we provided a varying range of data to the model and tested it, we came to know that increasing the data lead to an increase of 2.5% in the accuracy, resizing was also done on the frames to make it easier for the model to extract features and to load and execute the program faster. Confusion matrix and accuracy were utilized for the evaluation of the proposed model. We have also provided comparative analysis for the KTH and the Weizmann datasets. The confusion matrix is also formulated with the graph as the performance matrix for both datasets as shown in Fig. 6 (a), (b). Our proposed model's effectiveness is analyzed by comparing the highest accuracy of our model with the other methodologies on both datasets as shown in Table 2.

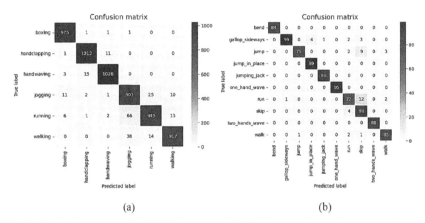

(a)   (b)

Fig. 6. Confusion matrix of (a) KTH dataset and (b) Weizmann dataset for proposed model

Table 2. Comparative analysis between obtained accuracy (%) for KTH and Weizmann dataset.

Author	Year	Methodology	KTH	Weizmann
[22]	2016	MI-ULBP	77.16	75.66
[23]	2019	Fusion of heterogeneous features + SMO + SVM	92.28	91.69
[24]	2020	SVM + ANN	87.57	86
[11]	2021	Distance Transform + Entropy Features + ANN	91.4	92.5
[25]	2022	Fuzzy Clustering Model	92.90	91.41
[26]	2022	CAE-DMD	-	91.1
[14]	2022	CNN + LSTM	96.24	93.39
[18]	2023	CLSTDN	90.1	-
-	**2023**	**Conv-AE + CNN**	**96.3**	**94.89**

5   Conclusions and Future Scope

This work mainly focuses on vision based HAR with the help of deep learning algorithms. We have employed deep learning through convolutional autoencoders to extract features from the input data. And then a convolutional neural network is implemented to classify the activities according to their respective classes. The proposed model improves the quality of extracted features which then provides better classification. On implementing these algorithms separately we realized that CNN performed quite efficiently in predicting the labeled dataset whereas AE provided better features. Therefore, we proposed a hybrid HAR model of AE and CNN algorithms for classifying the human activities. In this paper, Conv-AE is employed for the purposes of dimensionality reduction and feature extraction whereas CNN was used for the classification of the human activities. For the testing of the proposed model, vision based public benchmark datasets namely KTH and Weizmann were utilized. We have implemented proposed methodology with varying hyper-parameters from which we were able to obtain an accuracy of 96.3% and 94.89%for the KTH dataset and Weizmann dataset, respectively. From the given comparative analysis, our model was able to perform well in comparison to other state of the art methods and enhances the performances of HAR methods. A HAR system using deep learning techniques needs better feature representation, so fusion techniques for getting hybrid features with better resolution images can give us better results. Other deep learning techniques instead of CNN can be combined with the various feature descriptors to explore better accuracy results.

References

1. Basly, H., Ouarda, W., Sayadi, F.E., Ouni, B., Alimi, A.M.: CNN-SVM Learning Approach based Human Activity Recognition, pp. 271–281. ICISP, Springer (2020)
2. Bouchabou, D., Nguyen, S.M., Lohr, C., LeDuc, B., Kanellos, I.: A survey of human activity recognition in smart homes based on IoT sensors algorithms: taxonomies, challenges, and opportunities with deep learning. Sensors, MDPI **21**, 6037 (2021)
3. Zhang, S., et al.: Deep learning in human activity recognition with wearable sensors: a review on advances. Sensors, MDPI **4**, 1476 (2022)
4. Alo, U.R., Nweke, H.F., The, Y.W., Murtaza, G.: Smartphone motion sensor-based complex human activity identification using deep stacked autoencoder algorithm for enhanced smart healthcare system. Sensors, MDPI **20**, 6300 (2020)
5. Gu, F., Khoshelham, K., Valaee, S., Shang, J., Zhang, R.: Locomotion activity recognition using stacked denoising autoencoders. IEEE Internet of Things Journal, IEEE **5**, 2085–2093 (2018)
6. Sunny, J.T., et al.: Applications and challenges of human activity recognition using sensors in a smart environment. IJIRST Int. J. Innov. Res. Sci. Technol **2**, 50–57 (2015)
7. Kiruba, K., Shiloah, E.D., Sunil, R.R.C.: Hexagonal Volume Local Binary Pattern (H-VLBP) with Deep Stacked Autoencoder for Human Action Recognition. Cognitive Systems Research, Elsevier **58**, 71–93 (2019)
8. Gnouma, M., Ladjailia, A., Ejbali, R., Zaied, M.: Stacked sparse autoencoder and history of binary motion image for human activity recognition. Multimedia Tools and Applications, Springer **78**, 2157–2179 (2019)
9. Nigam, S., Singh, R., Singh, M.K., Singh, V.K.: Multiview human activity recognition using uniform rotation invariant local binary patterns. J. Ambient Intell. Humani. Comp. Springer, 1–19 (2022)
10. Song, X., Zhou, H., Liu, G.: Human behavior recognition based on multi-feature fusion of image. Cluster Computing, Springer **22**, 9113–9121 (2019)
11. Ramya, P., Rajeswari, R.: Human action recognition using distance transform and entropy based features. Multimedia Tools and Applications, Springer **80**, 8147–8173 (2021)
12. Mahmoud, R., Belgacem, S., Omri, M.N.: Towards an end-to-end Isolated and continuous deep gesture recognition process. Neural Computing and Applications, Springer **34**, 13713–13732 (2022)

13. Karuppannan, K., Darmanayagam, S.E., Cyril, S.R.R.: Human action recognition using fusion-based discriminative features and long short term memory classification. Concurrency and Computation: Practice and Experience, Wiley Online Library **34**, e7250 (2022)
14. Garg, A., Nigam, S., Singh, R.: Vision based Human Activity Recognition using Hybrid Deep Learning. CSI, IEEE, 1–6 (2022)
15. Singh, R., Nigam, S., Singh, A.K., Elhoseny, M.: Wavelets for Activity Recognition. Intelligent Wavelet Based Techniques for Advanced Multimedia Applications, Springer **10**, 109–121 (2020)
16. Dwivedi, N., Singh, D.K., Kushwaha, D.S.: A Novel Approach for Suspicious Activity Detection with Deep Learning. Multimedia Tools and Applications, pp. 1–24. Springer (2023)
17. Badhagouni, S.K., ViswanadhaRaju, S.: HBA optimized Efficient CNN in Human Activity Recognition. The Imaging Science Journal, Taylor & Francis **71**, 66–81 (2023)
18. Saif, A.S., Wollega, E.D., Kalevela, S.A.: Spatio-temporal features based human action recognition using convolutional long short-term deep neural network. Int. J. Adv. Comp. Sci. Appl. Sci. Info. (SAI) Organization Limited **14**, 66–81 (2023)
19. https://towardsdatascience.com/acomprehensive-guide-to-convolutional-neural-networks-the-eli5-way3bd2b1164a53/
20. Schuldt, C., Laptev, I., Caputo, B.: Recognizing Human Actions: A Local SVM Approach. ICPR, IEEE **3**, 32–36 (2004)
21. Blank, M., Gorelick, L., Shechtman, E., Irani, M., Basri, R.: Actions as Space-time Shapes. ICCV, IEEE **2**, 1395–1402 (2005)
22. Nigam, S., Khare, A.: Integration of moment invariants and uniform local binary patterns for human activity recognition in video sequences. Multimedia Tools and Applications, Springer **75**, 17303–17332 (2016)
23. Naveed, H., Khan, G.A.U., Siddiqi, A., Khan, M.U.G.: Human activity recognition using mixture of heterogeneous features and sequential minimal optimization. International Journal of Machine Learning and Cybernetics, Springer **10**, 2329–2340 (2019)
24. Nadeem, A., Jalal, A., Kim, K.: Human Actions Tracking and Recognition based on Body Parts Detection via Artificial Neural Network. ICACS, IEEE, pp. 1–6 (2020)
25. Song, B.: Application of Fuzzy Clustering Model in the Classification of Sports Training Movements. Computational Intelligence and Neuroscience, Hindawi, 2022 (2022)
26. Haq, I.U., Iwata, T., Kawahara, Y.: Dynamic mode decomposition via convolutional autoencoders for dynamics modeling in videos. Comput. Vis. Image Underst. **216**, 103355 (2022)

Deep Learning Approach for Enhanced Object Recognition and Assembly Guidance with Augmented Reality

Boon Giin Lee[1(✉)] ⓘ, Xiaoying Wang[1], Renzhi Han[1], Linjing Sun[1],
Matthew Pike[1], and Wan-Young Chung[2]

[1] School of Computer Science, University of Nottingham Ningbo China,
Ningbo 315100, China
{boon-giin.lee,scyxw7,renzhi.han,scyls1,matthew.pike}@nottingham.edu.cn
[2] Department of Artificial Convergence, Pukyong National University,
Busan 48513, Korea
wychung@pknu.ac.kr

Abstract. In an effort to enhance the efficiency and precision of manual part assembly in industrial settings, the development of software for assembly guidance becomes imperative. Augmented reality (AR) technology offers a means to provide visual instructions for assembly tasks, rendering the guidance more comprehensible. Nevertheless, a significant challenge lies in the technology's limited object detection capabilities, especially when distinguishing between similar assembled parts. This project proposes the utilization of deep learning neural networks to enhance the accuracy of object recognition within the AR guided assembly application. To achieve this objective, a dataset of assembly parts, known as the Visual Object Classes (VOC) dataset, was created. Data augmentation techniques were employed to expand this dataset, incorporating scale HSV (hue saturation value) transformations. Subsequently, deep learning models for the recognition of assembly parts were developed which were based on the Single Shot Multibox Detector (SSD) and the YOLOv7 detector. The models were trained and fine-tuned, targeting on the variations of the positions of detected parts. The effectiveness of this approach was evaluated using a case study involving an educational electronic blocks circuit science kit. The results demonstrated a high assembly part recognition accuracy of over 99% in mean average precision (MAP), along with favorable user testing outcomes. Consequently, the AR application was capable of offering high-quality guidance to users which holds promise for application in diverse scenarios and the resolution of real-world challenges.

Keywords: Augmented Reality · Assembly Tasks · Object Detection · Object Recognition

This work was supported in part by the Zhejiang Provincial Natural Science Foundation of China under Grant LQ21F020024, and in part by the Ningbo Science and Technology (S&T) Bureau through the Major S&T Program under Grant 2021Z037 and 2022Z080.

B. J. Choi et al. (Eds.): IHCI 2023, LNCS 14532, pp. 105–114, 2024.
https://doi.org/10.1007/978-3-031-53830-8_11

1 Introduction

An AR system enriches the real world with computer-generated virtual objects, seamlessly blending them into the physical environment. These virtual elements coexist with real-world surroundings, interact in real-time, and are accurately positioned relative to the real environment in three dimensions [1]. Leveraging these characteristics, numerous researchers have endeavored to integrate augmented reality into practical scenarios, such as enhancing digital education [2], refining spot-welding precision [3], and optimizing facility layout planning [4].

The prevailing techniques for object detection in augmented reality systems encompass marker-based and markerless algorithms [5]. Marker-based approaches utilize physical markers affixed to real-world objects for localization and tracking, while markerless methods rely on spatial geometry for object discrimination, dispensing with the need for physical markers. Although both approaches perform capably in a variety of augmented reality (AR) applications, they exhibit certain design limitations and cannot consistently deliver high object identification accuracy across all scenarios.

The integration of deep learning and computer vision to execute object detection tasks within augmented reality broadens its applicability to practical scenarios. Deep learning-based object detectors can be broadly classified into two categories: two-stage detectors and one-stage detectors. Two-stage detectors initially extract feature vectors from images, which are then employed as inputs for model training. In contrast, one-stage detectors classify regions of interest as either target objects or background within input images, offering significant speed advantages for real-time object detection applications. However, they tend to exhibit slightly lower prediction performance compared to two-stage detectors [6].

Joseph et al. [7] developed YOLO, a one-stage, real-time deep learning-based detector that treats object detection as a regression problem. It divides the image into predetermined matrix cells and uses a single neural network to predict probabilities and bounding boxes simultaneously. Despite its fast speed, earlier versions like YOLOv2 and YOLOv3 faced challenges in precise localization and detecting small and crowded objects [6]. However, the latest version, YOLOv7, boasts significant improvements, surpassing both SWIN-L Cascade-Mask R-CNN and ConvNeXt-XL Cascade-Mask R-CNN in speed and accuracy [8]. To address YOLO's limitations in detecting small objects, Lin et al. [9] introduced the Single Shot Multibox Detector (SSD) in 2016. Like YOLO, SSD also divides the input image into cells. However, SSD utilizes a range of anchors with different sizes, scales, and aspect ratios to optimize bounding boxes. This approach allows SSD to precisely detect small objects accurately. Furthermore, SSD performs detection exclusively on deeper network levels, in contrast to previous single-stage detectors that conducted experiments on multiple network layers with varying sizes.

This study seeks to address the challenge of slow object detection in two-stage object detectors by proposing the utilization of YOLOv7 for detecting objects of interest within input images. Additionally, the study integrates the SSD to enhance the precision of detecting smaller objects. The proposed guided system also provides instructions for correct or incorrect assembly steps, with the goal of improving user experiences.

2 Methodology

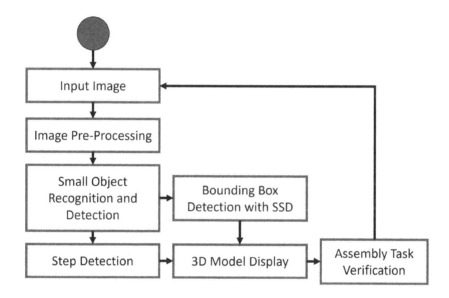

Fig. 1. System overview.

Figure 1 provides an overview of the proposed system. The initial pre-processing step primarily focuses on data augmentation to enhance the quality of input images. This involves various image transformations that allow the model to train on multiple images subjected to different lighting conditions, flips, and scales. These transformations help mitigate inaccuracies caused by camera movements and variations in ambient light. For more details on the transformations employed in this work, refer to Fig. 2. The SSD architecture utilizes VGG16 as the backbone network to extract multiple layers of feature maps and applies 3×3 convolution filters for object detection [9]. In YOLOv7, several enhancements are introduced, such as the use of Extended Efficient Layer Aggregation Networks (E-ELAN) to improve model efficiency and learning convergence, model scaling to optimize hardware usage, and planned re-parameterized convolution to maintain accuracy when combined with other architectures. Additionally, a coarse-to-fine lead head guided label assigner is designed as a deep supervision technique to enhance model performance [8].

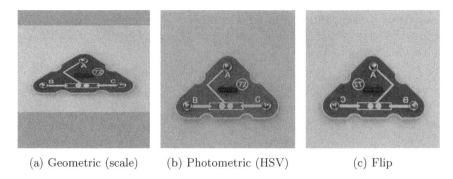

(a) Geometric (scale)   (b) Photometric (HSV)   (c) Flip

Fig. 2. The data augmentation involved in the study that include (a) geometric, (b) photometric, and (c) flip transformations.

2.1 Dataset Collection and Image Pre-processing

The educational electronic blocks circuit science kit components were captured under two scenarios: using an iPhone rear camera under a lamp, resulting in slightly sharpened images with high contrast and noticeable shadows, and with a different camera under specialized lighting equipment, producing softer colors with no apparent shadows. A total of nearly 500 images were captured from a nearly constant perspective. Seven target objects were identified, including five components required for the target assembly, an assembly board, and other assembly components not in use. These components are illustrated in Table 1.

2.2 Step Detection

A step detection algorithm is employed to ascertain the fulfillment of assembly step requirements. During this phase, the output from the object detector undergoes processing to extract step-related information, as illustrated in Fig. 3. The assembly step is marked as completed only when the assembly board is detected, and the required assembly part is detected inside the board at the specified position. The correctness of the part's placement is determined by calculating the relative position of the assembly part's center in relation to the assembly board's center and comparing it to a preset value. The step detection algorithm produces a 12-element Boolean list as output. The prototype is developed using Unity3D, integrated with Vuforia Engine, and runs on a mobile device.

3 Results and Discussion

3.1 Prototype

Figure 4a showcases the primary components of the user interface (UI): three hints, a 3D model, labels, bounding boxes, and three buttons. The first hint

Table 1. The components from the educational electronic blocks circuit science kit used for this study.

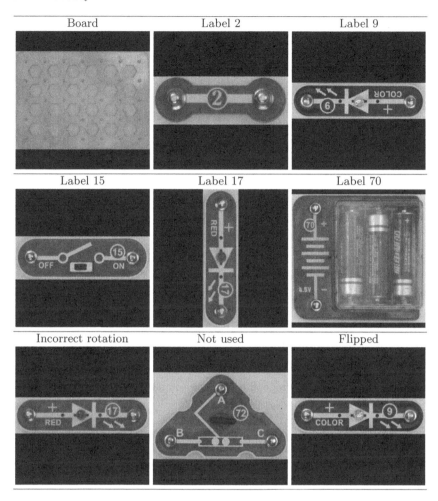

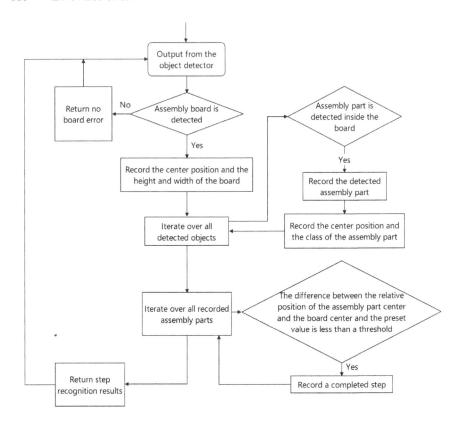

Fig. 3. Proposed step detection algorithm.

conveys the current step number, the second hint provides responses upon clicking the check button (with red text denoting incorrect assembly), and the third hint delivers current step instructions. The 3D model guides users on the placement of assembly components. Labels and bounding boxes display the predicted class label (9), confidence value (95%), and location of the detected assembly part, with green indicating necessary components. The left-hand preview button permits users to revert to the previous assembly step, while the right-hand next button enables progression to the next step when active. Clicking the check button in the middle yields current assembly results.

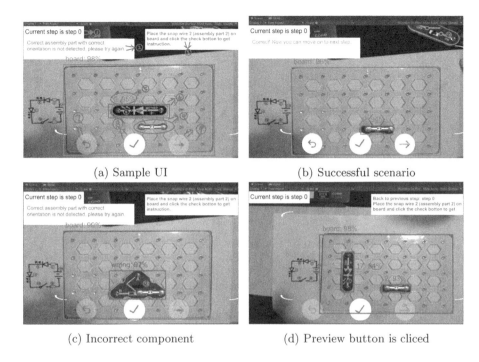

(a) Sample UI

(b) Successful scenario

(c) Incorrect component

(d) Preview button is cliced

Fig. 4. The sample UI of the developed prototype with different scenarios.

Figure 4b illustrates a successful assembly scenario. The green text in the second hint confirms correct assembly. In this state, preview and next buttons are interactive, and the current step instruction disappears. Users have the choice to continue assembly, return to the previous step, or remain in the current step. Figure 4c presents a situation where the correct assembly part is identified but placed incorrectly. The red text in the second hint notifies users of the assembly's incorrect placement. Upon clicking the preview button, as shown in Fig. 4d, the third hint informs users of their return to the previous step and displays the current step instruction.

3.2 SSD Performance

Following 300 epochs of SSD model training, the fine-tuned detector achieves assembly component detection with a mean average precision (MAP) of 99.39%. Notably, the average recall value is marginally lower than the average precision value, with the lowest recall observed for assembly part 17. This suggests that the deep learning model is more likely to miss detecting an object, especially assembly part 17, rather than incorrectly recognizing it. A possible explanation is the color variation in assembly part 17 under slightly changing lighting conditions. Table 2 provides precision, recall, F1-score, and mAP values for each class.

Table 2. Performance of SSD detector

Class	Precision	Recall	F1-score
15	94.25%	94.80%	0.95
17	100.00%	87.18%	0.93
2	99.25%	91.70%	0.95
70	99.02%	99.02%	0.99
9	100.00%	95.00%	0.97
Board	98.35%	100.00%	0.99
Incorrect Components	95.35%	91.99%	0.94
Mean mAP	99.39%		

3.3   YOLOv7 Performance

The fine-tuned YOLOv7 detector achieves a MAP of 99.68% in detecting assembly parts after 300 training epochs. Precision and recall values approach 100% for all classes. Instances of model misidentification or failure to detect assembly parts, such as parts 15, 2, or incorrect components, are rare. Table 3 presents precision, recall, and F1-score values for each class.

Table 3. Performance of YOLOv7 detector

Class	Precision	Recall	F1 Score
15	99.30%	99.30%	0.99
17	100.00%	100.00%	1.00
2	100.00%	98.78%	0.99
70	100.00%	100.00%	1.00
9	100.00%	100.00%	1.00
Board	100.00%	100.00%	1.00
Incorrect Components	99.63%	98.55%	0.99
Mean mAP	99.68%		

3.4   User Testing

User testing, involving 10 participants, was conducted to gather feedback on user experiences with the developed prototype. Results reveal that 90% of users find the prototype's latency acceptable. However, users reported that assembly component recognition was accurate in 70 of cases during actual usage, with occasional issues such as intermittent battery detection failures or the detector's inability to distinguish between very similar parts. These challenges appear to be related to factors like the battery's reflective glass cover and the lighting environment.

Regarding instructions, 90% of users found them readable and accurate. One user raised concerns about the uncommon term "diode" and the unclear button name "check" in the instructions. In terms of UI design, 80% of users expressed satisfaction, with two users providing suggestions for improvement based on their feedback. The System Usability Scale (SUS) questionnaire [10] revealed that 90% of users agreed that the prototype is user-friendly, integrates functions well, and provides trustworthy instructions, while one user expressed a neutral opinion.

4   Conclusion

This study has introduced the development of a guided assembly task prototype through an augmented reality (AR)-based mobile application. The prototype effectively positions 3D models to instruct users on the correct placement of essential components. The performance of the SSD detector, which enhances boundary box detection, and the YOLOv7 model for detecting small objects, surpasses 95% in various aspects, meeting the requirements for real-world applications. The prototype generally fulfills the criteria of accurately detecting assembly parts, offering clear and accurate instructions, and presenting augmented reality components to aid in assembly. Nevertheless, several factors, including lighting conditions, can influence detection accuracy. Further improvements can be pursued by collecting more images under diverse lighting conditions to enhance assembly component detection accuracy. Additionally, refining the data pre-processing stage can result in optimized, faster, and more precise small object detection and recognition performance.

References

1. Azuma, R., Baillot, Y., Behringer, R., Feiner, S., Julier, S., MacIntyre, B.: Recent advances in augmented reality. IEEE Comput. Graphics Appl. **21**(6), 34–47 (2001)
2. Lindner, C., Rienow, A., Jürgens, C.: Augmented reality applications as digital experiments for education - an example in the earth-moon system. Acta Astronaut. **161**, 66–74 (2019)
3. Doshi, A., Smith, R.T., Thomas, B.H., Bouras, C.: Use of projector based augmented reality to improve manual spot-welding precision and accuracy for automotive manufacturing. Int. J. Adv. Manuf. Technol. **89**(5–8), 1279–1293 (2017)
4. Jiang, S., Ong, S.K., Nee, A.Y.C.: An AR-based hybrid approach for facility layout planning and evaluation for existing shop floors. Int. J. Adv. Manuf. Technol. **72**(1–4), 457–473 (2014)
5. Katiyar, A., Kalra, K., Garg, C.: Marker based augmented reality. Adv. Comput. Sci. Inf. Technol. (ACSIT) **2**(5), 441–445 (2015). Cited by: 43
6. Vipul Sharma and Roohie Naaz Mir: A comprehensive and systematic look up into deep learning based object detection techniques: a review. Comput. Sci. Rev. **38**, 100301 (2020)
7. Redmon, J., Divvala, S., Girshick, R., Farhadi, A.: You only look once: unified, real-time object detection (2015)

8. Wang, C.-Y., Bochkovskiy, A., Liao, H.-Y.M.: YOLOv7: trainable bag-of-freebies sets new state-of-the-art for real-time object detectors (2022)
9. Liu, W., et al.: SSD: single shot multibox detector. In: Leibe, B., Matas, J., Sebe, N., Welling, M. (eds.) ECCV 2016. LNCS, vol. 9905, pp. 21–37. Springer, Cham (2016). https://doi.org/10.1007/978-3-319-46448-0_2
10. Brooke, J.: SUS: a quick and dirty usability scale. Usab. Eval. Ind. **189**, 11 (1995)

3D Facial Reconstruction from a Single Image Using a Hybrid Model Based on 3DMM and Deep Learning

Isha Deshmukh[✉], Vikas Tripathi, and Durgaprasad Gangodkar

Graphic Era (Deemed to be) University, Dehradun, India
{21062011,vikastripathi.cse,dr.gangodkar}@geu.ac.in

Abstract. A fundamental challenge in computer vision is accurately modelling 3D faces from images. It facilitates the creation of immersive virtual experiences, realistic facial animations, and reliable identity verification. This research introduces an innovative approach aimed at reconstructing intricate facial attributes, encompassing shape, pose, and expression from a single input image. The proposed methodology employs a fusion of two potent techniques: 3D Morphable Models (3DMMs) and advanced Deep Learning (DL) methodologies. By integrating DL into tasks like face detection, expression analysis, and landmark extraction, the framework excels in reconstructing realistic facial attributes from single images even in diverse environments. The framework achieves compelling results in reconstructing "in-the-wild" faces, exhibiting notable fidelity while preserving essential facial characteristics. Experimental evaluations confirm the effectiveness and robustness of our approach, confirming its adaptability across various scenarios. Our research contributes to the advancement of 3D face modelling techniques, addressing the challenges of accurate reconstruction and holding promise for applications in virtual reality, facial animation, medical, security, and biometrics.

Keywords: 3D Facial Reconstruction · 3D Morphable Model · Face Detection · Facial Landmark Detection · Face Model Fitting

1 Introduction

Facial reconstruction from images has evolved into a critical challenge in computer vision. The accurate modelling and reconstruction of the 3D shape, pose, and expression of a face from an image has garnered significant attention and found crucial applications in domains such as virtual reality, facial animation, medical, security, and biometrics [1–4]. The advancements in accurate and realistic 3D face modelling have paved the way for immersive virtual experiences, lifelike facial animations in movies and games, and reliable identity verification systems [5–7].

Image-based methods for facial reconstruction have played a pivotal role in driving progress in this field. These methods leverage the abundance of visual information available in images and enable the reconstruction process without requiring expensive and intrusive hardware setups. The development of accurate and effective image-based

B. J. Choi et al. (Eds.): IHCI 2023, LNCS 14532, pp. 115–126, 2024.
https://doi.org/10.1007/978-3-031-53830-8_12

reconstruction techniques has been considerably aided by the abundance of readily accessible image data, the rapid progress of Deep Learning (DL) techniques, and the easy availability of large-scale datasets.

The domain of 3D face modelling has made significant advancements lately, propelled by breakthroughs in DL techniques, particularly, Convolutional Neural Networks (CNN) for image processing [8]. This progress has been fueled by the need to overcome the limitations of traditional approaches that relied on handcrafted features and labour-intensive manual annotation. To achieve accurate facial reconstruction, researchers have explored various techniques and representations, aiming to capture the intricate details of human faces.

An extensively employed approach for 3D facial modelling is the utilization of 3D Morphable Models (3DMMs) [9], which offers a versatile and parametric representation of facial geometry and appearance [10]. 3DMMs provide a condensed yet comprehensive representation that can be utilised to reconstruct and modify 3D faces by capturing the variations in facial form, texture, and expressions within a low-dimensional space.

This paper introduces an innovative approach to 3D face modelling that combines the strengths of 3DMMs and integrates DL-based techniques for face detection, landmark extraction and expressions. The primary focus is on achieving a realistic reconstruction of facial shape, pose, and expression from a single input image, particularly in complex "in-the-wild" conditions. Our method aims to overcome the limitations and deliver a robust and efficient solution for facial reconstruction.

This paper makes several significant contributions:

- A unique approach that combines the flexibility and parametric representation of 3DMMs by leveraging the precision of DL-based approaches. This integration enables realistic and detailed reconstruction of facial geometry and appearance from an input image.
- The proposed method uses a single image for reconstruction that can be implemented in real-time systems as it is computationally less expensive and has more processing speed.
- The proposed method holds promise for various applications such as virtual reality, facial animation, and biometrics, where 3D facial modelling is crucial.

2   Related Work

The growth in 3D facial modelling and reconstruction has been particularly driven by the need to accurately capture the pose, shape, and expression of human faces. Previous studies have explored various approaches, with a particular emphasis on 3DMMs and image-based methods. This section provides an overview of the related work in these areas, highlighting the key contributions and limitations of each approach.

2.1   3DMM-Based Methods

3DMMs are widely used for 3D facial modelling, providing a parametric framework to capture facial geometry and appearance variations. Blanz and Vetter [9] pioneered the concept and showcased their effectiveness in reconstructing faces from 3D scans. These models encode shape, texture, and expression variations in a low-dimensional linear space, enabling efficient and compact representation.

Since then, researchers have made significant advancements in 3DMMs to enhance their accuracy and applicability. Booth et al. [11] introduced the Large-Scale Facial Model (LSFM), a comprehensive 3DMM that incorporates statistical information from a diverse human population. The model analyzes the high-dimensional facial manifold, revealing clustering patterns related to age and ethnicity. Although the LSFM shows promise for medical applications due to its sensitivity to subtle genetic variations, further research and validation in this domain are warranted.

Tran et al. [12] introduced a method that learns a nonlinear 3DMM model from a large set of unconstrained face images, eliminating the need for 3D scans. They employed weak supervision and leveraged a large collection of 2D images. Similarly, [13] utilized an encoder-decoder architecture to estimate projection, lighting, shape, and albedo parameters, resulting in a nonlinear 3DMM. However, the learned shape exhibits some noise, especially around the hair region. In [14], an approach to enhance the nonlinear 3D face morphable model by incorporating strong regularization and leveraging proxies for shape and albedo was presented. The method utilized a dual-pathway network architecture that balances global and local-based models. Nevertheless, the model may face challenges when dealing with extreme poses and lighting conditions.

Dai et al. [15] proposed the Liverpool-York Head Model (LYHM), a fully-automatic statistical approach for 3D shape modelling, enhancing correspondence accuracy and modelling ability. However, variations in lighting, expressions, or occlusions may impact texture mapping quality. Similarly, [16] introduced 3DMM-RF, a facial 3DMM combining deep generative networks and neural radiance fields for comprehensive rendering, yet challenges remain in accurately rendering occluded areas and flattened eye representation in the training data.

In [17], the authors introduced the SadTalker system to create stylized audio-driven animations of talking faces from single images. This approach involves generating 3D motion coefficients from audio and utilizing a unique 3D-aware face rendering method for animation. However, the emphasis of this method is primarily on lip movement and eye blinking, leading to generated videos having fixed emotions.

2.2 Image-Based Methods

These methods leverage the abundance of visual information available in images and enable the reconstruction process without requiring expensive and intrusive hardware setups. Recent advancements in DL techniques have revolutionized image-based facial reconstruction.

Jiang et al. [18] employed a coarse-to-fine optimization strategy for 3D face reconstruction, refining a bilinear face model with local corrective deformation fields. However, it is sensitive to face deviations from the training datasets, ambiguities in albedo and lighting estimation, and reliance on the quality of detected landmarks. In [19], the incorporation of expression analysis and supervised/unsupervised learning for proxy face geometry generation and facial detail synthesis was proposed. Their method excels in handling surface noise and preserving skin details, but it has limitations in accounting for occlusions, hard shadows, and low-resolution images.

Afzal et al. [3] utilized feature extraction and depth-based 3D reconstruction method. However, the method does not consider facial expressions, which limits its applicability in dynamic scenarios or facial recognition applications. On the other hand, [20] focused on high-fidelity facial texture reconstruction using GANs and DCNNs for single-image reconstruction. Their approach achieves impressive results but may face challenges with extreme expressions, challenging lighting conditions, limited data availability, and computational complexity, impacting its real-time performance and scalability.

In [21], AvatarMe, a method for reconstructing high-resolution realistic 3D faces through single "in-the-wild" images was proposed. The approach includes facial mesh reconstruction and head topology inference that allows for complete head models with textures. However, the training dataset contains insufficient cases of individuals from various ethnicities, potentially resulting in lower performance in reconstructing faces. In [22], a model utilizing a generative prior of 3D GAN and an encoder-decoder network was proposed that can be generalized to new identities efficiently. This addresses the limitation of personalized methods and expands practicality.

Approaches for the reconstruction from multi-view images were explored by [23, 24], and [25]. The approach of [23] combined traditional multi-view geometry with DL techniques, but it relies on high-quality 3D scans, limiting performance. A fast and accurate spatial-temporal stereo matching scheme using speckle pattern projection was proposed by [24], while [25] introduced a method for high-quality 3D head model recovery from a few multi-view portrait images. However, results depend on input image quality and computational demands may restrict real-time or resource-constrained applications. Obtaining sufficient high-quality images from different viewpoints can be challenging in practical or real-world settings.

3DMMs have offered a parametric representation for capturing facial variations, while image-based methods have leveraged DL techniques to extract information directly from images. However, several challenges remain, including the robustness to varying illumination and occlusion, handling large pose variations, and preserving fine-scale details in "in-the-wild" scenarios. The proposed method aims to address these challenges by leveraging the advantages of both 3DMM-based and image-based approaches, providing a more accurate and robust framework for 3D facial reconstruction.

3   Methodology

This section outlines the methodology employed for the proposed approach. The process involves several key steps, including initialization, face detection and landmark extraction, fitting process, and output generation. Figure 1 provides a visual representation of the methodology proposed in our research.

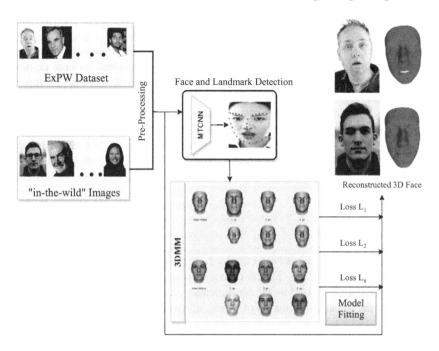

Fig. 1. Our facial reconstruction process includes face detection, landmark detection, and refinement of a 3D face model based on the input image. By optimizing the model to minimize discrepancies between projected and extracted landmarks, we achieve a realistic reconstruction. The final result is a composed image obtained by rendering and compositing the reconstructed face with the original image, enabling further analysis.

3.1 3D Morphable Model

For reconstruction using 3DMM, we utilize the popular Basel Face Model (BFM) 2009 [26]. The parameterization of each face involves angular meshes with 53,490 vertices.

$$S = S(\alpha, \beta) = \overline{S} + B_{id}\alpha + B_{exp}\beta \qquad (1)$$

$$T = T(\gamma) = \overline{T} + B_t\gamma \qquad (2)$$

In Eqs. (1) and (2), $\overline{S}$ and $\overline{T}$ represent the average face shape and texture, respectively. B_{id}, B_{exp}, and B_t are the PCA bases of identity, expression, and texture respectively. These bases are scaled with standard deviations. The coefficient vectors α, β, and γ are used to generate a 3D face.

The expression bases, B_{exp}, utilized in our method, as described in [27], consist of 53,215 vertices. To reduce dimensionality, a subset of these bases is selected, resulting in coefficient vectors $\alpha \in \mathbb{F}^{80}$, $\beta \in \mathbb{F}^{64}$ and $\gamma \in \mathbb{F}^{80}$ where $\mathbb{F}$ represents the field of real numbers. It is important to note that the cropped model used in our approach contains 35,709 vertices.

3.2 Camera Model

A perspective camera model is employed to record the 3D-2D projection geometry of the face. The camera model incorporates a focal length determined through empirical observations, enabling us to precisely represent the connection between the 3D face and its 2D projection.

The 3D pose of the face, denoted as θ, is expressed using a rotation matrix $R \in$ SO(3) (Special Orthogonal group in three dimensions) and a translation vector $t \in \mathbb{F}^3$ (three-dimensional space). These parameters, R and t, define the camera's orientation and position relative to the face. By applying this camera model, we can project the 3D facial information onto a 2D image plane, facilitating further analysis and processing of the face data.

3.3 Illumination Model

The illumination model used is based on the concept of spherical harmonics (SH) [28, 29] basis functions $H_b: \mathbb{F}^3 \to \mathbb{F}$. The colour C at a vertex with normal vector n and tangent vector t, parameterized by the coefficients γ, can be expressed as the dot product between t and the linear combination of spherical harmonic basis functions:

$$C(n, t|\gamma) = t \cdot (\gamma 1 * \Phi 1(n) + \gamma 2 * \Phi 2(n) + ... + \gamma B * \Phi B(n)) \tag{3}$$

In Eq. (3), $\Phi 1(n)$, $\Phi 2(n)$,..., $\Phi B(n)$ represent the spherical harmonic basis functions evaluated at the normal vector n. The coefficients $\gamma 1, \gamma 2, ..., \gamma B$ are the weights associated with each basis function.

3.4 Model Fitting

Model fitting is a crucial stage in the reconstruction process, as it seeks to optimize the parameters of the 3D face model for precise alignment with the face in the input image and detected landmarks.

Face and Landmark Detection. Before initiating the model fitting process, the input image undergoes a series of preprocessing steps. Initially, the face region is detected using multi-task Cascaded Convolutional Networks (MTCNN) [30]. Subsequently, 68 facial landmarks are extracted using the landmark detection model presented by [31].

Loss Functions. These functions are used to measure the discrepancy between the expected values and the actual data during the model-fitting process.

Photometric Loss. The resemblance between the rendered image created by the 3D model and the input image is determined by comparing their colour and texture. This comparison is performed using a skin-aware photometric loss, as described by [32] given by the Eq. (4):

$$\mathcal{L}_p(x) = \frac{\sum_{i \in \mathcal{M}} A_i \cdot \|I_i - I_i'\|_2}{\sum_{i \in \mathcal{M}} A_i} \tag{4}$$

In this equation, i represents the pixel index, $\mathcal{M}$ represents the projected face region, and A_i is the skin colour-based attention mask.

Reflectance Loss. We use the naive Bayes classifier of mixture models [33] to compute the skin-colour probability P_i for each pixel i in order to handle difficult and complicated facial appearances, such as occlusions like beards and makeup. This is shown in the Eq. (5) and (6):

$$A_i = \begin{cases} 1, & if P_i > 0.5 \\ P_i, & otherwise \end{cases} \tag{5}$$

Therefore, predicted reflectance loss is calculated by

$$\mathcal{L}_R(x) = \frac{1}{|S|} \cdot \sum_{i \in S} R_i'^2 \tag{6}$$

where, $|S|$ is the number of skin pixels and $R_i'^2$ is the difference between the predicted reflectance and the mean reflectance for pixel I.

Landmark Loss. It calculates the distance between the projected landmarks of the 3D model and the corresponding detected landmarks in the input image to ensure precise alignment. For landmark loss during the detection, we use Eq. (7):

$$\mathcal{L}_l(x) = \frac{1}{N} \sum_{n=1}^{N} \omega_n \|q_n - q_n'(x)\|^2 \tag{7}$$

Here, ω_n represents the manually assigned landmark weight for specific landmarks such as mouth and nose points.

Gamma Loss. The gamma loss encourages consistent gamma correction by measuring the deviation of gamma correction parameters from their mean value, as shown in Eq. (8):

$$\mathcal{L}_g(x) = \|\triangle \lambda\|^2 \tag{8}$$

where $\triangle \lambda$ is the difference between the gamma correction parameters and their mean value.

4   Results Analysis

Our experimental setup involved the ExPW dataset [34], which consists of approximately 91, 793 "in-the-wild" images with seven fundamental expression categories assigned to each face image; as well as other images found on the internet. The experimental setup included an Intel Core i7 processor, an NVIDIA RTX 3050 Ti graphics card, and 16GB of RAM. The implemented method combines DL and computer vision techniques for face fitting and 3D reconstruction. We utilized Python programming language and leveraged the open-source libraries OpenCV [35], Pytorch3D [36], and NumPy [37] for implementation.

We employed the MTCNN algorithm [30] to detect faces in images, resizing them to 224x224 pixels. For landmark detection, the face-alignment method [31] was utilized. The fitting process began with refining the BFM's pose (rotation and translation) to align with the detected face. Subsequently, the BFM was deformed to capture shape and expression details. Fitting was optimized using the Adam optimizer [38] to minimize the discrepancy between projected 3D landmarks of the BFM and extracted 3D landmarks from the image. The optimizer also minimized a combination of various loss terms, iteratively refining BFM parameters for minimizing overall loss. After fitting, optimized BFM parameters rendered a deformed face image. This image was composited with the original input, replacing the face region. The composed image, BFM coefficients, and mesh could be saved as output for diverse applications.

To assess the performance of our approach, we conducted comparisons with state-of-the-art approaches and relevant baseline methods. The evaluation encompasses both qualitative visual comparisons and quantitative analysis utilizing a variety of loss metrics. Figures 2 and 3 illustrate the outcomes of our approach juxtaposed with those of prominent state-of-the-art techniques. The visual comparisons vividly underscore the strengths of our method in capturing intricate facial details, expressions, and lifelike texture mapping. Across various test images, our method consistently generates more accurate and realistic 3D facial reconstructions, effectively preserving the nuances of individual appearances. Table 1 showcases a comprehensive summary of the computed losses across different types. This table presents representative values for each loss category, complemented by their corresponding mean and standard deviation. These metrics not only offer a concise snapshot of the experimental results but also provide insights into the dispersion and trends of the loss values.

While direct comparisons are limited by dataset variations and evaluation metrics, our method exhibits promising performance in terms of facial reconstruction and expression preservation. Our method offers several significant advantages. Firstly, it eliminates the need for manual marking of landmarks, which is required by many other methods. Additionally, it is computationally efficient as it only requires a single image for efficient 3D reconstruction. This efficiency is achieved through less expensive computations and faster processing speeds, enabling real-time implementation.

One aspect to consider is that the method may encounter challenges when dealing with occlusions, such as individuals wearing sunglasses, despite its ability to handle faces with spectacles. In such cases, the reconstructed 3D face may exhibit dark areas under the eyes, reflecting the colour of the sunglasses. Addressing these occlusion challenges and improving the generation of realistic facial features in such scenarios would be a valuable avenue for future enhancements. Furthermore, refining the model's ability to reproduce finer details, including wrinkles and eye tracking can contribute to achieving even greater realism in the reconstructed 3D faces.

Input Image   Tewari et al. [39]   Tran et al. [13]   Ours

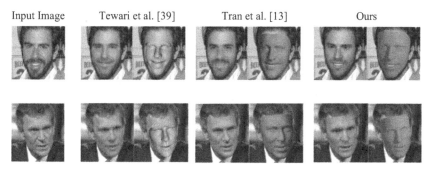

Fig. 2. The visual comparison of our outcomes with other innovative methods.

Table 1. Summary of the calculated losses for different loss types. It includes three representative values for each loss type to provide a concise representation of the results along with their mean and standard deviation. These values provide insights into the variations and distribution of the losses, offering a concise overview of the experimental results.

Loss Type	Measure 1	Measure 2	Measure 3	Mean	Standard Deviation
Landmark Loss $\mathcal{L}_l(x)$	0.000062	0.000098	0.000152	0.000104	0.000045
Photometric loss $\mathcal{L}_p(x)$	0.053055	0.095443	0.093362	0.080953	0.020273
Texture Loss $\mathcal{L}_T(x)$	0.006199	0.009445	0.007481	0.007375	0.001303

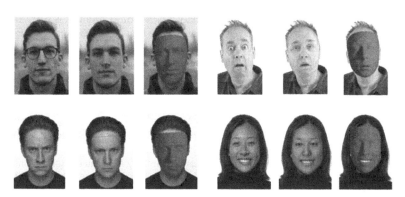

Fig. 3. The figure presents a compilation of reconstructed 3D faces utilizing our proposed method, highlighting its ability to generate realistic facial reconstructions.

5 Conclusion

In this research, we demonstrated a technique for generating a 3D facial model from a single input image. It incorporates face and landmarks detection models to precisely locate and extract the facial region. The approach enhances the quality of the reconstructed face by using photometric consistency constraints, local corrective deformation fields, and coarse-to-fine optimization. The use of a single image eliminates the need for multiple images or complex scanning setups, making the reconstruction process more practical and cost-effective. It further utilizes fitting process optimizations to minimize various loss functions, resulting in a refined and realistic 3D reconstruction, enabling further analysis and application possibilities.

Overall, the paper presents a powerful approach for 3D face reconstruction, with potential applications in computer graphics, virtual reality, facial animation, and biometrics. While the effectiveness relies on the performance of the detection models and input image quality, fine-tuning the optimization parameters can further enhance the accuracy and fidelity of the reconstruction. By providing a comprehensive solution for reconstructing 3D faces from a single image, this paper opens doors for advancements in facial modelling and realistic virtual representations.

References

1. Widanagamaachchi, W.N., Dharmaratne, A.T.: 3D Face Reconstruction from 2D Images. In: 2008 Digital Image Computing: Techniques and Applications, pp. 365–371. IEEE (2008). https://doi.org/10.1109/DICTA.2008.83
2. Zollhöfer, M., et al.: State of the Art on Monocular 3D Face Reconstruction, Tracking, and Applications. Computer Graphics Forum. **37**, 523–550 (2018). https://doi.org/10.1111/cgf.13382
3. Afzal, H.M.R., Luo, S., Afzal, M.K., Chaudhary, G., Khari, M., Kumar, S.A.P.: 3D Face Reconstruction From Single 2D Image Using Distinctive Features. IEEE Access. **8**, 180681–180689 (2020). https://doi.org/10.1109/ACCESS.2020.3028106
4. Diwakar, M., Kumar, P.: 3-D Shape Reconstruction Based CT Image Enhancement. In: Handbook of Multimedia Information Security: Techniques and Applications, pp. 413–419. Springer International Publishing, Cham (2019). https://doi.org/10.1007/978-3-030-15887-3_19
5. Uddin, M., Manickam, S., Ullah, H., Obaidat, M., Dandoush, A.: Unveiling the Metaverse: Exploring Emerging Trends, Multifaceted Perspectives, and Future Challenges. IEEE Access. 1–1 (2023). https://doi.org/10.1109/ACCESS.2023.3281303
6. Jha, J., et al.: Artificial intelligence and applications. In: 2023 1st International Conference on Intelligent Computing and Research Trends (ICRT), pp. 1–4. IEEE (2023). https://doi.org/10.1109/ICRT57042.2023.10146698
7. Sharma, H., Kumar, H., Gupta, A., Shah, M.A.: Computer Vision in Manufacturing: A Bibliometric Analysis and future research propositions. Presented at the (2023)
8. Khari, M., Garg, A.K., Gonzalez-Crespo, R., Verdú, E.: Gesture Recognition of RGB and RGB-D Static Images Using Convolutional Neural Networks. Int. J. Interact. Multi. Artifi. Intell. **5**, 22 (2019)
9. Blanz, V., Vetter, T.: A morphable model for the synthesis of 3D faces. In: Proceedings of the 26th annual conference on Computer graphics and interactive techniques, pp. 187–194 (1999)

10. Kittler, J., Huber, P., Feng, Z.-H., Hu, G., Christmas, W.: 3D Morphable Face Models and Their Applications. Presented at the (2016). https://doi.org/10.1007/978-3-319-41778-3_19
11. Booth, J., Roussos, A., Ponniah, A., Dunaway, D., Zafeiriou, S.: Large Scale 3D Morphable Models. Int. J. Comput. Vis. **126**, 233–254 (2018)
12. Tran, L., Liu, X.: Nonlinear 3D Face Morphable Model. In: 2018 IEEE/CVF Conference on Computer Vision and Pattern Recognition, pp. 7346–7355. IEEE (2018)
13. Tran, L., Liu, X.: On Learning 3D Face Morphable Model from In-the-wild Images. IEEE Trans. Pattern Anal. Mach. Intell. **43**, 157–171 (2019). https://doi.org/10.1109/TPAMI.2019. 2927975
14. Tran, L., Liu, F., Liu, X.: Towards High-Fidelity Nonlinear 3D Face Morphable Model. In: 2019 IEEE/CVF Conference on Computer Vision and Pattern Recognition (CVPR), pp. 1126–1135. IEEE (2019). https://doi.org/10.1109/CVPR.2019.00122
15. Dai, H., Pears, N., Smith, W., Duncan, C.: Statistical modeling of craniofacial shape and texture. Int. J. Comput. Vis. **128**, 547–571 (2020). https://doi.org/10.1007/s11263-019-012 60-7
16. Galanakis, S., Gecer, B., Lattas, A., Zafeiriou, S.: 3DMM-RF: convolutional radiance fields for 3D face modeling. In: 2023 IEEE/CVF Winter Conference on Applications of Computer Vision (WACV), pp. 3525–3536. IEEE (2023)
17. Zhang, W., et al.: SadTalker: Learning Realistic 3D Motion Coefficients for Stylized Audio-Driven Single Image Talking Face Animation. In: Proceedings of the IEEE/CVF Conference on Computer Vision and Pattern Recognition, pp. 8652–8661 (2023)
18. Jiang, L., Zhang, J., Deng, B., Li, H., Liu, L.: 3D face reconstruction with geometry details from a single image. IEEE Trans. Image Process. **27**, 4756–4770 (2018). https://doi.org/10. 1109/TIP.2018.2845697
19. Chen, A., Chen, Z., Zhang, G., Mitchell, K., Yu, J.: Photo-realistic facial details synthesis from single image. In: 2019 IEEE/CVF International Conference on Computer Vision (ICCV), pp. 9428–9438. IEEE (2019). https://doi.org/10.1109/ICCV.2019.00952
20. Gecer, B., Ploumpis, S., Kotsia, I., Zafeiriou, S.: GANFIT: generative adversarial network fitting for high fidelity 3D face reconstruction. In: 2019 IEEE/CVF Conference on Computer Vision and Pattern Recognition (CVPR), pp. 1155–1164. IEEE (2019). https://doi.org/10. 1109/CVPR.2019.00125
21. Lattas, A., et al.: AvatarMe: Realistically Renderable 3D Facial Reconstruction "In-the-Wild." In: 2020 IEEE/CVF Conference on Computer Vision and Pattern Recognition (CVPR), pp. 757–766. IEEE (2020). https://doi.org/10.1109/CVPR42600.2020.00084
22. Yu, W., et al.: NOFA: NeRF-based One-shot Facial Avatar Reconstruction. In: Special Interest Group on Computer Graphics and Interactive Techniques Conference Conference Proceedings, pp. 1–12. ACM, New York, NY, USA (2023)
23. Bai, Z., Cui, Z., Rahim, J.A., Liu, X., Tan, P.: Deep facial non-rigid multi-view stereo. In: 2020 IEEE/CVF Conference on Computer Vision and Pattern Recognition (CVPR), pp. 5849–5859. IEEE (2020). https://doi.org/10.1109/CVPR42600.2020.00589
24. Fu, K., Xie, Y., Jing, H., Zhu, J.: Fast spatial–temporal stereo matching for 3D face reconstruction under speckle pattern projection. Image Vis. Comput. **85**, 36–45 (2019). https://doi. org/10.1016/j.imavis.2019.02.007
25. Wang, X., Guo, Y., Yang, Z., Zhang, J.: Prior-Guided Multi-View 3D Head Reconstruction. IEEE Trans. Multimedia **24**, 4028–4040 (2022)
26. Paysan, P., Knothe, R., Amberg, B., Romdhani, S., Vetter, T.: A 3D face model for pose and illumination invariant face recognition. In: 2009 Sixth IEEE International Conference on Advanced Video and Signal Based Surveillance, pp. 296–301. IEEE (2009). https://doi.org/ 10.1109/AVSS.2009.58

27. Guo, Y., Zhang, J., Cai, J., Jiang, B., Zheng, J.: CNN-based real-time dense face reconstruction with inverse-rendered photo-realistic face images. IEEE Trans. Pattern Anal. Mach. Intell. **41**, 1294–1307 (2019). https://doi.org/10.1109/TPAMI.2018.2837742

28. Ramamoorthi, R., Hanrahan, P.: A signal-processing framework for inverse rendering. In: Proceedings of the 28th annual conference on Computer graphics and interactive techniques, pp. 117–128. ACM, New York, NY, USA (2001). https://doi.org/10.1145/383259.383271

29. Ramamoorthi, R., Hanrahan, P.: An efficient representation for irradiance environment maps. In: Proceedings of the 28th annual conference on Computer graphics and interactive techniques, pp. 497–500. ACM, New York, NY, USA (2001)

30. Schroff, F., Kalenichenko, D., Philbin, J.: FaceNet: a unified embedding for face recognition and clustering. In: 2015 IEEE Conference on Computer Vision and Pattern Recognition (CVPR), pp. 815–823. IEEE (2015). https://doi.org/10.1109/CVPR.2015.7298682

31. Bulat, A., Tzimiropoulos, G.: How far are we from solving the 2D & 3D face alignment problem? (and a Dataset of 230,000 3D Facial Landmarks). In: 2017 IEEE International Conference on Computer Vision (ICCV), pp. 1021–1030. IEEE (2017)

32. Deng, Y., et al.: Accurate 3D face reconstruction with weakly-supervised learning: from single image to image set. In: 2019 IEEE/CVF Conference on Computer Vision and Pattern Recognition Workshops (CVPRW), pp. 285–295. IEEE (2019). https://doi.org/10.1109/CVPRW.2019.00038

33. Jones, M.J., Rehg, J.M.: Statistical color models with application to skin detection. Int. J. Comput. Vis. **46**, 81–96 (2002). https://doi.org/10.1023/A:1013200319198

34. Hou, Z.-D., Kim, K.-H., Lee, D.-J., Zhang, G.-H.: Real-time markerless facial motion capture of personalized 3D real human research. Int. J. Inter. Broadcas. Comm. **14**, 129–135 (2022)

35. OpenCV: Open Source Computer Vision Library (2015)

36. Johnson, J., et al.: Accelerating 3D deep learning with PyTorch3D. In: SIGGRAPH Asia 2020 Courses, p. 1. ACM, New York, NY, USA (2020). https://doi.org/10.1145/3415263.3419160

37. Harris, C.R., et al.: Array programming with NumPy. Nature **585**, 357–362 (2020)

38. Kingma, D.P., Ba, J.: Adam: a method for Stochastic Optimization. CoRR. abs/1412.6980 (2014)

39. Tewari, A., et al.: MoFA: Model-Based Deep Convolutional Face Autoencoder for Unsupervised Monocular Reconstruction. In: 2017 IEEE International Conference on Computer Vision (ICCV), pp. 3735–3744. IEEE (2017)

Human Activity Recognition with a Time Distributed Deep Neural Network

Gunjan Pareek[1], Swati Nigam[1,2(✉)], Anshuman Shastri[2], and Rajiv Singh[1,2]

[1] Department of Computer Science, Banasthali Vidyapith, Rajasthan 304022, India
swatinigam.au@gmail.com
[2] Centre for Artificial Intelligence, Banasthali Vidyapith, Rajasthan 304022, India
anshumanshastri@banasthali.in

Abstract. Human activity recognition (HAR) is necessary in numerous domains, including medicine, sports, and security. This research offers a method to improve HAR performance by using a temporally distributed integration of convolutional neural networks (CNN) and long short-term memory (LSTM). The proposed model combines the advantages of CNN and LSTM networks to obtain temporal and spatial details from sensor data. The model efficiently learns and captures the sequential dependencies in the data by scattering the LSTM layers over time. The proposed method outperforms baseline CNN, LSTM, and existing models, as shown by experimental results on benchmark datasets UCI-Sensor and Opportunity-Sensor dataset and achieved an accuracy of 97% and 96%, respectively. The results open up new paths for real-time applications and research development by demonstrating the promise of the temporally distributed CNN-LSTM model for improving the robustness and accuracy of human activity recognition from sensor data.

Keywords: Sensor Data · Action Identification · Convolution Neural Network · Time Distributed Feature Extraction · Long Short-Term Memory

1 Introduction

Healthcare monitoring, sports analysis, and human-computer interaction (HCI) are just a few of the areas where human activity recognition (HAR) is finding increasing use [1–3]. Recognizing human behavior accurately from sensor data in real-time is crucial for delivering individualized and contextualized support. Recent years have seen encouraging results from deep learning models in HAR, since they can automatically generate discriminative characteristics from raw sensor data.

Owing to their different strengths in capturing temporal and spatial correlations, deep learning architectures such as convolutional neural networks (CNN) [6] and long short-term memory (LSTM) [4] have been extensively used. However, there are limits to using either CNN or LSTM models. While CNN models are more inclined towards spatial data than temporal dynamics, LSTM models have difficulty capturing long-term relationships.

B. J. Choi et al. (Eds.): IHCI 2023, LNCS 14532, pp. 127–136, 2024.
https://doi.org/10.1007/978-3-031-53830-8_13

In order to alleviate the existing limitations, we offer a technique to improve HAR performance by applying a time-distributed CNN-LSTM model to sensor data. To extract both temporal and spatial characteristics from sensor input, the temporally distributed CNN-LSTM network associates the improvements of CNN and LSTM architectures. To better recognize activity patterns across time, the proposed model uses a time-distributed LSTM to capture the sequential dependencies in the data. However, the model can gather important information across several sensor channels since the CNN layers focus on extracting spatial characteristics from sensor input.

The aim of this study is to assess the effectiveness of the proposed time-distributed CNN-LSTM model in enhancing HAR, relative to the conventional CNN and LSTM models. We test the model's efficiency using publically available datasets. We expect the suggested technique to greatly enhance the accuracy and reliability of human activity detection from sensor data by harnessing the combined strength of CNN and LSTM architectures.

The remainder of the paper is structured as follows: In Sect. 2, we examine relevant work in the area of sensor data and the limits of existing methods. The proposed temporally distributed CNN-LSTM model is described in depth, including its architecture and training method, in Sect. 3. The experimental design, including datasets, measures for success, and implementation specifics, is presented in Sect. 4. The research finishes with Sect. 5, in which the contributions are summarized.

2   Related Works

There has been a lot of work done on HAR. Researchers have investigated a wide range of approaches to improve the robustness and precision of action recognition systems. Here, we summarize recent research that has improved HAR using sensor data.

Representational analysis of neural networks for HAR using transfer learning is described by An et al. [1]. To compare and contrast the neural network representations learned for various activity identification tasks, they proposed a transfer learning strategy. The results show that the suggested strategy is useful for increasing recognition accuracy with little additional training time.

Ismail et al. [2] offer AUTO-HAR, an automated CNN architecture design-based HAR system. They present a system that mechanically generates an activity-recognition-optimized CNN structure. The recognition performance is enhanced due to the framework's excellent accuracy and flexibility across datasets.

A storyline evaluation of HAR in an AI frame is provided by Gupta et al. [4]. This study compiles and assesses many techniques, equipment, and datasets that have been requested for the problem of human activity identification. It gives an overview of the state-of-the-art techniques and talks about the difficulties and potential future developments in this area.

Gupta et al. [6] offer a method for HAR based on deep learning and the information gathered from wearable sensors. In particular, convolutional and recurrent neural networks, two types of deep learning models, are investigated for their potential. Findings show that the suggested method is efficient in obtaining high accuracy for activity recognition.

A transfer learning strategy for human behaviors employing a cascade neural network architecture is proposed by Du et al. [11]. The approach takes the lessons acquired from one activity recognition job and applies them to another similar one. This research shows that the cascade neural network design is superior at identifying commonalities across different types of motion.

Wang et al. [13] provide a comprehensive overview of deep learning for HAR based on sensor data. Their work summarizes many deep learning models and approaches that have been applied to the problem of activity recognition. It reviews the previous developments and talks about the difficulties and potential future paths.

For HAR using wearable sensors, CNN is proposed by Rueda et al. [14]. The research probes several CNN designs and delves into the merging of sensor data from various parts of the body. Findings prove that CNN can reliably identify actions from data collected by sensors worn on the body.

A multi-layer parallel LSTM network for HAR using smartphone sensors is presented by Yu et al. [15]. In order to extract both three-dimensional and sequential characteristics from sensor input, the network design makes use of parallel LSTM layers. The experimental findings demonstrate the effectiveness of the proposed network in performing activity recognition tasks. A few other methods are described in [15–17].

3   The Proposed Method

The block diagram of the proposed method for improving human activity identification using a temporally distributed CNN-LSTM model using sensor data is shown in Fig. 1. Each component of the block diagram is described here.

3.1   Input Dataset

The study uses two datasets, namely UCI-Sensor [2] and Opportunity-Sensor [5], as input data. These datasets contain sensor readings captured during various human activities.

3.2   Data Pre-processing

The input data undergoes pre-processing steps, including null removal and normalization. Null removal involves handling missing or incomplete data, while normalization ensures that the data is scaled and standardized for better model performance.

3.3   Time Distributed Frame Conversion

The pre-processed data is then converted into time-distributed frames. This step involves splitting the data into smaller frames based on a specific time step and the total number of sensor channels. This enables the model to capture temporal dynamics and extract features from the data.

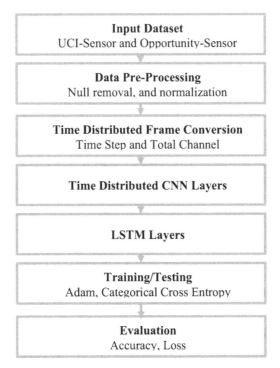

Fig. 1. Block diagram of the proposed time distributed CNN-LSTM model.

3.4 Time Distributed CNN Layers

Convolutional neural network (CNN) layers play a crucial role in handling the time-distributed frames. These CNN layers are designed to enable the model to identify significant patterns and structures by extracting spatial attributes from the input sensor data. A typical convolutional layer consists of numerous convolution kernels or filters.

Let us designate the number of convolution kernels as K. Each individual kernel is tasked with capturing distinct features, thereby generating a corresponding feature matrix. When employing K convolution kernels, the convolutional operation's output would consist of K feature matrices, which can be illustrated as:

$$Zk = f(WK*X + b) \tag{1}$$

In this given context, let X denote the input data with dimensions $m \times n$. The Kth convolution kernel with dimensions $k_1 \times k_2$ is represented by W_K, and the bias is denoted by 'b'. The convolution operation is depicted by ' $*$ '. The dimension of the Kth feature matrix Z_k depends on the chosen stride and padding method during the convolution operation. For instance, when using a stride of $(1,1)$ and no padding, the size of Z_k becomes $(m - k_1 + 1) \times (n - k_2 + 1)$. The function f signifies the selected nonlinear activation function, applied to the output of the convolutional layer. Common activation functions include sigmoid, tanh, and ReLU.

3.5 LSTM Layers

The layers get the results from the CNN layers. Temporal dependencies in the data may be captured and learned by the LSTM layers. The network's ability to learn and anticipate future activity sequences is greatly enhanced by the addition of LSTM layers. LSTM utilizes three gates to manage the information flow within the network. The forget gate (ft) regulates the extent to which the previous state ($ct - 1$) is preserved. The input gate (it) decides whether the current input should be employed to update the LSTM's information. The output gate (ot) dictates the specific segments of the current cell state that should be conveyed to the subsequent layer for further iteration.

$$ft = \sigma(W(f)xt + V(f)ht - 1 + bf) \tag{2}$$

$$it = \sigma(W(i)xt + V(i)ht - 1 + bi) \tag{3}$$

$$ot = \sigma(W(o)xt + V(o)ht - 1 + bo) \tag{4}$$

$$ct = ft \otimes ct - 1 + it \otimes \tanh(W(c)xt + V(c)ht - 1 + bc) \tag{5}$$

$$ht = ot \otimes \tanh(ct) \tag{6}$$

Here, xt represents the input data fed into the memory cell during training, while ht signifies the output within each cell. Additionally, W, V, and b denote the weight matrix and biases correspondingly. The function σ refers to the sigmoid activation, which governs the significance of the message being propagated, and $\otimes$ indicates the dot product operation.

3.6 Training and Testing

Loss function "categorical cross-entropy" and "Adam" as an optimizer are used during training and testing. During training, the model uses the annotated data to fine-tune its settings and becomes better at identifying people at work.

3.7 Evaluation

Metrics like accuracy and loss are used to assess the trained model's performance. The accuracy and loss metrics gauge the model's effectiveness in categorizing human behaviors by measuring its precision and accuracy, respectively. The model's overall performance and its capacity to reliably distinguish various actions may be depicted from these assessment indicators.

4 Experimental Results and Discussion

4.1 UCI Sensor Dataset [2] Results

Six basic human activities—walking, sitting, standing, laying down, walking upstairs and downstairs are represented in the UCI-HAR [2] machine learning repository dataset. The information was collected from 30 people (aged 19 to 48) using an Android mobile device (Galaxy S2) equipped with inertial sensors. This dataset also includes transitions between other types of stationary postures, such as standing to sit, sitting to stand, lying to sit, laying to stand, and standing to laying.

The accuracy and loss calculated for each epoch for the proposed CNN-LSTM model are shown in Fig. 2. The confusion matrix for the proposed method is shown in Fig. 3 for six activities, and classification report is shown in Fig. 4 for the UCI-Sensor dataset. A comparison with the state of the art [1, 2, 4, 6], and baseline CNN and LSTM models is shown in Table 1. From this comparative analysis, one can conclude that the proposed model performs better.

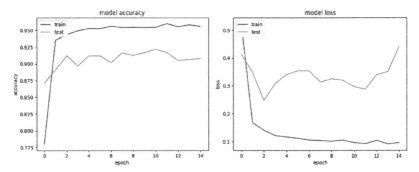

Fig. 2. Accuracy-loss plot for the proposed CNN-LSTM model.

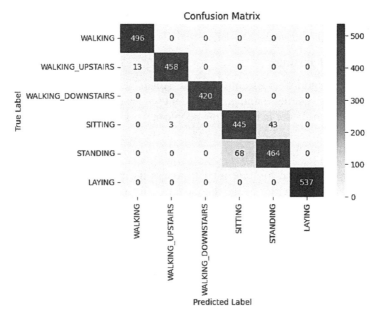

Fig. 3. Confusion matrix for the proposed CNN-LSTM model.

```
              precision    recall  f1-score   support

         0       0.97      1.00      0.99       496
         1       0.99      0.97      0.98       471
         2       1.00      1.00      1.00       420
         3       0.87      0.91      0.89       491
         4       0.92      0.87      0.89       532
         5       1.00      1.00      1.00       537

  accuracy                          0.96      2947
 macro avg       0.96      0.96      0.96      2947
weighted avg     0.96      0.96      0.96      2947
```

Fig. 4. Classification report for the proposed CNN-LSTM model.

Table 1. UCI-sensor dataset comparative analysis.

Model	Accuracy (%)	Precision (%)	Recall (%)	F1-Score (%)
Time-Series CNN [1]	93.09	91.10	92.10	92.10
Parallel LSTM [2]	94.34	93.86	93.34	93.80
Feature Learning CNN [4]	67.51	66.80	66.78	67.35
Auto-Har [6]	94.80	94.65	94.70	95
Baseline CNN	74	75	73	73
Baseline LSTM	43	43	42	38
Time Distributed CNN-LSTM	**96**	**96**	**96**	**96**

4.2 OPPORTUNITY Sensor Dataset Results

Standing, laying down, walking, and navigating the stairwell are only some of the six basic human actions included in the Opportunity [5] machine learning repository dataset. Thirty people, ranging in age from 19 to 48, were surveyed using Android smartphones (Samsung Galaxy S II) equipped with inertial sensors. This dataset also includes transitions between other static postures, such as sitting, standing, lying, laying, sitting, lying, and standing.

The accuracy and loss calculated for each epoch for the proposed CNN-LSTM model are shown in Fig. 5. The confusion matrix for the proposed method is shown in Fig. 6 for six activities and classification report is shown in Fig. 7 for OPPORTUNITY-Sensor dataset. A comparison with the state of the art [11, 13–15], and baseline CNN and LSTM models is shown in Table 2. From this comparative analysis, one can conclude that the proposed model performs better.

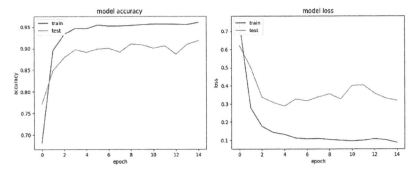

Fig. 5. Accuracy-loss plot for the proposed CNN-LSTM model.

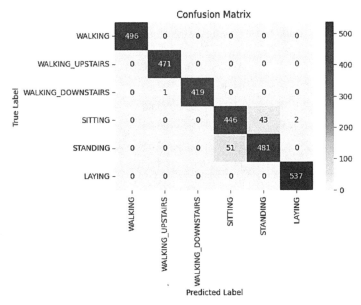

Fig. 6. Confusion matrix for the proposed CNN-LSTM model.

```
              precision    recall  f1-score   support

           0       1.00      1.00      1.00       496
           1       1.00      1.00      1.00       471
           2       1.00      1.00      1.00       420
           3       0.90      0.91      0.90       491
           4       0.92      0.90      0.91       532
           5       1.00      1.00      1.00       537

    accuracy                           0.97      2947
   macro avg       0.97      0.97      0.97      2947
weighted avg       0.97      0.97      0.97      2947
```

Fig. 7. Classification report for the proposed CNN-LSTM model.

Table 2. Opportunity dataset comparative analysis.

Model	Accuracy (%)	Precision (%)	Recall (%)	F1-Score (%)
Hybrid M [11]	46.68	46.75	46.70	47
b-LSTM-S [13]	92.70	92.45	92.10	92.90
InnoHAR [14]	94.60	94.20	94.20	94.80
CNN [15]	93.70	93.70	93.70	93.70
CNN	73	74	72	72
LSTM	38	34	35	27
Time Distributed CNN-LSTM	**97**	**97**	**97**	**97**

5   Conclusions

This research shows that a time-distributed CNN-LSTM model using sensor data significantly improves the performance of human activity recognition. The proposed model outperforms baseline CNN and LSTM, and other existing models, as shown by experimental results on the UCI-Sensor dataset and the Opportunity-Sensor dataset. The temporally distributed CNN-LSTM model achieved 97% accuracy for the Opportunity-Sensor dataset and 96% accuracy for the UCI-Sensor dataset across the board. These results demonstrate the value of integrating CNN and LSTM architectures to better capture temporal and spatial characteristics, which in turn enhances the accuracy and reliability of human activity classification from sensor data. Improving the effectiveness and scalability of the proposed model may require more investigation into broadening the assessment to other datasets and investigating optimization strategies.

References

1. An, S., Bhat, G., Gumussoy, S., Ogras, U.: Transfer learning for human activity recognition using representational analysis of neural networks. ACM Transactions on Computing for Healthcare 4(1), 1–21 (2023)
2. Ismail, W.N., Alsalamah, H.A., Hassan, M.M., Mohamed, E.: AUTO-HAR: An adaptive human activity recognition framework using an automated CNN architecture design. Heliyon 9(2), e13636 (2023). https://doi.org/10.1016/j.heliyon.2023.e13636
3. Nigam, S., Singh, R., Misra, A.K.: A review of computational approaches for human behavior detection. Archives of Computational Methods in Engineering 26, 831–863 (2019)
4. Gupta, N., Gupta, S.K., Pathak, R.K., Jain, V., Rashidi, P., Suri, J.S.: Human activity recognition in artificial intelligence framework: a narrative review. Artif. Intell. Rev. 55(6), 4755–4808 (2022)
5. Ciliberto, M., Fortes Rey, V., Calatroni, A., Lukowicz, P., Roggen, D.: Opportunity++: A multimodal dataset for video- and wearable, object, and ambient sensors-based human activity recognition. Frontiers in Computer Science 3, 1–7 (2021). https://doi.org/10.3389/fcomp.2021.792065
6. Gupta, S.: Deep learning based human activity recognition (HAR) using wearable sensor data. Int. J. Info. Manage. Data Insights 1(2), 100046 (2021). https://doi.org/10.1016/j.jjimei.2021.100046

7. Lv, T., Wang, X., Jin, L., Xiao, Y., Song, M.: Margin-based deep learning networks for human activity recognition. Sensors **20**(7), 1871 (2020)
8. Cruciani, F., et al.: Feature learning for human activity recognition using convolutional neural networks: a case study for inertial measurement unit and audio data. CCF Trans. Pervasive Comp. Interac. **2**(1), 18–32 (2020)
9. Shuvo, M.M.H., Ahmed, N., Nouduri, K., Palaniappan, K.: A hybrid approach for human activity recognition with support vector machine and 1d convolutional neural network. A hybrid approach for human activity recognition with support vector machine and 1D convolutional neural network. In: 2020 IEEE Applied Imagery Pattern Recognition Workshop (AIPR), pp. 1–5. IEEE, Washington DC, USA (2020)
10. Nematallah, H., Rajan, S.: Comparative study of time series-based human activity recognition using convolutional neural networks. In: 2020 IEEE International Instrumentation and Measurement Technology Conference (I2MTC), pp. 1–6. IEEE, Dubrovnik, Croatia (2020)
11. Du, X., Farrahi, K., Niranjan, M.: Transfer learning across human activities using a cascade neural network architecture. In: 2019 ACM International Symposium on Wearable Computers, pp. 35–44. London United Kingdom (2019)
12. Xu, C., Chai, D., He, J., Zhang, X., Duan, S.: InnoHAR: A deep neural network for complex human activity recognition. IEEE Access **7**, 9893–9902 (2019)
13. Wang, J., Chen, Y., Hao, S., Peng, X., Hu, L.: Deep learning for sensor-based activity recognition: a survey. Pattern Recogn. Lett. **119**, 3–11 (2019)
14. Rueda, F.M., Grzeszick, R., Fink, G.A., Feldhorst, S., Ten Hompel, M.: Convolutional neural networks for human activity recognition using body-worn sensors. Informatics **5**(2), 1–17 (2018)
15. Yu, T., Chen, J., Yan, N., Liu, X.: A multi-layer parallel LSTM network for human activity recognition with smartphone sensors. In: 2018 10th International conference on wireless communications and signal processing (WCSP), pp. 1–6. IEEE, Hangzhou, Zhejiang, China (2018)
16. Hammerla, N.Y., Halloran, S., Plötz, T.: Deep, convolutional, and recurrent models for human activity recognition using wearables. In: 25th International Joint Conference on Artificial Intelligence (IJCAI), pp. 1533–1540. New York, USA (2016)
17. Nigam, S., Singh, R., Singh, M.K., Singh, V.K.: Multiview human activity recognition using uniform rotation invariant local binary patterns. J. Ambient. Intell. Humaniz. Comput. **14**(5), 4707–4725 (2023)

Intelligent Systems

Artificial Neural Network to Estimate Deterministic Indices in Control Loop Performance Monitoring

John A. Gómez-Múnera[1] (ID), Luis Díaz-Charris[2](✉) (ID), and Javier Jiménez-Cabas[2] (ID)

[1] Corporación de Estudiantes y Profesionales de Marinilla (CORUM), Marinilla, Antioquia, Colombia
[2] Universidad de la Costa, Barranquilla, Atlántico, Colombia
ldiaz28@cuc.edu.co

Abstract. In many industrial processes, the control systems are the most critical components. Evaluate performance and robustness of a control loops is an important task to maintain the health of a control system and an efficiency in the process. In the area of Control-Loop Performance Monitoring (CPM), there are two groups of indices to evaluate the performance of the control loops: stochastic and deterministic. Using stochastic indices, a control engineer can calculate the performance indices of a control loop with the data in normal operation and a minimum knowledge of the process; but the problem is that to do a performance analysis is so hard, due it is necessary an advanced knowledge about the interpretation. Instead, an interpretation or analysis of deterministic indices is simpler; however, the problem with this approach is that an invasive monitoring of the plant is required to calculate the indices. In this paper, it is proposed to use an Artificial Neural Network to estimate deterministic indices, considering as input the stochastic indices and some process information, taking advantage of the fact that data collection for stochastic indices is simpler.

Keywords: Artificial Neural Network · Control Loop Performance Monitoring · Deterministic Indices · Stochastic Indices

1 Introduction

The primary objective of control systems revolves around maximizing profits through the transformation of raw materials into finished products, reducing production costs, and adhering to various standards, including product quality, operational conditions a restrictions, safety measures, regulations and standards. As a result, the development, adjustment, and application of control strategies take place in the initial stage to address control issues. While PID-type controllers are widely utilized in various industries, research has also explored optimal control for linear problems [1–4] and trajectory linearization for nonlinear issues [5].

Upon successful implementation, the initial phase aims for a well-functioning control system. However, changes in materials, operational strategies, and plant conditions can lead to reduced performance in the control system over time. Even well-designed control loops may encounter issues, necessitating strategy adjustments or redesigning due to factors like the degradation of components (e.g., sensors and actuators) operating within the loop [6, 7].

Consequently, the second phase in addressing control problems involves monitoring control loops to swiftly detect any decline in performance [8, 9]. Industries dealing with process management face mounting pressure to enhance product quality, meet delivery schedules, improve productivity, and adhere to environmental standards, urging them to operate their facilities at maximum efficiency. Hence, the demand for consistently high-performing control systems. As a result, control systems are increasingly recognized as valuable assets requiring regular automated maintenance, monitoring, and evaluation. Presently, these tasks fall under the domain of Control Performance Monitoring (CPM), gaining significant attention from both academic and industrial sectors in recent decades.

Evaluating control loop performance metrics falls into two categories: stochastic and deterministic. Stochastic indicators require minimal process knowledge and data collection during regular operations for real-time performance evaluation, but they encounter challenges regarding scale and range issues. On the other hand, deterministic indicators offer easy interpretation, yet their estimation necessitates invasive plant tests, making them less practical. Is there a way to merge these two group of techniques? To answer this question, it is suggested creating a model that estimates deterministic indicators while integrating stochastic indicators and specific process information as model inputs.

The key features of control systems or models often demand an understanding that might not always be quickly interpretable by operators in industrial processes. The proposed strategies aim to facilitate easier and quicker interpretations by anyone involved in such processes. Therefore, this work contributes to the realm of human-machine interaction, offering a strategy that streamlines decision-making for operators with general knowledge of process control.

The structure of this paper is: Sect. 2 talks about a background on control performance monitoring and the utilization of neural networks to build an inferential model. Section 3 touch on the methodology carried out in this work. In Sect. 4, the proposed methodology is evaluated using an FOPDT process. Finally, the paper concludes with some key insights.

2   Background

The Control Performance Monitoring techniques are applied by Harris [12], across Minimum Variance Index (MVI) for achieving superior control advantages. This study caused a boom in the study of controllers and how they correlate with the performance of production processes. This growth stems from the demand for accurate and effective control systems. Consequently, numerous research endeavors have focused on exploring, creating, and overseeing control loops within feedback systems. This has led to the development of tools or frameworks like those proposed by Moudgalya [13], which systematically assess the MVI, automating the identification and diagnosis of reasons for inadequate system performance and offering remedies to enhance control performance monitoring [14].

The process of evaluating a control system's performance involves acquiring performance metrics that juxtapose the process's capability under ideal conditions against its actual performance during data collection. Engaging with CPM can result in heightened control performance. The principal purpose is to help optimize the control throughout the lifespan of a system, irrespective of variability in operating conditions. Nevertheless, achieving ideal process control is contingent upon the proper functioning of all system components.

2.1 Control Performance Monitoring

2.2 CPM Performance Indices

Performance metrics must be capable of detecting deficiencies in model tuning and aging, irrespective of disturbances or a range of set-point values within a controller process. Metrics require computing data obtained from the normal operation of the process. While some metrics need to be non-invasive, others require invasive testing of the process.

It is crucial that performance metrics remain realistic and calculable within physical constraints. Moreover, they should offer insights into the causes of poor performance in control systems and measure performance enhancements resulting from controller adjustments.

Typically, within the framework of CPM, a Controller Performance Index (CPI) takes the form of Eq. (1).

$$\eta = \frac{J_{des}}{J_{act}}, \tag{1}$$

J_{des} represents the intended value for a specific performance criterion, often variance, while J_{act} denotes the observed value of the criterion, acquired from measured data within the plant.

Stochastic Indices. Minimum Variance Control (MVC) is a control system that incorporates an online parameter estimator for a linearized model within its structure. This model characterizes the behavior of the nonlinear process around an operational point. MVC aims to minimize the output variance. To achieve this, various indices have been established. One of the most recognized and commonly used is the harris index [17], formulated in Eq. (1). This index compares the system's output variance, σ_{MV}^2, with the minimum variance σ_{MV}^2 obtained from a time series model estimated using operational data.

While the expansion of the Harris index to MIMO systems (multiple inputs and multiple outputs) has been achieved [18], the incorporation of the interaction or equivalence matrix proves to be pivotal. Determining this matrix doesn't solely rely on understanding time delays but also on the data obtained in closed-loop scenarios [19–21]. A practical approach to addressing the MIMO architecture and Control Performance Assessment (CPA) involves commencing the process through time series analysis for control loops with a Single Input and Single Output (SISO). This method independently estimates the output variable of the process for each y_i output [22], commonly employing AR/ARMA models. From these models, the reaction to a process impulse is computed, as demonstrated by Jelali in [15], where the initial τ terms of the response remain unaffected by the process model or the controller, relying solely on the characteristics of the disturbance impacting the process.

Additional stochastic indices, as defined in [26], are derived from the autocorrelation function. The primary index, termed *AcorSI*, is represented as follows:

$$AcorSI = \frac{\rho_\tau - CI}{\theta_{cross} - \tau},\tag{2}$$

This index represents the ratio between the autocorrelation value at the process delay time $(\rho\tau)$ minus the Confidence Interval (CI) and the variance between the process delay value (τ) and the delay or lag value just before the curve reaches the confidence interval θ_{cross}. The subsequent index is termed *AcorAr* (Eq. 3).

$$AcorAr = \int \begin{cases} |\rho\kappa| - CI, si|\rho\kappa| > CI \\ 0, si|\rho\kappa| < CI \end{cases} dlag,\tag{3}$$

This index signifies the area outside the confidence interval of the autocorrelation function (ACF) curve.

Deterministic Indices. Deterministic indices rely on the relationship between closed and open-loop rise time R_{tr}, closed and open-loop settling time S_{tr}, gain margin (GM), phase margin (MP), maximum sensitivity (MS), and other factors. As mentioned in [27], assessing driver performance using the MVC-based approach proved to be challenging in terms of interpretation and lacked the ability to evaluate the impact of deterministic changes in a closed-loop system. This led to the introduction of alternative indices that necessitate precise models of the process and the controller. In [28], it is demonstrated that deterministic indices offer a more accurate assessment of loop performance when compared to stochastic methods. However, it's worth noting that real-time quantification of deterministic indices is expensive due to the need for intrusive testing.

For stable systems, managing loop performance involves the use of classical parameters that characterize dynamic systems. In such cases, it becomes essential to calculate the rise time (R_t) required for the response to transition from 5 to 95% of its final value.

2.3 Machine Learning

Selecting an appropriate model is a pivotal aspect of machine learning, involving steps such as postulation, identification, estimation, diagnosis, and verification. Following these steps, the model becomes ready for deployment in a production environment.

Neural Networks. Known as artificial neural networks (ANNs), have found extensive application in areas allowing for prediction or classification due to their inherent capability for non-linear modeling without presumptions about statistical distributions. Instead, these models are adaptively generated based on data, aiming to mirror numerical values that emulate the significance of specific physical systems.

Neural Networks Models. The Inferential Model, a non-linear model, estimates deterministic indices from stochastic ones. In certain scenarios, operational changes are necessary to calculate and quantify the stochastic index.

Construction involves the use of neural networks with three layers: the input layer, the hidden layer, and the output layer [29]. The model's output is established through linear combinations of inputs within a non-linear fixed-base function, where adaptive parameters serve as the coefficients in this combination. The output is represented in Eq. (4).

$$yt = \alpha_0 + \sum_{j=1}^{q} \alpha_j g \left(\beta 0j + \sum_{i=1}^{p} \beta_{ij} y_t - i \right) + \varepsilon_t, \forall t \tag{4}$$

with $y_{(t-1)} (i = 1, 2, ..., p)$ as p inputs and y as the output, where integers p and q represent the number of inputs and hidden nodes or neurons, respectively. The $\alpha_j (j = 0, 1, 2, ..., q)$ and $\beta_{ij} (i = 0, 1, 2, ..., p; j = 0, 1, 2, ..., q)$ denote the weights of connections, while ε_t signifies a random change. Typically, constants α_0, β_{0j} are referred to as the bias term [30]. The term g represents the activation function, dictating the behavior of the node.

The estimation of weights for neuron connections is carried out using least-squares methods, with the most recognized approaches in literature being the backpropagation algorithms or the generalized delta rule [31].

Various metrics frequently employed for regression problems include the Mean Absolute Error (MAE):

$$MAE = \frac{1}{n} \sum_{i=1}^{n} |y_i - \widehat{y}_i|, \tag{5}$$

Offering a straightforward interpretation while maintaining units consistent with the output.

The Mean Squared Error (MSE) is applied during the training of the model with the aim of minimizing it. Utilizing the squared residuals, it increases sensitivity to substantial errors and outliers compared to the Mean Absolute Error (MAE).

$$MSE = \frac{1}{n} \sum_{i=1}^{n} \left(y_i - \widehat{y}_i \right)^2 \tag{6}$$

The Root Mean Square Error (RMSE) similarly accentuates notable errors and outliers (akin to MSE) while preserving consistency with the response units.

$$RMSE = \sqrt{\frac{1}{n} \sum_{i=1}^{n} \left(y_i - \widehat{y}_i \right)^2} \tag{7}$$

R^2 represents the relative variance in the total error during model fitting, providing a value within the range of 0 to 1. When a model fits the data effectively, resulting in minimal error. Conversely, when a model poorly fits the data, resulting in substantial error, R^2 tends toward 0.

$$R^2 = \frac{\sum_{i=1}^{n} (y_i - \overline{y})^2 - \sum_{i=1}^{n} \left(y_i - \widehat{y}_i \right)^2}{\sum_{i=1}^{n} (y_i - \overline{y})^2} \tag{8}$$

2.4 Control-Loop Performance Assessment Whit Machine Learning

Advancements in control-loop performance assessment (CPA) or control-loop performance monitoring (CPM), have reached a pivotal juncture, marked by the addition of machine learning strategies in this field. The current state-of-the-art aims to explore and analyze the forefront of CPA and your relationship with machine learning techniques, shedding light on the recent developments, trends, and implications of the use of these technologies.

In [29], showing a machine learning classification system applied in control performance assessment. The system proposed [29] is dedicated to a wide class of proportional-integral-derivative (PID) control industrial loops; it is capable to distinguish between acceptable and poor performance of the control loops. In [29], 30 deterministic features were taken to generate two datasets: training and validation. The training dataset was labeled with OK and NOK, indicating that a row of data was a good or bad performance of a PID-based control loop. In [31] an iterative learning control (ILC) strategy was evaluated by CPA technics. In this study, the authors proposed a novelty method to evaluate ILC loops, because ILC strategy is little studied in the field of CPA. The difference between [29, 30] and the work presented in this paper, is that in the second one, the algorithms are inputs based on stochastic indices and the output is the deterministic indices.

A rule-based method of pattern recognition for performance monitoring of a proportional-integral-derivative (PID) controller is presented in [32]. In this work, the authors start from 3 training datasets, obtained from three typical responses of control systems. The evidential K-Nearest Neighbour (EKNN) classifier was used to recognize the underlying patterns of an online PID controller. The authors simulate the process of changing system control performance by altering the damping factor parameter. Additionally, they employ the EKNN algorithm to calculate the mass function and determine whether the system's pattern has migrated or shows a tendency to migrate based on the size of the mass functions.

3 Methodology

The inferential model proposed is obtained from the training of a superficial neural network, which has as input parameters deterministic indices and parameters of the system identified through a transfer function of order one plus dead time. The system is the one shown in Fig. 1.

To obtain deterministic indices of a control system, they normally need to vary the operating points of the control system to be analyzed. The strategy proposed below allows you to obtain these indices without having to make these variations.

A data set was generated that served as a reference to train the neural network through simulations with MATLAB and SIMULINK. The parameters of the first-order model shown in Fig. 1 were varied, and a rise time relationship between open loop and closed loop was determined as a design parameter when a proportional-integrative (PI) type controller was applied.

To find the PI controller that adjusted the values of the constants according to the requirements, variations of the constants were made, and the closed loop time was calculated to see if it met the criteria of the design rise time relationship. This process allowed the construction of a dataset that can cover a wide range of types of control processes that have a transfer function with first-order characteristics plus dead time. In this way, for a new analysis of a system, only the identification and recording of the operating points at which the process is working would be necessary.

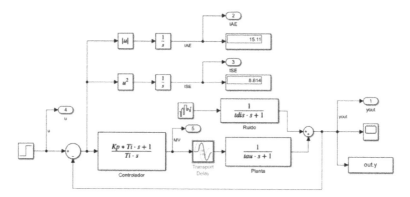

Fig. 1. Control-loop for simulation: First Order Plus Dead Time process (FOPDT).

4 Results

The stochastic indices used as input to the inferential model were: τ, deli, AcorAr CV_{AR}, CV_{AR}. A simulation of a First Order Plus Dead Time process (FOPDT) was conducted, wherein τ represents the process's time constant, deli was computed according to the method outlined in [25], $AcorAr$ was calculated using Eq. (3), and step changes in set-point were employed to quantify CV_{AR}, MV_{AR} and R_{tr}.

Afterwards, the data undergoes pre-processing, which involves arranging the inputs and outputs in a table, cleaning, balancing, and reducing the dataset. Following this stage, an inferential model is constructed using a three-layer neural network. The initial layer requires the inputs, the second layer involves a hidden layer where the number of neurons is adjusted, and ultimately, the output (R_{tr}) is acquired in the third layer.

4.1 The Model with Machine Learning

In the implementation of the Machine Learning model, neural networks were developed using MATLAB, utilizing a dataset created by altering both plant and disturbance parameters. This process resulted in a total of 21,760 data points, derived from various combinations reflecting changes in the stochastic indices. After data pre-processing, a dataset containing 3916 elements was established, with 3500 data points dedicated to training the neural network, and the remaining values are utilized for network validation.

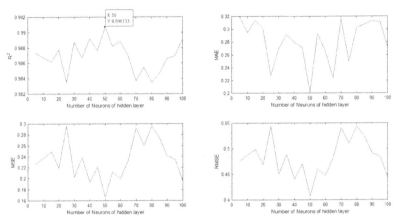

Fig. 2. Metrics of the learning model.

The 3500 data points were segmented into three subsets, allocating 70% for training, 15% for validation, and another 15% for testing the machine learning model. In this methodology, a three-layer network was employed. Initially, the number of neurons in the hidden layer was iteratively increased by five until reaching 100. The subsequent process involved using two distinct activation functions-one for the sigmoid and another for the hyperbolic tangent. This analysis focused on evaluating the value of R^2 and selecting coefficient of determination values closest to 1 (Table 1).

Table 2 displays the metrics derived from the 416 data points that were not utilized in training the machine learning model (Figs. 2 and 3).

Table 1. Metrics for new data

Activation Function	MAE	MSE	RMSE	$R2$
Sigmoidal	0.23999	0.19034	0.43628	0.98939
Hyperbolic tangent	0.27501	0.2273	0.47676	0.98733

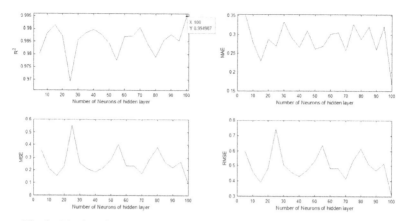

Fig. 3. Metrics of the Hyperbolic tangent activation function learning model.

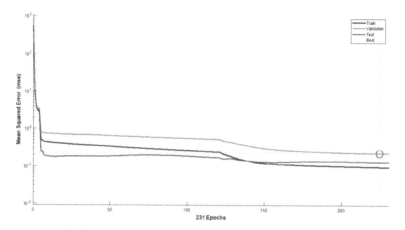

Fig. 4. Cost functional with a sigmoidal activation function.

Table 2. Variation of process parameters to design the machine learning model.

Parameter	Value
τ	$[5, 10 : 10 : 100, \quad 120 : 20 : 200]$
θ	$[1 : 2 : 21, \quad 25 : 5 : 50]$
τ_d	$[30 \ 50 \ 100 \ 200]$
R_{tr}	$[0.1 : 0.1 : 1, \quad 1.5 : 0.5 : 6]$

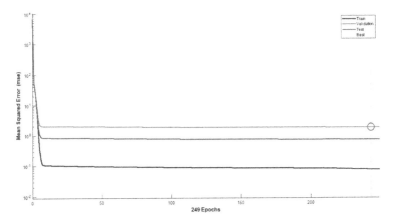

Fig. 5. Cost functional with Hyperbolic tangent activation function.

5 Conclusions

Assessing the performance of control loops typically relies on calculating stochastic indices (like the minimum variance index), which often fail to yield conclusive insights. Hence, deterministic indices are favored due to their ease of interpretation, providing decision-making criteria for corrective actions. However, the drawback of deterministic indices lies in the necessity for set-point alterations or invasive system tests to derive them. Hence, the proposed approach involves building models using machine learning techniques to predict these indices, allowing for their evaluation in industrial environments.

The model, constructed using neural networks, requires input of stochastic indices and process characteristics. It comprises an input layer, a hidden layer, and an output layer. Variations in the number of neurons within the hidden layer were explored to derive performance metrics. The R^2 metric served as a benchmark for selecting the model that best suited the data. Additionally, two activation functions were considered in the hidden layer for comparison purposes.

Figures 4 and 5 initially display notably high training and validation errors, necessitating the calculation of new functions to determine hyperparameters that reduce the error. This error diminishes as the network undergoes training until it reaches a point where the error either increases or reaches a tolerance level. The optimal value of the network's hyperparameters, resulting in the lowest error with the validation set data, is then chosen. Finally, the test set is utilized to verify the outcomes derived from using the network. Notably, an increased number of neurons in the hidden layer provides the network with more adaptability, offering additional parameters for optimization. However, an excessively large hidden layer may lead to under-characterization of the problem.

References

1. Hazil, O., et al.: A robust model predictive control for a photovoltaic pumping system subject to actuator saturation nonlinearity. Sustainability **15**(5), 4493 (2023)
2. Dai, T., Sznaier, M.: Data-driven quadratic stabilization and LQR control of LTI systems. Automatica **153**, 111041 (2023)
3. Jiménez-Cabas, J., Meléndez-Pertuz, F., Ovallos-Gazabon, D., Vélez-Zapata, J., Castellanos, H.E., Cárdenas, C.A., Sánchez, J.F., Jimenez, G., Mora, C., Sanz, F.A. and Collazos, C.A.: Robust control of an evaporator through algebraic Riccati equations and DK iteration. In: Computational science and its applications–ICCSA 2019: 19th international conference, Saint Petersburg, Russia, July 1–4, Proceedings, Part II 19, pp. 731–742. Springer International Publishing (2019)
4. Abdollahzadeh, M., Pourgholi, M.: Adaptive dynamic programming discrete-time LQR optimal control on electromagnetic levitation system with a H∞ Kalman filter. Int. J. Dyn. Control, 1–16 (2023)
5. Gómez, J.A., Rivadeneira, P.S., Costanza, V.: A cost reduction procedure for control-restricted nonlinear systems. IREACO **10**, 1–24 (2017)
6. Huang, X., Song, Y.: Distributed and performance guaranteed robust control for uncertain MIMO nonlinear systems with controllability relaxation. IEEE Trans. Autom. Control **68**(4), 2460–2467 (2022)
7. Borrero-Salazar, A.A., Cardenas-Cabrera, J.M., Barros-Gutierrez, D.A., Jiménez-Cabas, J.A.: A comparison study of MPC strategies based on minimum variance control index performance (2019)
8. Cardenas-Cabrera, J., et al.: Model predictive control strategies performance evaluation over a pipeline transportation system. J. Control Sci. Eng. **2019**, 1–11 (2019)
9. Jiménez-Cabas, J., Manrique-Morelos, F., Meléndez-Pertuz, F., Torres-Carvajal, A., Cárdenas-Cabrera, J., Collazos-Morales, C., González, R.E.: Development of a tool for control loop performance assessment. In: Computational science and its applications–ICCSA 2020: 20th international conference, Cagliari, Italy, July 1–4, 2020, Proceedings, Part II 20, pp. 239–254. Springer International Publishing (2020)
10. Harris, T.J.: Assessment of control loop performance. Canadian J. Chem. Eng. **67**(5), 856–861 (1989)
11. Moudgalya, K.M.: CL 692-Digital Control (2007)
12. Jelali, M.: An overview of control performance assessment technology and industrial applications. Control. Eng. Pract. **14**(5), 441–466 (2006)
13. Jelali, M.: Control performance management in industrial automation: assessment, diagnosis and improvement of control loop performance (2012)
14. Zeroual, A., Harrou, F., Dairi, A., Sun, Y.: Deep learning methods for forecasting COVID-19 time-series data: a comparative study. Chaos Solitons Fractals **140**, 110121 (2020)
15. Desborough, L., Harris, T.: Performance assessment measures for univariate feedforward/feedback control. Canadian J. Chem. Eng. **71**(4), 605–616 (1993)
16. Hajizadeh, I., Samadi, S., Sevil, M., Rashid, M., Cinar, A.: Performance assessment and modification of an adaptive model predictive control for automated insulin delivery by a multivariable artificial pancreas. Ind. Eng. Chem. Res. **58**(26), 11506–11520 (2019)
17. Huang, B., Shah, S.L.: Performance assessment of control loops: theory and applications. Springer Science & Business Media (1999)
18. Del Portal, S.R., Braccia, L., Luppi, P., Zumoffen, D.: Modeling-on-demand-based multivariable control performance monitoring. Comput. Chem. Eng. **168**, 108061 (2022)
19. Ettaleb, L.: Control loop performance assessment and oscillation detection (Doctoral dissertation, University of British Columbia) (1999)

20. Akhbari, A., Rahimi, M., Khooban, M.H.: Various control strategies performance assessment of the DFIG wind turbine connected to a DC grid. IET Electr. Power Appl. **17**(5), 687–708 (2023)
21. Qin, S.J.: Control performance monitoring—a review and assessment. Comput. Chem. Eng. **23**(2), 173–186 (1998)
22. Farenzena, M.: Novel methodologies for assessment and diagnostics in control loop management (2008)
23. Qamsane, Y., Phillips, J.R., Savaglio, C., Warner, D., James, S.C., Barton, K.: Open process automation-and digital twin-based performance monitoring of a process manufacturing system. IEEE Access **10**, 60823–60835 (2022)
24. Bezergianni, S., Georgakis, C.: Controller performance assessment based on minimum and open-loop output variance. Control. Eng. Pract. **8**(7), 791–797 (2000)
25. Wang, J., Lu, S., Wang, S.H., Zhang, Y.D.: A review on extreme learning machine. Multimedia Tools Appl. **81**(29), 41611–41660 (2022)
26. Guo, X., Li, W. J., Qiao, J.F.: A self-organizing modular neural network based on empirical mode decomposition with sliding window for time series prediction. Appl. Soft Comput. 110559 (2023)
27. Fieguth, P.: An introduction to pattern recognition and machine learning. Springer Nature (2022). Zhang, G.P.: A neural network ensemble method with jittered training data for time series forecasting. Inf. Sci. (Ny)., **177**(23), pp. 5329–5346 (2007)
28. Bishop, C.M.: Pattern recognition and machine learning. Springer (2006)
29. Grelewicz, P., Khuat, T.T., Czeczot, J., Klopot, T., Nowak, P., Gabrys, B.: Application of machine learning to performance assessment for a class of PID-based control systems. IEEE Trans. Syst. Man, Cybern. Syst. **53**(7) (2023)
30. Wang, Y., Zhang, H., Wei, S., Zhou, D., Huang, B.: Control performance assessment for ILC-controlled batch processes in a 2-D system framework. IEEE Trans. Syst., Man, Cybern. Syst. **48**(9) (2018)
31. Wang, Y., Zhang, H., Wei, S., Zhou, D., Huang, B.: Control performance assessment for ILC-controlled batch processes in a 2-D system framework. IEEE Trans. Syst., Man, Cybern.: Syst. **48**(9), 1493–1504 (2017)
32. Xu, M., Wang, P.: Evidential KNN-based performance monitoring method for PID control system. In: 2020 5th international conference on mechanical, control and computer engineering (ICMCCE)

Interference Mitigation in Multi-radar Environment Using LSTM-Based Recurrent Neural Network

Hum Nath Parajuli[1], Galymzhan Bakhtiyarov[1], Bikash Nakarmi[2], and Ikechi Augustine Ukaegbu[1(✉)]

[1] School of Engineering and Digital Sciences, Nazarbayev University, Astana, Kazakhstan
`{humnath.parajuli,galymzhan.bakhtiyarov,`
`ikechi.ukaegbu}@nu.edu.kz`
[2] Key Laboratory of Radar Imaging and Microwave Photonics, Nanjing University of
Aeronautics and Astronautics, Nanjing, China
`bikash@nuaa.edu.cn`

Abstract. External disturbances, such as interference, have a significant impact on the functionality of radio detection and ranging (radar) systems, which are employed for the identification, ranging, and imaging of target objects. As radar systems are increasingly adopted across various sectors for different applications, it is essential to handle interference issues appropriately to mitigate false detections, poor signal-to-noise ratio (SNR), and reduced resolution. In the current paper, we introduce a Long Short-Term Memory (LSTM)—based multi-layer recurrent neural network (RNN) to tackle interference problems in a multi-radar setting. In the simulation, a 4-layered LSTM-RNN is trained with 50 different chirp rate interference signals. The efficiency of the introduced interference mitigation technique is evaluated by testing the randomly selected coherently interfered signals, non-coherently interfered signals, and a combination of both on the trained model. The LSTM-RNN effectively suppresses ghost targets in the range profile in the case of coherently interfered signals. Furthermore, the LSTM-RNN enhances the signal-to-interference noise ratio (SINR) by > 18dB in all cases. Thus, the proposed LSTM-RNN offers a promising solution to improve the accuracy and reliability of radar operation in multi-radar environments.

Keywords: Radar · interference · deep learning · recurrent neural network · LSTM

1 Introduction

In today's context, radio detection and ranging systems (radars) are extensively utilized across various domains, ranging from remote sensing, defense, agriculture, industry, and automotive to consumer electronics [1–3]. With the growing number of radar deployments, signals originating from external radars often referred to as interferers, spoofers or intruders can introduce disturbances to the legitimate radar, also known as victim

© The Author(s), under exclusive license to Springer Nature Switzerland AG 2024
B. J. Choi et al. (Eds.): IHCI 2023, LNCS 14532, pp. 151–161, 2024.
https://doi.org/10.1007/978-3-031-53830-8_15

radar (VR). One significant disturbance is the signal arising from the interferer radar (IR) which is operating at the same or nearby frequency band as the VR. The IR signal can significantly impact the performance of the VR by adding external noise, introducing ghost signals, and deteriorating the waveform of the VR. In a multi-radar scenario, these degradations can lead to a poor detection capability of the VR and impose threats to the safe and secure operation of radar systems. Consequently, effective interference handling becomes crucial.

To address the effect of external interference/intrusion in radar systems, various methods have been proposed, such as the development of noise waveform-based radars [4, 5], interference-tolerant waveforms [6, 7], wavelet denoising techniques [8], adaptive beamforming techniques [9, 10], and more. Additionally, recently deep learning (DL)-based interference mitigation techniques have gained attention to the radar research community, as these techniques can reconstruct the echo signal by suppressing the interference and can be applied regardless of the VR and IR signal types [11–15]. Among different DL techniques, convolution neural networks (CNN) based interference mitigation techniques utilize the range-Doppler map technique, which minimizes the interference after the range-Doppler measurement [11, 12]. In general, the IR signals corrupt the VR signal and create temporal gaps in the time domain signal. Consequently, recurrent neural network (RNN) based solutions are effective in addressing this kind of temporal problem. Most previously reported RNN-based techniques involve detecting the interference corrupted segments in the signal, followed by interference mitigation steps [13–15]. However, interference detection may not be robust if the interference corrupts the signal smoothly or uniformly without creating substantial gaps. Therefore, a robust interference mitigation mechanism is always favored, which can mitigate the interference irrespective of the interference type.

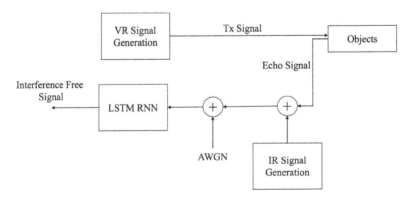

Fig. 1. Overall functional block diagram.

Figure 1 presents the overall functional diagram. To demonstrate the robustness of the proposed scheme, in the simulation environment, we generate a VR linear frequency modulated (LFM) transmission signal (Tx Signal) with a bandwidth of 4GHz (4GHz–8GHz). We used this signal to detect 2 objects located 3m away from the radar sensor, with a 4cm separation between each object. In the current paper, we introduce and present a robust LSTM-based RNN (LSTM-RNN) to suppress the interference signals from the received echo. For the training purpose, the received echo serves as the reference signal to the 4-layered LSTM-RNN. Each object was detected using 50 different IR LFM signals with varying chirp rates, which were subsequently used as an input signal to train the LSTM-RNN. In the testing phase, randomly interfered signals are evaluated. The interference-suppressed echo signal is retrieved by using the previously trained model. In general, IR signals can be categorized as coherent (having the same chirp rate as the VR) and incoherent (having a different chirp rate than the VR). The performance of the proposed LSTM-RNN is evaluated by calculating the range profiles for scenarios involving coherently interfered VR signals, incoherently interfered VR signals, and a combination of both. The proposed 4-layered LSTM-RNN structure mitigates coherent interference cases by suppressing the ghost target in the range profile. Furthermore, it enhances the signal-to-interference noise ratio (SINR) by > 18dB in all cases, thereby indicating its robustness in mitigating interference including different chirp rate interference signals.

2   Signal Model and Interference Effect Analysis

Among various types of radar waveforms, LFM is the most common one which can be represented as [1]

$$T_x(t) = A_{tx} \cos\left(\left(2\pi\left(f_c - \frac{B}{2}\right)t + \pi\alpha_{tx}t^2\right)\right), \; for(0 \leq t \leq T), \tag{1}$$

where A_{tx} is the amplitude of the transmitted signal, f_c is the center frequency, B is the operating bandwidth, α_{tx} is the chirp rate of the signal which is equal to $\frac{B}{T}$, and T is the time period. For the time period of $0 \leq t \leq T$, the frequency of the transmitted signal linearly increases from $f_c - \frac{B}{2}$ to $f_c + \frac{B}{2}$. When the transmitted signal is reflected from the target of K number of stationary objects, the received signal, $R_x(t)$ is represented as

$$R_x(t) = \sum_{k=1}^{K} A_{Rk} \cos\left(\left(2\pi\left(f_c - \frac{B}{2}\right)(t - \tau_{dk}) + \pi\alpha_{tx}(t - \tau_{dk})^2\right)\right), \tag{2}$$

where A_{Rk} and τ_{dk} represent the amplitude of the signal reflected from the kth objects and the time delay resulted by the relative range between the radar receiver and the kth target object, respectively. The LFM signal's range resolution is its ability to distinguish the two target entities' separation distance, which is proportional to the frequency shift. Hence, the range resolution can be expressed as $\Delta R = \frac{c}{2B}$, and is a function of the bandwidth.

Next, we consider the case where the interference radar (IR) LFM signal, which acts on the receiver, can be expressed as,

$$R_I(t) = \sum_{y=1}^{Y} A_{Iy} \cos\left(\left(2\pi\left(f_c - \frac{B_{Iy}}{2}\right)(t - \tau_{Iy}) + \pi\alpha_{Iy}(t - \tau_{Iy})^2\right)\right), \tag{3}$$

where A_{Iy} is the amplitude of the yth IR signal, α_{Iy} is the chirp rate of the IR signal, and τ_{Iy} is the delay of the IR signal to the victim radar (VR) signal. The chirp rate of the

IR signal is given by $\alpha_{Iy} = \frac{B_{Iy}}{T_{Iy}}$, where B_{Iy} and T_{Iy} are bandwidth and the time period of yth IR signal, respectively. When the chirp rate of the VR signal and IR signal are equal, i.e., $\alpha_{Iy} = \alpha_{tx}$ it is known as a coherent interference, and when they are different, $\alpha_{Iy} \neq \alpha_{tx}$, we considered it as a noncoherent interference. In the presence of IR signal, whether it is coherent or non-coherent, the output of the receiver is the mixture of the echo signal from the transmitted signal and the IR signal. For simplification, we consider a single VR and IR source. The beating frequency due to the IR signal $R_I(t)$, after the de-chirping is given by,

$$f_{bi} = (\alpha_{Iy} - \alpha_{tx})t + \alpha_{Iy}\tau_{Iy} \tag{4}$$

Figure 2 illustrates the LFM VR signal and the de-chirping of the signal at the receiver in the appearance of an IR signal with the same/various chirp rate. In Fig. 2 (a-i), the IR signal has the same chirp rate as that of the VR, i.e., $\alpha_{Iy} = \alpha_{tx}$, resulting in two constant beat-frequencies f_{bo} and f_{bi}, observed at the receiver's output as they fall inside the bandwidth of the receiver as shown in Fig. 2a-ii. These two constant beat frequencies will result in two ranges, giving the information of the existence of two target objects, one of which is a ghost target object, as illustrated in the range profile of Fig. 2a-iii.

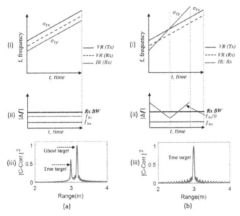

Fig. 2. Demonstration of interference impact **a** coherent interference, **b** non-coherent interference, (i) echo with interference, (ii) produced beat signals in the receiver, (iii) range profile observation.

In Fig. 2b-i, the chirp rates are $\alpha_{Iy} \neq \alpha_{tx}$, leading to the observation of a constant beat frequency, f_{bo}, caused by the actual target object, along with a varying beat frequency $f_{bi}(t)$ as shown in Fig. 2b-ii. The varying beat frequency within the receiver bandwidth adds additional undesired frequency bins in the received signal and adds additional noise floor in the range profile as illustrated in Fig. 2b-iii. Both the interferences, whether coherent, Fig. 2a, or non-coherent, Fig. 2b, severely degrade the performance of the VR's detection capability in a multi-radar environment either by introducing ghost targets or by significantly reducing the SNR. The illustration of Fig. 2 is for a case of 1 object detection and 1 interference signal acting in the received echo. However, in practical scenarios, multiple object detection should be carried out along with the various interference signals. Various approaches can be implemented to minimize the effect of IR signals in a VR signal while operating in a multi-radar environment. In this manuscript, our focus is on analyzing and mitigating interference through an LSTM-RNN based DL technique, which can provide similar output performance for detecting target objects in both interference and non-interference scenarios, in terms of range resolution, SNR, and detection capability.

3   LSTM-RNN Architecture

In the training process of the LSTM-RNN we assume that the echo signal without the interference is known. Therefore, the desired purpose of the training is to develop the model that correctly identifies the interference in the echo and successfully removes it. The LSTM-RNN based training model consists of the control mechanism with three gates regulating the passage of data within the cell allowing the network to learn long-range dependencies more effectively. Additionally, it avoids the exploding and vanishing gradient problem for the large number of input data samples. Thus, the LSTM-RNN can significantly enhance the ability of radar systems to detect and remove the interference signals at the receiver side and recover the corrupted samples.

As illustrated in Fig. 3a, the corrupted sample (x_n), can be recovered based on the preceding samples $x_1, x_2, ..., x_{n-1}$. Figure 3b illustrates the LSTM unit, which consist of the Forget Gate (Γ_F), the Input Gate (Γ_I), , and the Output Gate (Γ_{out}), and can be described as

$$\Gamma_F = \sigma(x_n \cdot \omega_0 + h_{n-1} \cdot \omega_1 + b_1), \tag{5}$$

$$\Gamma_I = \sigma(x_n \cdot \omega_2 + h_{n-1} \cdot \omega_3 + b_2) \cdot tanh(x_n \cdot \omega_4 + h_{n-1} \cdot \omega_5 + b_3), \tag{6}$$

$$\Gamma_{out} = \sigma(x_n \cdot \omega_6 + h_{n-1} \cdot \omega_7 + b_4), \tag{7}$$

where σ is the sigmoid activation function, $tanh$ is the hyperbolic tangent activation function, x_n is the input sample, h_{n-1} is the previous value of the short-term memory, ω_0—ω_7 are corresponding weights, and b_1—b_4 are corresponding biases. The Forget Gate is the first stage in a LSTM unit. This stage determines the percentage of the previous long-term memory to remember. As can be seen in (5), the Forget Gate uses the sigmoid activation function which turns any input into a value within the range of 0 to 1. If the output of the function is 0, the previous long-term memory will be completely forgotten. On the other hand, if the output is equal to 1 the long-term memory remains unchanged. The next stage of the unit is the Input Gate. This gate contains both sigmoid and hyperbolic tangent activation functions, as illustrated in (6). The $tanh$ function part of the Input Gate combines the input and the previous short-term memory to determine a potential long-term memory. The σ function part determines the percentage of the potential long-term memory to add to the current long-term memory. Overall, this part of the LSTM unit updates the current long-term memory. The final part of an LSTM unit is the Output Gate. This stage, in turn, updates the short-term memory by passing the updated long-term memory to the $tanh$ activation function and using the σ function.

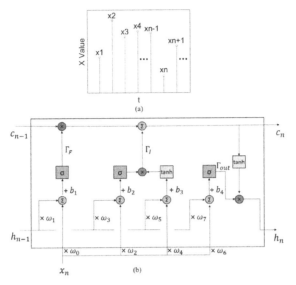

Fig. 3. LSTM **a** LSTM principle, **b** LSTM unit structure.

Thus, the new short-term memory is the output of the entire LSTM unit. By utilizing these Gates, the new long-term and short-term memories are calculated using the following equations:

$$c_n = c_{n-1} \cdot \Gamma_F + \Gamma_I, \tag{8}$$

$$h_n = \Gamma_{out} \cdot \tanh(c_n) \tag{9}$$

4 Methodology, Results and Discussions

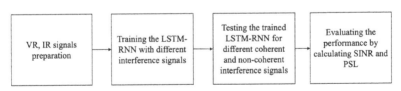

Fig. 4. Steps of investigation.

Figure 4 demonstrates the steps of investigation. Overall, after preparing the VR and IR signals, we train the LSTM-RNN with various interference signals and test the trained network with different coherent and non-coherent interference signals. The last step in our investigation is the evaluation of the system performance by calculating SINR and peak to sidelobe level (PSL).

Table 1 illustrates the parameters for VR and IR signals. The VR signal is generated with a 1μs time period and a 4GHz bandwidth. The VR signal is used to detect objects, and the received echo serves as the output label for the LSTM-RNN. The AWGN noise equivalent to the SNR of 20dB is added to the received echo. To train the LSTM-RNN, the input signal is obtained by applying various chirp rate LFM interference signals to the echo signal. In the training set, the echo signal is mixed with 50 different interference signals with varying chirp rates ranging from 1GHz/μs to 10GHz/μs. All the signals prior to applying to LSTM-RNN, were normalized using zero mean and unit variance normalization functions. For the training step, the mean square error (MSE) values in each iteration were minimized applying the Adam optimization algorithm [16] with a starting learning rate of 0.01. For the testing purpose, we apply different interference signals than the training phase to ensure the trained model works properly for various interference signals with different parameters. Used training parameters and hyper-parameters are provided in Table 2.

Table 1. Parameters of VR and IR waveforms

Parameters	Values for VR	Values of IR
Center frequency	6GHz	0GHz–8GHz
Time period	1μs	1μs
Bandwidth	4GHz	1GHz–6GHz
Chirp rate	4GHz/μs	1GHz/μs–10GHz/ μs
Sampling freq	20GHz	20GHz
No. of signals	1	50

Table 2. LSTM-RNN training parameters and hyper-parameters

Parameters	Values
Sample length	20000
No. of layers	4
No. of hidden units	50
Learning rate	0.01
Learning rate drop factor	0.2
Validation freq	10
Min. Batch size	32

With the use of 4 layers and a learning rate of 0.01, the optimum number of hidden units (50) was obtained by simulating the LSTM-RNN and selecting the hidden unit number that resulted in the minimum MSE. The optimization function in the training process minimizes the loss function, which calculates the error between the LSTM-RNN output (echo with interference suppression) and the expected output (echo without interference). In the radar systems, external interference generally causes two major problems (a) increment in noise floor and (b) appearance of ghost targets in the range profile. SINR and PSL values provide quantitative measure of the radar detection performance by identifying these problems.

To evaluate the performance with and without interference mitigation, the SINR calculation was performed, which is given as

$$SINR = 10 \, log \left\{ \frac{|S|^2}{|\tilde{S} - S|^2} \right\}, \tag{10}$$

where, $\tilde{S}$ is the interference-suppressed echo signal, and S is the echo signal without the interference. In radar, range resolution determines the minimum separation distance between objects that can be distinguished by the radar waveform. This is determined by calculating the delay between the transmitted and received echo. In this paper, we calculate the range resolution by cross-correlating the transmitted signal with the received echo. The PSL power in the range profile is evaluated by determining the difference in the cross-correlated power ($|C - Corr.|^2$) between the detected object's peak with the sidelobe peak, denoted as (PSL).

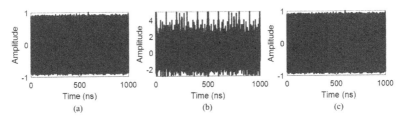

Fig. 5. Interference mitigation **a** Echo signal, **b** Echo signal with interference, and **c** interference removed echo signal.

Figure 5a–c shows examples of the amplitude-time diagrams of the echo signal (label), interference corrupted signal (input), and the output signal after the interference mitigation, respectively. As illustrated in Fig. 5b, the interference corrupts the signal in the time domain.

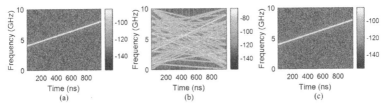

Fig. 6. Training process. **a** Echo signal, **b** Echo signal with various interference, and **c** interference removed echo signal.

Figure 6 shows the frequency-time diagrams at various stages in the training phase. Figure 6a depicts the echo signal (label), whereas Fig. 6b demonstrates the interference corrupted signal (input), and Fig. 6c represents the output signal after the interference mitigation.

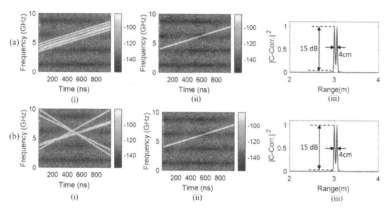

Fig. 7. Interference mitigation with the trained model. **a** Coherent interferences, **b** incoherent and mixed interferences. (i) Spectrogram before interference mitigation, (ii) spectrogram after interference mitigation, and (iii) range profile

Figure 7 shows the range profile results for different test scenarios. The coherent and non-coherent types of interference signals are added to the victim echo signals for the test purposes. In Fig. 7, Fig. 7i represents the interfered test signal input to the trained LSTM-RNN, Fig. 7ii represents the recovered test signal with the previously trained model and Fig. 7iii shows the range profile acquired using interference-recovered signal.

Figure 7a is the case for coherent interference, while Fig. 7b corresponds to the case of both coherent and non-coherent interferences. As can be seen, in the case of coherent interference (Fig. 7a), the ghost target does not appear in the range profile. For both cases, a PSL > 15dB is obtained. The SINR was calculated using (10). With 25 runs, it is confirmed that the presented LSTM-RNN network can recover the test signal with a SINR of > 18dB in the presence of AWGN noise equivalent to the SNR of 20dB.

5   Conclusions

In this paper, we presented LSTM-RNN based interference mitigation method that can be applied to mitigate both coherent and non-coherent interferences in the multi-radar setup. The LSTM-RNN is trained with 50 different chirp rates of interference signals. The trained model is then used to recover the interference corrupted signal in the testing phase. With 25 tests, it is confirmed that the presented LSTM network can recover the echo signal, with the SINR > 18dB. Additionally, with the recovered signal, the PSL > 15dB is obtained for all test cases. These results indicate that the presented LSTM-RNN can be used to effectively mitigate the interference in multi-radar environments. The interference mitigation in multi-radar environment holds vital importance in the radar application area such as autonomous vehicle. We plan to further investigate the effectiveness of the proposed LSTM-RNN in experimental data. Additionally, with experimental data we will investigate the optimum values of parameters and hyper-parameters of the proposed LSTM-RNN by considering sufficient number of interference signals for training.

Acknowledgement. This work was supported by the Nazarbayev University (NU) Collaborative Research Project (grant no. 11022021CRP15070), Faculty Development Competitive Research Grant (grant no. 021220FD0451), and the Ministry of Education and Science of the Republic of Kazakhstan (AP14871109, AP13068587).

References

1. Parker, M.: Radar Basics, Digital Signal Processing 101. 2nd edn., pp. 231–240. Newnes (2017)
2. Gurbuz, Z., Griffiths, H., Charlish, A., Rangaswamy, M., Greco, M., Bell, K.: An overview of cognitive radar: past, present, and future. In: IEEE aerospace and electronic systems magazine, **34**(12), 6–18 (2019)
3. Bilik, I., Longman, O., Villeval, S., Tabrikian, J.: The rise of radar for autonomous vehicles: signal processing solutions and future research directions. IEEE Signal Process. Magazine **36**(5), 20–31 (2019)
4. Xu, Z., Shi, Q.: Interference mitigation for automotive radar using orthogonal noise waveforms. IEEE Geosci. Remote Sensing Lett. **15**(1), 137–141 (2018)
5. Galati, G., Pavan, G., Wasserzier, C.: Signal design and processing for noise radar. EURASIP J. Adv. Signal Process **52** (2022)
6. Skaria, S., Al-Hourani, A., Evans, R., Sithamparanathan, K., Parampalli, U.: Interference mitigation in automotive radars using pseudo-random cyclic orthogonal sequences. Sensors **19**(20), 4459 (2019)
7. Uysal, F.: Phase-coded FMCW automotive radar: system design and interference mitigation. IEEE Trans. Veh. Technol. **69**(1), 270–281 (2020)
8. Lee, S., Lee, J., Kim, S.: Mutual interference suppression using wavelet denoising in automotive FMCW radar systems. IEEE Trans. Intell. Transp. Syst. **22**(2), 887–897 (2021)
9. Choi, Y.: Adaptive nulling beamformer for rejection of coherent and noncoherent interferences. Signal Process. **92**(2), 607–610 (2012)
10. Lee, T., Lin, T.: Coherent interference suppression with complementally transformed aadaptive beamformer. IEEE Trans. Antennas Propag. **46**(5), 609–617 (1998)
11. Rock, J., Toth, M., Messner, E., Meissner, P., Pernkopf, F.: Complex signal denoising and interference mitigation for automotive radar using convolutional neural networks. In: 22th International conference on information fusion (FUSION), pp. 1–8, Ottawa, ON, Canada (2019). https://doi.org/10.23919/FUSION43075.2019.9011164
12. Fuchs, J., Dubey, A., Lübke, M., Weigel, R., Lurz, F.: Automotive radar interference mitigation using a convolutional autoencoder. IEEE Int. Radar Conf. (RADAR), pp. 315–320, Washington, DC, USA (2020). https://doi.org/10.1109/RADAR42522.2020.9114641
13. Mun, J., Ha, S., Lee, J.: Automotive radar signal interference mitigation using RNN with self attention. In: ICASSP IEEE international conference on acoustics, speech and signal processing (ICASSP), pp. 3802–3806, Barcelona, Spain (2020). https://doi.org/10.1109/ICASSP40776.2020.9053013

14. Hille, J., Auge, D., Grassmann, C., Knoll, A.: FMCW radar2radar interference detection with a recurrent neural network. In: IEEE radar conference (RadarConf22), pp. 1–6, New York City, NY, USA (2022). https://doi.org/10.1109/RadarConf2248738.2022.9764236
15. Rameez, M., Javadi, S., Dahl, M., Pettersson, M.: Signal reconstruction using Bi-LSTM for automotive radar interference mitigation. In: 18th European radar conference (EuRAD), pp. 74–77, London, United Kingdom (2022). https://doi.org/10.23919/EuRAD50154.2022.9784516
16. Kingma, Diederik, P., Ba, J.: Adam: A method for stochastic optimization. arXiv preprint arXiv:1412.6980 (2014)

Centrifugal Pump Health Condition Identification Based on Novel Multi-filter Processed Scalograms and CNN

Zahoor Ahmad[1], Muhammad Farooq Siddique[1], Niamat Ullah[1], Jaeyoung Kim[2], and Jong-Myon Kim[1,2(✉)]

[1] Department of Electrical, Electronics, and Computer Engineering, University of Ulsan, Ulsan 44610, South Korea
jmkim07@ulsan.ac.kr
[2] PD Technology Co. Ltd, Ulsan 44610, South Korea

Abstract. This paper proposes a fault diagnosis method for centrifugal pumps (CP) based on multi-filter processed scalograms (MFS) and convolutional neural networks (CNN). Deep learning (DL) based autonomous Health-sensitive features extraction from continuous wavelet transform (CWT) scalograms are popular adoption for the health diagnosis of centrifugal pumps. However, vibration signals (VS) acquired from the centrifugal pump consist of fault-related impulses and unwanted macrostructural noise which can affect the autonomous Health-sensitive features extraction capabilities of the deep learning models. To overcome this concern, novel multi-filter processed scalograms are introduced. The new multi-filter processed scalograms enhance the fault-related color intensity variations and remove the unwanted noise from the scalograms using Gaussian and Laplacian image filters. The proposed techniques identified the ongoing health condition of the centrifugal pump by extracting fault-related information from the multi-filter processed scalograms and classifying them into their respective classes using convolutional neural networks. The proposed method resulted in higher classification accuracy as compared to the existing method when it was applied to a real-world centrifugal pump vibration signals dataset.

Keywords: Multi-filter processed scalograms · Fault diagnosis · Centrifugal pump

1 Introduction

CPs play a crucial role in various aspects of business operations. Unexpected failures in CPs can result in extended periods of downtime, financial losses, costly repairs, and potential hazards to worker safety [1]. It is vital to promptly identify and diagnose faults to ensure the extended functionality of centrifugal pumps [2].

The amplitude of the VS serves as a valuable indicator for detecting mechanical faults in the CP arising from the mechanical seal and impeller. Time-domain features

© The Author(s), under exclusive license to Springer Nature Switzerland AG 2024
B. J. Choi et al. (Eds.): IHCI 2023, LNCS 14532, pp. 162–170, 2024.
https://doi.org/10.1007/978-3-031-53830-8_16

effectively identify emerging faults within the VS. However, their utility diminishes when dealing with severe faults due to the inherent variability in fault severity [3–6].

To address this issue, the frequency spectrum emerges as a more adept tool for pinpointing faults of varying degrees of severity, supported by the use of Frequency-Domain features for CP fault diagnosis. The VS obtained from the CP under a faulty health state is characterized by its complexity and nonstationary nature [7]. While spectrum analysis is optimal for stationary signals, non-stationary signals necessitate a different approach [8]. Time and multiresolution domain transformations, offering multi-resolution analysis suited for these dynamic signals. Empirical mode decomposition, an adaptive decomposition method, has found efficient application in diagnosing faults in rotating machinery. Nevertheless, EMD faces challenges like mode mixing and extreme interpolation, rendering it less attractive for VS analysis [9–12]. For analyzing CP's non-stationary transients, variational mode decomposition (VMD) and CWT emerge as favorable choices. In this context, The critical aspect is the selection of the fundamental wavelet, which profoundly influences the distinctiveness of the features extracted. This choice calls for a nuanced blend of domain expertise and exhaustive empirical exploration to ensure its suitability for the given diagnostic task [13]. To address the above-mentioned concerns, in this work, the paper introduces innovative MFSs that effectively enhance fault-related color intensity variations while eliminating undesirable noise in the CWT scalograms using Gaussian [14] and Laplacian [15] image filters.

The CP fault diagnosis system involves two key steps: extracting fault-related features from the VS and classifying the CP working conditions based on these extracted features. DL methods are preferred over traditional machine learning techniques because they can effectively analyze intricate data and autonomously derive meaningful discriminant information for pattern recognition tasks [16, 17]. Prominent DL methods used for fault diagnosis include neural auto-encoders, deep belief networks, CNN, and recurrent neural networks. CNN, in particular, mitigates overfitting risks, offers low computational complexity through weight sharing, and employs local representative fields, and special domain subsampling [18, 19]. Furthermore, CNNs have showcased their proficiency in effectively recognizing patterns in fault diagnosis situations related to bearings, CPs, and pipelines [12, 20–23]. For this reason in this paper, the proposed method uses CNN to identify the ongoing health condition of CPs by extracting crucial fault-related information from the MFS.

The arrangement of this study unfolds across the subsequent segments: In Sect. 2, the experimental testbed used in this study is described. Section 3 elucidates the details of the proposed framework. Sect. 4 describes the results and discussion. The conclusion and future direction are presented in Sect. 5.

2 Experimental Setup

For experimental purposes, a test rig has been created, comprising various components: a CP (PMT-4008) powered by a 5.5 kW motor, a control panel featuring an ON/OFF switch, speed control, flowrate control, temperature control, water supply control, and display screens. Additionally, it includes pressure gauges, transparent pipes, and two tanks, namely the main tank and buffer tank. To ensure an efficient CP suction head, a water tank has been placed at an elevated position. The test rig setup, along with a schematic representation, is displayed in Fig. 1. Once the primary setup was established, the test rig was set in motion to circulate water within a closed loop. Vibration data from the CP were gathered while maintaining a constant speed of 1733 rpm. This data was collected using four accelerometers, with two affixed to the pump casing using adhesive, while the other two were positioned close to the mechanical seal and near the impeller. Each sensor recorded the pump's vibrations through its own dedicated channel. The recorded VS was subsequently directed to a signal monitoring unit. Within this unit, the signal underwent digitization via a National Instruments 9234 device. Data was collected over a duration of 300 s, with a sampling frequency of 25.6 kHz. In total, 1200 sets of samples were gathered, and each set had a sample length of 25,600 data points. These measurements were obtained from the CP under various operational conditions such as normal and defective operating conditions. The faults considered in this study are mechanical seal scratch defect (MSS-D), mechanical seal hole defect (MSH-D), and impeller defect (IF) The description of the whole dataset is shown in Table 1.

Table 1. CP dataset.

CP condition	Defect specification			VS samples
	Defect length (mm)	Defect diameter (mm)	Defect depth (mm)	
Normal	–	–	–	300
MSH	–	2.8	2.8	300
MSS	10	2.5	2.8	300
IF	18	2.5	2.8	300

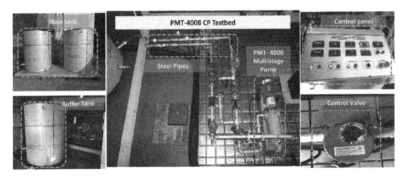

Fig. 1. Experimental testbed for data acquisition

3   Proposed Framework

The proposed approach is depicted graphically in Fig. 2. The method comprises the following steps:

Step 1: VSs are collected from the CP.

Step 2: The acquired VSs are represented in their respective time-frequency scalograms using CWT. The scalogram images illustrate variations in energy levels across various time-frequency scales through the application of distinct color intensities.

Step 3: VSs Vibration signals acquired from the centrifugal pump encompass fault-related impulses as well as unwanted macrostructural vibration noise. This noise can potentially impact the autonomous health-sensitive feature extraction capabilities of the CNN. To extract discernible health-sensitive features, the current step involves the processing of CWT scalograms to derive new MFSs.

The process of generating MFSs comprises several crucial stages. Initially, the proposed method employs a Gaussian filter on the CWT scalogram to achieve smoother results and effectively mitigate any noise interference. Subsequently, a Laplacian filter is applied as an edge detector to the CWT scalogram. This enhances the accuracy of edge detection within the CWT scalogram, ultimately leading to the extraction of MFSs.

Step.4: To identify the ongoing health conditions of the CP, fault-related information from the MFS is extracted and classified into their respective classes using CNN in this step. The CNN model used in this study is presented in Table 2 which consists of three convolutional layers followed by max-pooling layers, designed to extract intricate features from input data. These layers utilize the ReLU activation function for introducing non-linearity. After feature extraction, the flattened data is processed through three densely connected layers, each with ReLU activation, gradually reducing dimensionality and capturing higher-level patterns. The final output layer employs the softmax activation function, enabling the model to make multi-class predictions, making it well-suited for tasks like image classification. Overall, this architecture excels at feature extraction and pattern recognition, facilitating the accurate classification of diverse objects or categories.

Table 2. CNN for feature extraction and classification:

Type of layers	Output	Param#	Activation function
Conv2D	[None, 62, 62, 32]	320	ReLU
MaxPool2D	(None, 31, 31, 32)	0	–
convo2D_1 (Convo2D)	(None, 29, 29, 64)	18496	ReLU
max_pooling2D_1 (MaxPooling2D)	(None, 14, 14, 64)	0	–
convo2D_2 (Convo2D)	(None, 12, 12, 128)	73856	ReLU
max_pooling2D_2 (MaxPooling2D)	(None, 6, 6, 128)	0	–
flatten (Flatten)	(None, 4608)	0	ReLU
dense (Dense)	(None, 512)	2359808	–
dense_1 (Dense)	(None, 256)	131328	ReLU
dense_2 (Dense)	(None, 128)	32896	–
dense_3 (Dense)	(None, 4)	516	softmax

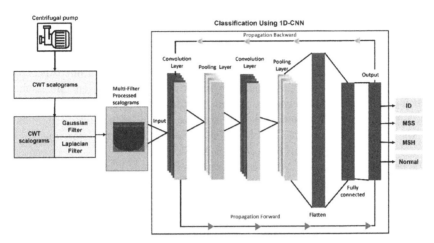

Fig. 2. Graphical flow of the proposed method

4 Results and Performance Evaluation

This study utilized a k-fold cross-validation (k = 3) approach, to gauge the effectiveness of the proposed methodology. To enhance result reliability, these experiments underwent 15 iterations. In each iteration, 200 samples were selected at random for model training per class, while the remaining samples were designated for model testing. Readers can find a comprehensive dataset description in Table 1.

A comparison is made between the proposed method and the reference technique proposed by Gu et al. [17]. For evaluation, matrices such as average accuracy (AA), precision (Pm), Recall (Rm), and F-1 score are calculated from the confusion matrices presented in Fig. 3. These matrices are computed using Eqs. 1,2,3, and 4.

$$AA = \frac{\sum_{k}^{K} TP_k}{N} \times 100, \tag{1}$$

$$P_m = \frac{1}{K}\left(\sum_{k=1}^{K} \frac{TP_k}{TP_k + FP_k}\right) \times 100, \tag{2}$$

$$R_m = \frac{1}{K}\left(\sum_{k=1}^{K} \frac{TP_k}{TP_k + FN_k}\right) \times 100, \tag{3}$$

$$F1 - score = 2 \times (\Pr ec_m \times \text{Rec}_m)/(\Pr ec_m + \text{Rec}_m), \tag{4}$$

The true positive and negatives are represented by *TP* and *TN*, False negatives of the classifier is represented by *FN*, while *N* denotes the total count of samples within each class.

The CP health conditions classification results obtained from the proposed and reference comparison method are presented in Table 3. The proposed approach achieved a higher AA of 96.4% compared to the reference method with Pm of 96.50%, Rm of 96.25, and F-1 score of 96.50. The higher AA of the proposed method can be elaborated as follows. The VSs acquired from the CP consist of fault-related impulses and unwanted macrostructural vibration noise which can affect the autonomous Health-sensitive features extraction capabilities of the CNN. To overcome this concern, novel MFSs are used for CP health state classification. The new MFS enhances the fault-related color intensity variations and removes the unwanted noise from the scalograms using Gaussian and Laplacian image filters. For a classification model, the accuracy of classification is directly proportional to the quality of input features. As can be seen from Fig. 4 the features extracted by CNN from the MFS are highly discriminant which is the key reason for the higher AA of the proposed method. A slight feature space overlap between the MSH and MSS can be noticed in Fig. 4. To increase the classification accuracy, in the future, signal filtering can be used to increase the discriminancy of MFS for different classes.

The reference method Gu et al. [17] decomposed VSs obtained from the rotating machinery using VMD and selected the informative IMF of VMD. Scalograms of the selected IMF are created using CWT from which features are extracted using CNN. Instead of a softmax layer for classification, the reference method used SVM for this task. This method is selected for comparison due to its correlation with the steps involved in the proposed technique. Furthermore, both techniques utilized VSs for the diagnosis of mechanical faults in machines. To make the comparison fair, instead of SVM, the softmax layer is used for the classification task. After applying the steps presented in [17] to our dataset, an AA of 89.25% was obtained which is lower than the proposed method, as presented in Table 3. This underperformance can be elaborated as follows. The CWT scalograms illustrate variations in energy levels across various time-frequency scales through the application of distinct color intensities. These color intensities help the CNN to extract discriminant features for the health state identification of the CP. However, VSs acquired from the CP are heavily affected by macrostructural vibration noise. Thus it is important to further preprocess the CWT scalograms prior to feature extraction.

Table 3. Performance comparison of the proposed method with Gu et al. [12].

Approaches	Performance metrics			Accuracy (Average)%
	Precision	Recall	F-1 Score	
Proposed	96.50	96.25	96.50	96.40%
Gu et al. [17]	89.25	89.60	89.25	89.00%

Based on the fault diagnosis capability of the proposed method for CPs, it can be concluded that the framework is suitable for diagnosing CP faults. The main advantage of the proposed framework lies in its fundamental concept, which involves preprocessing VS and selecting health-sensitive features based on their ability to improve classification accuracy. As can be seen from Fig. 4 the features extracted by CNN from the traditional CWT scalograms are not discriminant enough to represent the health state of the CP which is the key reason for the higher AA of the proposed method.

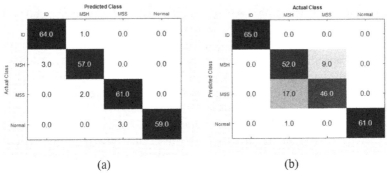

Fig. 3. Confusion matrices (a) Proposed (b) Gu et al. [17]

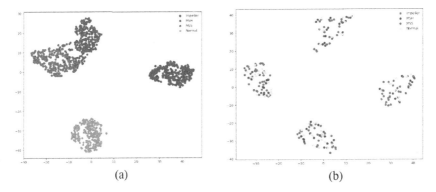

Fig. 4. Feature space (a) Proposed (b) Gu et al. [17]

5 Conclusion

This paper proposed a fault diagnosis method for centrifugal pumps using multi-filter processed scalograms and convolutional neural networks. Traditional approaches using continuous wavelet transform scalograms struggle with unwanted noise in vibration signals from pumps, affecting diagnostic accuracy. The proposed method overcomes this challenge by using the new multi-filter processed scalograms that effectively enhance fault-related color intensity variations while eliminating undesirable noise using Gaussian and Laplacian image filters. It effectively identifies pump health conditions and outperforms existing methods when tested on real-world data with an average accuracy of 98.4%, offering significant potential for improving maintenance and operational reliability in industries relying on centrifugal pumps. In the future, the proposed method will be applied to diagnose cavitation-related faults in the centrifugal pumps.

Acknowledgements. This work was supported by the National IT Industry Promotion Agency (NIPA), grant funded by the Korean government Ministry of Science and ICT (MSIT), Grant No. S0721–23-1011, for development of a smart mixed reality technology for improving the pipe installation and inspection processes in the offshore structure fabrication. This work was also supported by the Technology Infrastructure Program funded by the Ministry of SMEs and Startups(MSS, Korea).

References

1. Ahmad, Z., Rai, A., Hasan, M.J., Kim, C.H., Kim, J.M.: A novel framework for centrifugal pump fault diagnosis by selecting fault characteristic coefficients of walsh transform and cosine linear discriminant analysis. IEEE Access **9**, 150128–150141 (2021). https://doi.org/10.1109/ACCESS.2021.3124903
2. Ahmad, Z., Rai, A., Maliuk, A.S., Kim, J.M.: Discriminant feature extraction for centrifugal pump fault diagnosis. IEEE Access **8**, 165512–165528 (2020). https://doi.org/10.1109/ACCESS.2020.3022770
3. Zhang, X., Zhao, B., Lin, Y.: Machine learning based bearing fault diagnosis using the case western reserve university data: a review. IEEE Access **9**. Institute of Electrical and Electronics Engineers Inc., pp. 155598–155608, (2021). https://doi.org/10.1109/ACCESS.2021.3128669
4. Ahmad, S., Ahmad, Z., Kim, J.M.: A centrifugal pump fault diagnosis framework based on supervised contrastive learning. Sensors **22**(17) (2022). https://doi.org/10.3390/s22176448
5. Chen, L., Wei, L., Wang, Y., Wang, J., Li, W.: Monitoring and predictive maintenance of centrifugal pumps based on smart sensors. Sensors **22**(6) (2022) https://doi.org/10.3390/s22062106
6. Dong, L., Chen, Z., Hua, R., Hu, S., Fan, C., Xiao, X.: Research on diagnosis method of centrifugal pump rotor faults based on IPSO-VMD and RVM. Nuclear Eng. Technol. **55**(3), 827–838 (2023). https://doi.org/10.1016/j.net.2022.10.045
7. Ahmad, Z., Nguyen, T.K., Ahmad, S., Nguyen, C.D.,Kim, J.M.: Multistage centrifugal pump fault diagnosis using informative ratio principal component analysis. Sensors **22**(1), (2022). https://doi.org/10.3390/s22010179
8. Rapuano, S., Harris, F.J.: IEEE instrumentation & measurement magazine an introduction to FFT and time domain windows part 11 in a series of tutorials in instrumentation and measurement (2007)
9. Hou, Y., Wu, P., Wu, D.: An operating condition information-guided iterative variational mode decomposition method based on Mahalanobis distance criterion for surge characteristic frequency extraction of the centrifugal compressor. Mech Syst Signal Process **186** (2023) https://doi.org/10.1016/j.ymssp.2022.109836
10. Dai, C., Hu, S., Zhang, Y., Chen, Z., Dong, L.: Cavitation state identification of centrifugal pump based on CEEMD-DRSN. Nucl. Eng. Technol. (2023). https://doi.org/10.1016/j.net.2023.01.009
11. Nguyen, T.K., Ahmad, Z., Kim, J.M.: A deep-learning-based health indicator constructor using kullback–leibler divergence for predicting the remaining useful life of concrete structures. Sensors **22**(10) (2022) https://doi.org/10.3390/s22103687
12. Aguilera, J.J. et al.: A review of common faults in large-scale heat pumps. Renew. Sustain. Energ. Rev. **168**. Elsevier Ltd, Oct. 01, 2022. https://doi.org/10.1016/j.rser.2022.112826
13. Yang, Y., Zheng, H., Li, Y., Xu, M., Chen, Y.: A fault diagnosis scheme for rotating machinery using hierarchical symbolic analysis and convolutional neural network. ISA Trans. **91**, 235–252 (2019). https://doi.org/10.1016/j.isatra.2019.01.018
14. Gupta, S.B.: A hybrid image denoising method based on discrete wavelet transformation with pre-gaussian filtering. Indian J. Sci. Technol. **15**(43), 2317–2324 (2022). https://doi.org/10.17485/IJST/v15i43.1570
15. Ullah, N., Ahmed, Z., Kim, J.M.: Pipeline leakage detection using acoustic emission and machine learning algorithms. Sensors **23**(6) 2023. https://doi.org/10.3390/s23063226

16. Yafouz, A., Ahmed, A.N., Zaini, N., Sherif, M., Sefelnasr, A., El-Shafie, A.: Hybrid deep learning model for ozone concentration prediction: comprehensive evaluation and comparison with various machine and deep learning algorithms. Eng. Appl. Comput. Fluid Mech. **15**(1), 902–933 (2021). https://doi.org/10.1080/19942060.2021.1926328
17. Gu, J., Peng, Y., Lu, H., Chang, X., Chen, G.: A novel fault diagnosis method of rotating machinery via VMD, CWT and improved CNN. Measurement (Lond) **200** (2022). https://doi.org/10.1016/j.measurement.2022.111635
18. Hasan, M.J., Rai, A., Ahmad, Z., Kim, J.M.: A fault diagnosis framework for centrifugal pumps by scalogram-based imaging and deep learning. IEEE Access **9**, 58052–58066 (2021). https://doi.org/10.1109/ACCESS.2021.3072854
19. Ahmad, S., Ahmad, Z., Kim, C.H., Kim, J.M.: A method for pipeline leak detection based on acoustic imaging and deep learning. Sensors **22**(4) (2022). https://doi.org/10.3390/s22 041562
20. Nguyen, T.K., Ahmad, Z., Kim, J.M.: Leak localization on cylinder tank bottom using acoustic emission. Sensors **23**(1) (2023). https://doi.org/10.3390/s23010027
21. Saeed, U., Lee, Y.D., Jan, S.U., Koo, I.: CAFD: Context-aware fault diagnostic scheme towards sensor faults utilizing machine learning. Sensors (Switzerland) **21**(2), 1–15 (2021). https://doi.org/10.3390/s21020617
22. Li, G., Chen, L., Liu, J., Fang, X.: Comparative study on deep transfer learning strategies for cross-system and cross-operation-condition building energy systems fault diagnosis. Energy **263** (2023). https://doi.org/10.1016/j.energy.2022.125943
23. Siddique, M.F., Ahmad, Z., Kim, J.M.: Pipeline leak diagnosis based on leak-augmented scalograms and deep learning. Eng. Appl. Comput. Fluid Mech. **17**(1) 2023. https://doi.org/10.1080/19942060.2023.2225577

A Closer Look at Attacks on Lightweight Cryptosystems: Threats and Countermeasures

Khusanboy Kodirov, Hoon-Jae Lee, and Young Sil Lee[(⊠)]

Computer Engineering, Dongseo University, Busan 47011, Korea
hjlee@gdsu.dongseo.ac.krs, lys0113@dongseo.ac.kr

Abstract. Cryptosystems are fundamental to securing digital communication and information exchange in our interconnected world. However, as technology advances, so does the sophistication of malicious actors seeking to compromise these cryptographic mechanisms. A branch of cryptosystems called lightweight cryptography plays a pivotal role in ensuring secure communication and data protection in resource-constrained devices, such as IoT sensors and embedded systems. This paper provides an in-depth exploration of the attack vectors that lightweight cryptosystems face and introduces novel technical countermeasures aimed at bolstering their security. The research in this paper is positioned at the forefront of lightweight cryptography, aiming to address current and emerging threats.

Keywords: Cryptosystems · Cryptanalysis · Lightweight Encryption · Attacks

1 Introduction

From the ages, human beings had two inherent needs. First, to communicate and share information, and second, to communicate selectively. These two needs led people to create the art of coding the messages so that only authorized and intended personnel could access the information. Even if the scrambled secret messages fell into the hands of unintended people, they could not decipher the message and extract any hidden information. During 500 to 600 BC, Romans used a mono-alphabetic substitution cipher known as Caesar Shift Cipher which relied on shifting the letters of a message by some agreed amount (Fig. 1).

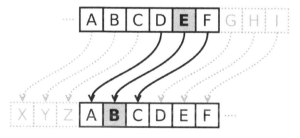

Fig. 1. Caesar Shift Cipher

B. J. Choi et al. (Eds.): IHCI 2023, LNCS 14532, pp. 171–176, 2024.
https://doi.org/10.1007/978-3-031-53830-8_17

In the 20th and 21st centuries, many encryption standards have been developed by computer scientists and mathematicians. Nowadays, in almost every aspect of human life, there is a need to transfer information secretly. As a result, it has become essential to protect useful information from cyber-attacks. This paper discusses some of the most common attacks on cryptosystems by each category and the necessary countermeasures to ensure the information being communicated is as secure as possible.

1.1 Lightweight Cryptography

Lightweight cryptography plays a pivotal role in ensuring secure communication and data protection in resource-constrained devices, such as IoT sensors and embedded systems. This paper provides an in-depth exploration of the attack vectors that lightweight cryptosystems face and introduces novel technical countermeasures aimed at bolstering their security. The research in this paper is positioned at the forefront of lightweight cryptography, aiming to address current and emerging threats.

2   Related Work

A significant amount of related work has been conducted in the field of securing lightweight encryption. Researchers have focused on various aspects of lightweight encryption, including threat analysis, vulnerabilities, and countermeasures. Here are some key areas of related work:

2.1   Side-Channel Analysis and Countermeasures:

There is a substantial body of research on side-channel attacks against lightweight encryption algorithms. Countermeasures like masking, blinding, and secure implementations have been proposed to mitigate these attacks (Fig. 2).

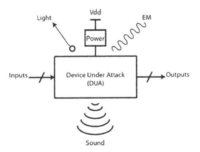

Fig. 2. Side channel attack

Frequently employed side-channel inputs encompass elements like power supply voltage, temperature, ambient light, and other primary signals not directly associated with the cryptographic module. These attacks involve a combination of monitoring side-channel outputs, manipulating side-channel inputs, observing primary outputs, and tampering with primary inputs. These activities are accompanied by progressively intricate analytical methods aimed at uncovering confidential data from the cryptographic system.

2.2 Light Lightweight Cryptographic Algorithm Design:

Researchers have developed and analyzed lightweight cryptographic algorithms designed specifically for resource-constrained devices. These algorithms aim to strike a balance between security and efficiency.

3   Types of Attacks on Cryptosystems

3.1 Passive Attacks

In order to obtain unauthorized access to the information, a passive attack is carried out. For instance, when an attacker intercepts or eavesdrops on the communication channel, it is regarded as a passive attack (Fig. 3).

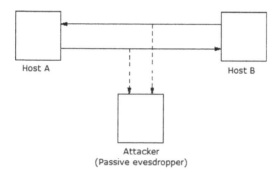

Host A   Host B

Attacker
(Passive evesdropper)

Fig. 3. Passive Attack

It is called a passive attack in nature since the attacker neither affects the information being transferred nor disrupts the communication channel. It can be seen as stealing information. As information theft may go unnoticed by the owner, this type of attack can be more dangerous.

3.2 Active Attacks

Contrary to a passive attack, an active attack involves altering the information in some way by conducting a certain process on the data (Fig. 4).

| Modifying the information in an unauthorized manner. | Initiating unauthorized transmission of information. | Unauthorized deletion of data. | Denial of access to information for legitimate users (denial of service). |

Fig. 4. Active Attack

4 Cryptographic Attacks

The main intention of an attacker is usually to break a cryptosystem and extract the plaintext from the ciphertext. As for the symmetric key encryption, the attacker only needs the secret key to obtain the plaintext, as in most cases, the algorithm itself is already in the public domain. For this, he (or she) takes maximum effort to find the secret key used in that particular communication channel. A cryptosystem is considered broken or compromised once the attacker successfully obtains the secret encryption key. Let us look at some of the most common attacks carried out on cryptosystems by each category .

4.1 Ciphertext Only Attacks (COA)

It involves a scenario where the attacker possesses a collection of ciphertexts but lacks access to their corresponding plaintexts. Success in COA is achieved when one can deduce the corresponding plaintext from the provided ciphertext set. In some cases, this type of attack may even reveal the encryption key. It's worth noting that contemporary cryptosystems are designed with robust defenses against ciphertext-only attacks to enhance security.

4.2 Known Plaintext Attack (KPA)

In this approach, the attacker possesses knowledge of the plaintext for certain portions of the ciphertext. The objective is to decipher the remaining ciphertext with the assistance of this known information. Achieving this can involve techniques such as identifying the encryption key or employing alternative methods. A prominent illustration of such an attack is seen in linear cryptanalysis when applied to block ciphers.

4.3 Dictionary Attack

This type of attack comes in various forms, but they all revolve around creating a 'dictionary.' In its most straightforward form, the attacker constructs a dictionary containing pairs of ciphertexts and their associated plaintexts that they have gathered over time. When faced with ciphertext in the future, the attacker consults this dictionary to identify the corresponding plaintext.

4.4 Brute Force Attack (BFA)

In this approach, the attacker ascertains the key by systematically testing every conceivable key. For instance, if the key consists of 8 bits, there are a total of 256 possible keys ($2^8 = 256$). Armed with knowledge of the ciphertext and the encryption algorithm, the attacker proceeds to test all 256 keys individually in an attempt to decrypt the data. However, if the key is lengthy, this method would require significant time to complete the attack due to the sheer number of potential keys (Fig. 5).

4.5 Man in Middle Attack (MIM)

This attack primarily focuses on public key cryptosystems that employ a key exchange process prior to initiating communication.

- Host A seeks to establish communication with host B and, consequently, requests the public key belonging to B.
- However, an assailant intercepts this request and substitutes their own public key.
- As a result, anything that host A transmits to host B becomes accessible to the attacker.
- To sustain the communication, the attacker re-encrypts the data with their public key after intercepting and reading it and then forwards it to B.
- The attacker disguises their public key as if it were A's public key, causing B to accept it as if it were originating from A.

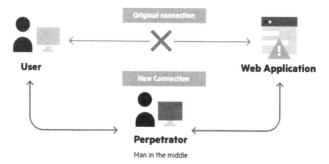

Fig. 5. Man in Middle Attack (MIM)

5 Countermeasures for Lightweight Encryption

5.1 Fault Injection Attacks and Protections:

Lightweight devices are vulnerable to fault injection attacks. Researchers have investigated fault attacks and proposed methods to protect lightweight cryptographic algorithms, such as error-detection codes and fault tolerance techniques.

5.2 Post-quantum Lightweight Cryptography:

With the advent of quantum computing, researchers are exploring lightweight cryptographic primitives that can resist quantum attacks. This includes lattice-based and code-based cryptography suitable for resource-constrained devices.

5.3 Energy-Efficient Cryptography:

Energy-efficient cryptographic algorithms are developed for low-power devices, considering the unique constraints of these systems. This work focuses on achieving security while minimizing energy consumption.

5.4 Machine Learning and Lightweight Cryptography:

Recent research has explored the application of machine learning techniques to enhance the security of lightweight cryptographic algorithms by identifying patterns and anomalies in data.

6 Conclusion

In conclusion, the world of cryptosystems is one where innovation and security continually collide with evolving threats. The paper concludes by summarizing the key findings and emphasizing the importance of addressing threats to lightweight cryptosystems with innovative countermeasures. The research presented in this paper is a testament to the ongoing development of lightweight cryptography, ensuring the security and privacy of resource-constrained devices. However, as the landscape of cyber threats continues to evolve, so too do the countermeasures and defense mechanisms designed to sfight against these attacks. The arsenal of countermeasures is diverse and dynamic, from employing state-of-the-art encryption algorithms and key management practices to fostering a culture of security awareness and regulatory compliance.

Acknowledgment. This research was supported by Korea Institute of Marine Science & Technology Promotion (KIMST) funded by the Ministry of Oceans and Fisheries (20210650).

References

1. Smith, J.: Attacks on cryptosystems: vulnerabilities and countermeasures. Int. J. Cybersecurity **12**(3), 401–420 (2023)
2. Johnson, A.L.: Cryptanalysis techniques and their role in cryptosystem security. J. Inf. Secur. Res. **7**(1), 56–72 (2023)
3. Brown, E.R., Garcia, M.P.: Enhancing cryptosystem resilience: a comprehensive review of countermeasures. Cybersecurity J. **15**(2), 189–210 (2023)
4. Chen, Q., Kim, S.H.: Social engineering attacks on cryptosystems: strategies and mitigation. J. Comput. Secur. **25**(4), 509–528 (2023)
5. Williams, R.T., Lee, C.H.: Quantum threats to cryptosystems: challenges and defenses. Cryptology Today **9**(4), 315–333 (2023)

A Prototype of IoT Medication Management System for Improved Adherence

Hyunbin Yun[1], Heekyung Chae[2], Bong Jun Choi[3], and Dhananjay Singh[4(✉)]

[1] Department of Nuclear and Quantum Engineering, Korea Advanced Institute of Science and Technology, Daejeon, South Korea
[2] Department of Electronics Engineering, Hankuk University of Foreign Studies, Yongin, South Korea
[3] School of Computer Science and Engineering and School of Electronic Engineering, Soongsil University, Seoul, Korea
[4] ReSENSE Lab, School of Ptrofessional Studies, Saint Louis University, Saint Louis, MO, USA
dhananjay.singh@slu.edu

Abstract. In response to the growing trend of non-face-to-face medical care, we have developed an Innovative Pharmaceutical IoT product to improve drug compliance among patients. This system allows doctors to remotely monitor and support their patients, helping to ensure that they are taking their medication as prescribed. By using this system, doctors can provide timely reminders to patients to take their medication and can also track their medication use to ensure that they are adhering to their treatment plan. In this way, the Innovative Pharmaceutical IoT product can help to improve the health outcomes of patients by ensuring that they receive the full benefits of their prescribed treatment.

1 Introduction

Proper treatment is essential for ensuring that patients receive the care they need to manage their health conditions and achieve optimal health outcomes. This includes receiving an accurate diagnosis from a doctor, as well as receiving the appropriate medication to manage their condition. It is important for patients especially older ones to follow their prescribed treatment regimen [1], including taking the correct dosage of their medicine at the prescribed times. This practice, known as drug compliance, is essential for ensuring that patients receive the full benefits of their medication [2]. However, drug compliance can be a challenge for many patients. The World Health Organization (WHO) said only 50% of patients with chronic diseases in developed countries comply with prescribed treatments [3]. There are a variety of factors that can impact a patient's ability to follow their treatment plan, including forgetfulness, difficulty remembering to take their medicine, and difficulty accessing their medication [4]. Also, Medication compliance is a major concern as it prevents hospitalization and increase deaths to medication errors [5]. Even proper medication can lead to decreased efficacy when patients do not adhere to their prescribed schedule and amount of medicine [6]. In order to address these challenges and improve drug compliance, researchers have been conducting studies to identify effective interventions and strategies. Also, recent review studies have identified various devices for monitoring and improving drug compliance [7].

B. J. Choi et al. (Eds.): IHCI 2023, LNCS 14532, pp. 177–190, 2024.
https://doi.org/10.1007/978-3-031-53830-8_18

With the increasing use of electronic devices, such as smartphones and tablets, people are spending more time-consuming digital media and engaging in social media activities alone in their rooms. This shift in behavior can lead to feelings of low self-esteem and depression, particularly when individuals compare themselves to others on social media platforms [8]. According to the World Economic Forum (WEF), depression is currently the most common mental illness, affecting nearly 4% of the world's population (Fig. 1).

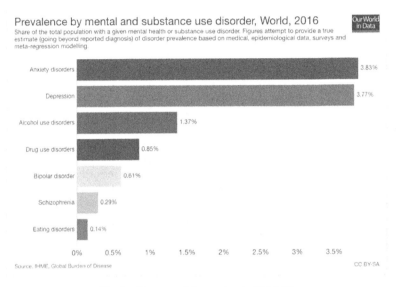

Fig. 1. The rate of depression in 2016 [9]

The COVID-19 pandemic has further exacerbated the issue of depression, as people have had to spend more time at home and have had fewer opportunities for social interaction [10]. This isolation and reduced socialization can lead to an increase in depression, as shown in Fig. 2 from the WEF. The data indicates that the rate of depression increased with age during the pandemic, and that the rate of increase was higher among women than men. Additionally, individuals who were already struggling with emotional issues, such as cancer or trauma, may be more vulnerable to depression as they are unable to return to their normal daily routines.

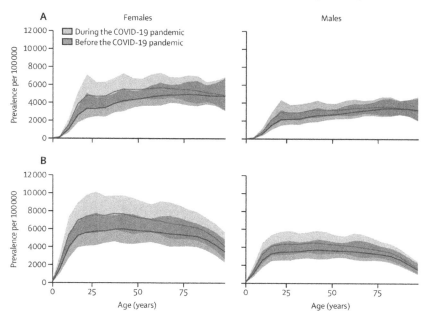

Fig. 2. Global prevalence of major depressive disorder (A) and anxiety disorders (B) before and after adjustment for (ie, during) the COVID-19 pandemic, 2020, by age and sex [11]

The global prevalence of major depressive disorder (A) and anxiety disorders (B) was measured before and during the COVID-19 pandemic in 2020. The data shows that the rates of these mental health conditions increased with age during the pandemic, and that the rate of increase was higher among women than men. The data also indicates that the prevalence of these disorders increased during the COVID-19 pandemic compared to before the pandemic.

One consequence of depression is that patients may not take their prescribed medication as directed by their doctors. This lack of drug compliance can hinder their recovery and overall health outcomes. It is important for individuals experiencing depression to seek professional help and follow their treatment plans in order to manage their condition and improve their quality of life. With the recent development of various electronic devices, people tend to spend more time watching video or Social Network Service alone in their rooms than outside activities with a lot of communication. He is discouraged by his appearance compared to others reflected in Social Network Service, and his self-esteem decreases, resulting in depressed feelings [12]. There is a study with similar device with good results. A total of 58 participants were recruited, of which 55% (32/58) were female with a mean age of 66.36 (SD 11.28; range 48–90) years. Eleven caregiver participants were recruited, of whom 91% (10/11) were female. The average monthly adherence over 6 months was 98% (SD 3.1%; range 76.5–100%). The average System Usability score was 85.74 (n = 47; SD 12.7; range 47.5–100). Of the 46 participants who provided data, 44 (96%) rated the product as easy, 43 (93%) as simple to use, and 43 (93%) were satisfied with the product. Caregiver burden prior to and following smart medication dispenser use for 6 months was found to be statistically significantly different (P < 0.001; CI 2.11–5.98) [13] (Fig. 3).

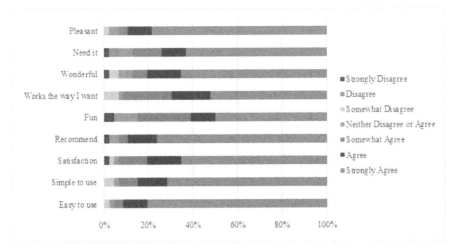

Fig. 3. Usefulness, satisfaction, and ease of use (USE) questionnaire response breakdown. (https://www.ncbi.nlm.nih.gov/pmc/articles/PMC9164090/figure/figure2/)

2   The Design Methodology of an Innovative Pharmaceutical IoT Medication Product

The designed medicine box is equipped with various technologies that allow it to manage and dispense medications efficiently. These technologies include the Arduino Uno board, Bluetooth module, temperature and humidity sensors, and a load cell sensor. The reason for using arduino is that it is efficient to collect data, motors and sensors are easy to control, and they can be used in conjunction with the application by utilizing the MIT App Inventor [14, 15]. The medicine box is divided into two layers, with the upper layer used for storing medication and the lower layer used for dispensing medication when needed.

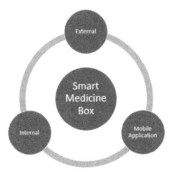

Fig. 4. Structure of Innovative Pharmaceutical IoT medication product

The medicine box also has several developed applications, including a user information system, a system for tracking prescribed drugs, a system for tracking daily recommended dosage, and a humidity and temperature control system. These technologies and applications work together to ensure that the medicine box is able to accurately dispense the correct medications to the user at the appropriate times. The structure of the medicine box is shown in Fig. 4.

2.1 External Technology of the Product

The smart medicine container is equipped with various sensors that allow it to monitor and dispense medication accurately. These sensors include Bluetooth sensors that connect electronic devices and medicine, temperature and humidity sensors that track the temperature and humidity of the medicine, and load cell sensors that measure the weight of the medication to ensure it is normal and to track whether the patient has taken it. The Bluetooth module HM-10 is connected to the Arduino Uno board and works with a developed application to facilitate communication between the medicine container and external devices [16] (Table. 1).

Table 1. Sensors and functions

Sensors	Functions
Bluetooth module	Connects the application and the box
DC motor	Carry the medicine from the 2nd floor to the 1st floor of the box
Loadcell	Carry whether the user has taken the medicine or not
Infrared sensor	Check only one pill comes out of the box
Temperature and humidity sensor	Check the condition of the box

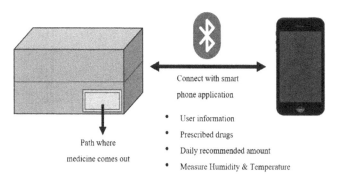

Fig. 5. External structure of the product

The temperature and humidity sensors are used to periodically check the condition of the medicine, and the data from these sensors is input into the application through the Arduino board to track the temperature and humidity of the medicine. The load cell sensors detect the weight of the medication to ensure it is dispensed normally, and if the medication is not taken, a notification is sent through Bluetooth communication to the patient's mobile phone reminding them to take their medicine. The external structure of the medicine container and the process of monitoring and dispensing medication are shown in Figs. 5 and 6, respectively.

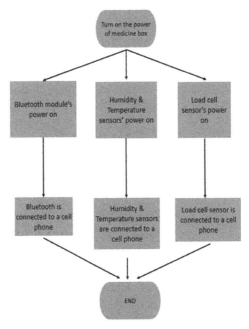

Fig. 6. Flowchart of the External technology of the product

2.2 Internal Technology of the Product

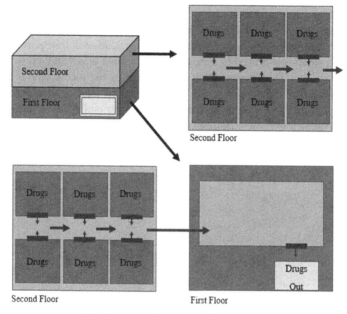

Fig. 7. Internal structure of the product

The smart medicine box is designed to store and dispense medications in a convenient and efficient manner. The interior of the smart medicine shown in Fig. 7 where the box is divided into two layers, with the upper layer used for storing medication and the lower layer used for dispensing medication when needed. When it is time for a dose of medication to be taken, the medication is released from the storage container on the upper layer and transported to the lower layer. This process allows the medication to be easily accessed and dispensed as needed. The smart medicine box may also be equipped with various technologies, such as Bluetooth sensors, temperature and humidity sensors, and load cell sensors, to facilitate communication with external devices and to ensure that the medication is dispensed accurately.

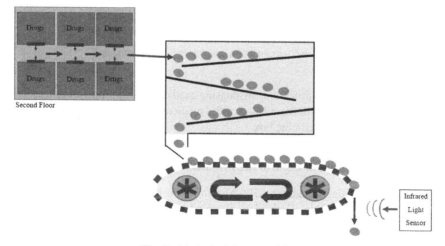

Fig. 8. Method of the second floor

The process of dispensing medication from the smart medicine box involves the use of a slope and a motor-powered conveyor belt to transport the medication from the upper storage layer to the lower dispensing layer. As the medication is transported along the conveyor belt, a small jaw is used to separate the individual doses and allow them to be dispensed one by one. The conveyor belt then transports the medication to a location where it is detected by a load cell sensor and an infrared light sensor. These sensors are used to ensure that the correct amount of medication is dispensed and to halt the transport of the medication when the dispense process is complete. By using this system, the smart medicine box is able to accurately dispense the correct doses of medication to the user at the appropriate times that shown in Fig. 8.

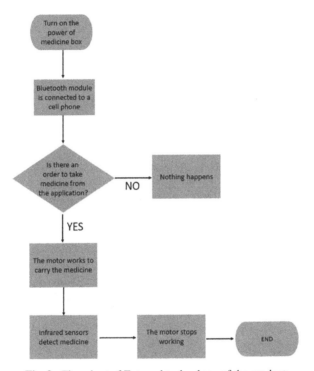

Fig. 9. Flowchart of External technology of the product

Figure 9 illustrates the process of dispensing medication from the smart medicine box. The process begins with the release of the medication from the storage container on the upper layer of the medicine box.

2.3 Mobile Application to Control the Product

The mobile application is made by MIT App inventor. It is connected to the Arduino board via a Bluetooth module, allowing users to book appointments with their doctors and receive treatment recommendations and prescriptions. When a prescription is received, the doctor can set the time and dosage of the medication to be taken. The smart medicine box is programmed to release the appropriate medication at the designated times, and a load cell sensor is used to detect whether the medication has been taken. If the sensor does not detect the weight of the medication, a notification is sent to the patient's cell phone reminding them to take their medication. If the patient ignores this notification and does not take their medication for an extended period of time, a separate notification is sent to the doctor so that they can directly assist the patient in taking their medication.

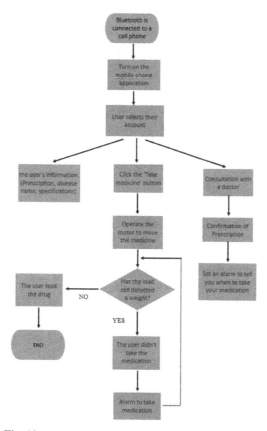

Fig. 10. Flowchart of to process the mobile application

In the event that the medicine does not dispense automatically due to a communication error in the box, there is also a feature that allows users to manually dispense their medication. This can prevent potential problems caused by communication errors. In addition, the temperature and humidity of the medicine container are monitored using temperature and humidity sensors, and this information is regularly reported to the doctor to ensure that the medicine is stored in optimal conditions [17]. Users can create individual accounts and receive personalized treatment recommendations and prescriptions, allowing them to take the correct medications at the appropriate times. The process of using the mobile application and the smart medicine box is shown in Fig. 10.

3   Prototype Development of an Innovative Pharmaceutical

The prototype we developed incorporates several key technologies. These include a Bluetooth module and a temperature and humidity sensor for external functions, and a motor and infrared sensor for internal functions. When a command is triggered from a mobile phone, the motor operates and the medicine container moves along a conveyor belt. When the medicine is detected by the infrared sensor as it moves along the conveyor belt, the motor stops and the medicine is dispensed from the outlet of the container.

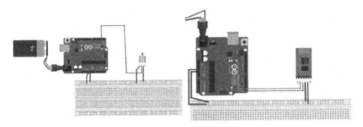

Fig. 11. Circuit of an external with sensors of temperature, humidity, and Bluetooth

The mobile application associated with the prototype includes a function for dispensing medication and a function for monitoring temperature and humidity. The circuit diagrams in Fig. 11 show the implementation of the external functions, including the temperature and humidity sensor on the left and the Bluetooth module on the right. The temperature and humidity sensor require a voltage of 5 V and is connected to the input/output. The Bluetooth module requires a voltage of 3.3 V and is also connected to the input/output. These sensors are connected to different breadboards due to the differences in their voltage requirements.

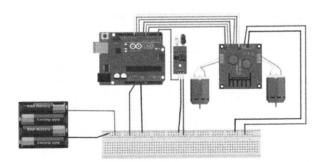

Fig. 12. Circuit of the internal with infrared sensor, DC motors, motor driver

Figure 12 shows the circuit used to implement the internal functions of the prototype. This circuit includes infrared sensors for detecting medication, two DC motors, and a motor driver for controlling the motors. The infrared sensors use a voltage of 5 V and are connected to the input/output. The motor driver is used to control the speed of the two motors, which allows a single dose of medication to be dispensed. In order to control the two motors, additional voltage is required, and batteries are connected separately to provide this power. Overall, this circuit allows the prototype to accurately dispense medication and monitor the movement of the medicine container (Fig. 13).

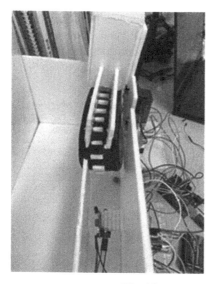

Fig. 13. The prototype of the product

In addition to these features, the medicine box is equipped with temperature and humidity sensors to ensure that the medication is stored in optimal conditions. The box also has a manual dispense function to prevent problems caused by communication errors. The mobile application associated with the product allows users to dispense medication and monitor the temperature and humidity of the medicine container. Overall, the product is intended to help patients take their medication on time and in the correct dosage, and to ensure that their medication is stored and dispensed safely (Table. 2).

Table 2. The functions of the application

Application functions
Create an Account
Select user's account
Login for identification
a description of a medicine case
Allows users to choose whether to take the medication
a sign that the medicine is coming out
Notify the user should take medicine
User took the medication
User didn't take medication (warning)
Check the condition of the medicine box

The medicine box is connected to a mobile application through a Bluetooth module, allowing users to book appointments with their doctors and receive treatment recommendations and prescriptions. When a prescription is received, the doctor can set the time and dosage of the medication to be taken. The medicine box is programmed to release the appropriate medication at the designated times, and a load cell sensor is used to detect whether the medication has been taken. If the sensor does not detect the weight of the medication, a notification is sent to the patient's cell phone reminding them to take their medication. If the patient ignores this notification and does not take their medication for an extended period, a separate notification is sent to the doctor so that they can directly assist the patient in taking their medication (Fig. 14).

Fig. 14. The application screen of the product

The detailed features of the app are as follows. You will sign up first. A maximum of 4 members can be registered as a prototype. After signing up, you can log in and your account. The doctor determines the prescribed medication and how many times to take them. By filling in the information, the patient will know the function of the medication and the number of times to take it. The patient will be able to take the medicine at a fixed time. If you take the medicine, the app will show you that you are taking the medicine, and if you are not taking it, the app will show you that you have not taken it. Nevertheless, if you don't take the medicine for more than a certain period of time, your doctor will be notified, and the doctor may ask you to take the medicine yourself. Additionally, you can view the temperature and humidity of the box in real time to maintain the drug's condition.

These functions enable immediate interaction between doctors and patients. Doctors can give patients the necessary prescriptions in real time, and patients can interact in real time when they need consultation with doctors. In addition, physicians can continue to monitor whether the patient has followed the prescription and, if necessary, notify the patient so that the medication can be taken at a fixed time. This can solve the problem of drug compliance.

4 Final Discussion

The smart medicine box is designed to improve drug compliance among patients suffering from depression and other conditions. It is an economical choice, as it can be used by multiple individuals in a single household, allowing multiple accounts to be set up and used. The medicine box can hold various types of medications and can dispense a set number of pills at a time. It also includes a weight sensor to detect whether a patient has taken their medication, providing immediate feedback to both the patient and their doctor. This feature can be especially useful for patients who are unable to visit the hospital in person, as it allows them to receive treatment and support remotely through the mobile phone application. In the future, the smart medicine box has the potential to be used by patients with a wide range of medical conditions. Its ability to connect to a mobile phone application makes it convenient for users to track their medication usage and for doctors to monitor their patients' drug use. This data can be valuable in the treatment of patients, helping doctors to ensure that their patients are receiving the appropriate care and treatment.

References

1. Lee, S., Jeong, K.-H., Lee, S., Park, H.: A study on types of medication adherence in hypertension among older patients and influencing factors. Healthcare 10(11), 2322 (2022). https://doi.org/10.3390/healthcare10112322
2. Aremu, T.O., Oluwole, O.S., Adeyinka, K.O., Schommer, J.C.: Medication adherence and compliance: recipe for improving patient outcomes. Pharmacy 10(5), 106 (2022). https://doi.org/10.3390/pharmacy10050106
3. Brown, M.T., Bussell, J.K.: Medication adherence: WHO cares? Mayo Clin. Proc. 86, 304–314. Medication adherence: WHO cares?—PubMed (nih.gov) (2011)
4. Martin, L.R., Williams, S.L., Haskard, K.B., Dimatteo, M.R.: The challenge of patient adherence. Ther. Clin. Risk Manag. 1(3), 189–99. PMID: 18360559; PMCID: PMC1661624 (2005)
5. Manmohan, T., Sreenivas, G., Sastry, V.V., Rani, E.S., Indira, K., Ushasree, T.: Drug compliance and adherence to treatment. J. Evol. Med. Dental Sci. 1(3). https://www.researchgate.net/publication/274881427_DRUG_COMPLIANCE_AND_ADHERENCE_TO_TREATMENT (2012)
6. Karagiannis, D., Mitsis, K., Nikita, K.S.: Development of a low-power IoMT portable pillbox for medication adherence improvement and remote treatment adjustment. Sensors 22(15), 5818 (2022). https://doi.org/10.3390/s22155818
7. Aldeer, M., Javanmard, M., Martin, R.P.: A review of medication adherence monitoring technologies. Appl. Syst. Innov. 1, 14 (2018). https://doi.org/10.3390/asi1020014
8. Kim, Y-J., Jang, H. M., Lee, Y., Lee, D., Kim, D-J.: Effects of internet and smartphone addictions on depression and anxiety based on propensity score matching analysis. Int. J. Environ. Res. Public Health 15(5), 859 (2018). https://doi.org/10.3390/ijerph15050859.
9. Fleming, S.: This is the world's biggest mental health problem—and you might not have heard of it. World Economic Forum. https://www.weforum.org/agenda/2019/01/this-is-the-worlds-biggest-mental-health-problem/ (2019)
10. Troisi, A.: Mental health challenges during the COVID-19 pandemic. J. Clin. Med. 12(3), 1213 (2023). https://doi.org/10.3390/jcm12031213

11. Global prevalence and burden of depressive and anxiety disorders in 204 countries and territories in 2020 due to the COVID-19 pandemic. The Lancet. https://doi.org/10.1016/S0140-6736(21)02143-7

12. Sarla, G.S.: Excessive use of electronic gadgets: health effects. Egypt J Intern Med **31**, 408–411 (2019). https://doi.org/10.4103/ejim.ejim_56_19

13. Patel, T., Ivo, J., Pitre, T., Faisal, S., Antunes, K., Oda, K.: An in-home medication dispensing system to support medication adherence for patients with chronic conditions in the community setting: prospective observational pilot study. JMIR Form Res. **6**(5), e34906 (2022). https://doi.org/10.2196/34906. PMID: 35587371; PMCID: PMC9164090

14. Muc, A., Bielecka, A., Iwaszkiewicz, J.: The use of wireless networks and mobile applications in the control of electric machines. Appl. Sci. **13**(1), 532 (2023). https://doi.org/10.3390/app13010532

15. Al-kahtani, M. S., Khan, F., Taekeun, W.: Application of internet of things and sensors in healthcare. Sensors **22**(15), 5738 (2022). https://doi.org/10.3390/s22155738

16. Sharkawy, A.-N., Hasanin, M., Sharf, M., Mohamed, M., Elsheikh, A.: Development of smart home applications based on arduino and android platforms: an experimental work. Automation **3**, 579–595 (2022). https://doi.org/10.3390/automation3040029

17. Horhota, S.T., Burgio, J., Lonski, L.: Effect of storage at specified temperature and humidity on properties of three directly compressible tablet formulations. J. Pharm. Sci. **65**(12), 1746–1749, ISSN 0022-3549. (1976) https://doi.org/10.1002/jps.2600651213

Mobile Computing and Ubiquitous Interactions

Navigating the Complexities of 60 GHz 5G Wireless Communication Systems: Challenges and Strategies

Sultan Maken[1], Koblan Kuanysh[1], Ikechi A. Ukaegbu[1(✉)], and Dhananjay Singh[2]

[1] Electrical and Computer Engineering Department, School of Engineering and Digital Sciences, Nazarbayev University, Qabanbay Batyr Ave 53 Akmola, Astana 010000, Kazakhstan
`ikechi.ukaegbu@nu.edu.kz`
[2] School of Professional Studies, Saint Louis University, 1 N. Grand Blvd. St, Louis, MO 63103, USA

Abstract. The increasing road traffic demands innovative safety solutions, and researchers are exploring V2V communication via 5G massive MIMO at 60 GHz using roadside lamp-based base stations. This approach, with its numerous antennas, offers benefits like improved spectral efficiency, higher throughput, expanded coverage, energy efficiency, and reduced latency. However, deploying massive MIMO in mmWave technology poses challenges, particularly in selecting a suitable channel estimation algorithm. This research paper addresses the channel estimation issue and introduces a sparsity adaptive algorithm that balances accuracy and computational complexity effectively. The algorithm optimizes channel estimation in 60 GHz scenarios relevant for vehicular communications by leveraging channel sparsity, reducing complexity while maintaining accuracy. It is well-suited for real-time vehicular communication applications. To validate the proposed algorithm's efficiency, the paper presents a comprehensive survey of existing channel estimation methods. A comparison table discusses various scenarios, particularly those in the 60 GHz range, providing insights into the algorithm's performance in different settings. These results hold promise for enhancing V2V communication systems, contributing to safer and more efficient road networks amid growing vehicular density.

Keywords: V2V communication · massive MIMO system · 5G · 60 GHz · channel estimation algorithm

1 Introduction

The rise of 5G technology meets the demand for faster data rates, low latency, and reliable wireless communication, particularly in the automotive sector. 5G is instrumental in enhancing Intelligent Transportation Systems (ITS), even in non-line-of-sight (NLOS) scenarios, fostering road safety and V2V (Vehicle-to-Vehicle) communication for improved safety and transportation efficiency [1]. In wireless communication, the growing emphasis on V2V employs massive MIMO (Multiple-Input Multiple-Output)

B. J. Choi et al. (Eds.): IHCI 2023, LNCS 14532, pp. 193–204, 2024.
https://doi.org/10.1007/978-3-031-53830-8_19

technology as a key component of the 5G system. This technology is highly recognized for its capacity to effectively manage the increasing traffic demands associated with vehicular communications in 5G wireless networks [2]. Massive MIMO, equipped with numerous antennas, significantly enhances spectral and energy efficiency in wireless networks, thereby improving overall performance for 5G and future iterations [3]. The integration of antennas, radios, and spectrum within massive MIMO technology is instrumental in boosting network capacity and speed, making it a crucial technology for upcoming wireless standards [4]. It surpasses point-to-point MIMO by more than tenfold in spectral efficiency while employing simpler signal processing [5]. Massive MIMO's array gain is indispensable for intelligent sensing systems in 5G and beyond. In summary, the benefits of massive MIMO encompass high spectral efficiency, communication reliability, simplified signal processing, energy efficiency, and favorable propagation [6]. Furthermore, millimeter wave (mmWave) communication in the 60 GHz range has emerged as a promising solution for V2X communication, offering increased transmission speed and capacity [7]. The 60 GHz mmWave frequency is gaining traction in mobile cellular communication due to cost-effective technology, high-data-rate services, and 5G deployment with small cells and low-powered base stations (BSs) [8]. This frequency range supports data rates of up to 7 Gbit/s and provides ample bandwidth while minimizing interference [9]. The 60 GHz frequency band is a crucial part of the millimeter-wave spectrum, ideal for high-bandwidth wireless communication. It offers wider bandwidth compared to 4G technology and better protection against interference, making it suitable for both stationary and mobile applications [10]. 5G wireless architecture emphasizes energy-efficient hardware and communication methods within this frequency range [11]. This research paper aims to address the intricacies of 60 GHz 5G wireless systems, examining challenges and proposing effective strategies and technologies to overcome them. It will shed light on the key features of the 60 GHz frequency band and how they affect the deployment of 5G wireless communication. Solutions will primarily revolve around channel estimation algorithms.

2   Weaknesses of the 60 GHz Massive MIMO System

Studies in wireless communication show that the 28 GHz and 38 GHz frequency bands experience minimal rain attenuation and oxygen absorption effects at 200-m distances [12]. However, the 60 GHz frequency exhibits significant rain attenuation and oxygen absorption under similar conditions, with air absorption rates of 15 to 30 dB/km and enhanced path loss of 20 to 40 dB [13]. At 60 GHz, the ability of electromagnetic waves to diffract around obstacles is weak, making the links susceptible to obstruction due to their short wavelength [12]. Multipath interference is common at 60 GHz due to factors like surface reflections, object scattering, and diverse media propagation [13]. Historically, millimeter-wave frequencies in wireless communication were costly and mainly used for non-consumer and government applications [14]. In massive MIMO, more antennas lead to increased pilot overhead and computational complexity, potentially reducing performance [15]. Circuit power consumption and antenna spacing considerations, which are typically at least one wavelength apart, can impact the efficiency of massive MIMO systems [16]. Hardware imperfections and channel estimation errors can limit the benefits of numerous base station antennas. Non-orthogonal pilots in multi-cell systems can lead to detrimental inter-cell interference, which can be mitigated with proper channel estimation [17]. Implementation and accurate channel estimation with many antennas can be challenging due to high costs and complexity [18].

3 Proposed Algorithm

Notations: Vectors and matrices are denoted, respectively, by lowercase and uppercase characters; $(\cdot)^{-1}$, $(\cdot)^{T}$, $(\cdot)^{H}$ represent a matrix's inverse, a transpose, and a conjugate transpose, respectively.

3.1 Channel Model of Sparse Multipath

The base station (BS) employs MIMO systems with Q transmitting antennas to transmit OFDM signals. Each individual antenna transmits an OFDM signal of length L, and a subset of P carriers (where $0 < P < L$) is selected as the pilot to estimate the channel. The channel length is denoted as C. Each transmit antenna, indexed as i (ranging from 1 to Q), follows a unique pilot pattern, denoted as $t(i)$. The pilot patterns for different antennas, $t(i)$ and $t(j)$, are disjoint $t(i) \cap^t (j) = \emptyset$, except when i is not equal to j. At the reception end, after the transmission, the pilot signal associated with each antenna is received and represented as $r(t(i))$, which can be denoted as $r^{(i)}$. The fundamental channel model can be expressed as follows:

$$r^{(i)} = G^{(i)}F^{(i)}c^{(i)} + n^{(i)}, i = 1, 2, \ldots, Q \tag{1}$$

In this context, the matrix $F^{(i)}$ represents a sub-matrix with dimensions $P \times C$, , derived from a Fourier matrix associated with a discrete Fourier transform (DFT) matrix of size $L \times L$. The elements of this sub-matrix are selected from the pilot lines, while the columns are limited to the first C columns. On the other hand, $G^{(i)}$ is a diagonal array represented by the pilot pattern $t(i)$ for antenna i. It effectively diagonalizes the matrix and allows for individual processing of the pilot patterns. Additionally, the term $n^{(i)}$ denotes Gaussian white noise with a variance of σ^2 and a mean of 0, contributing to the overall noise in the system. To jth antenna the corresponding channel impulse response (CIR) is $c^{(i)} = [c^{(i)}(1), c^{(i)}(2), \ldots, c^{(i)}(C)]^T$. Equation (1) now includes the following extra equation after changing $H^{(i)} = G^{(i)}F^{(i)}$:

$$r^{(i)} = H^{(i)}c^{(i)} + n^{(i)}, i = 1, 2, \ldots, Q \tag{2}$$

Here, the equation $supp\{c^{(i)}\} = \{l : |c^{(i)}(l)| > p_{th}, 1 \leq l \leq C\}$ represents the index set that defines the support for the ith sub-channel. The support set consists of indices for which the absolute value of $c^{(i)}$ exceeds a specific noise threshold p_{th}. . Recent studies have revealed that in massive MIMO systems, the size of the antenna array at the base station (BS) can be reduced as the transmission distance increases. Furthermore, there is consistency and a shared channel delay among the delay dispersion properties of sub-channels across various sending and receiving antennas. The delay model further indicates that the sub-channels' sparse support set remains unchanged across different users and transmitting antennas [19]:

$$supp\{c^{(i)}\} = supp\{c^{(j)}\}, i \neq j \tag{3}$$

3.2 SAMP Algorithm

Estimation of Sparseness
The problem of finding the minimum l_0 norm can be solved by applying compressed sensing techniques to the channel estimation problem.

$$\hat{c} = argmin||c||_0, subjectto||r - Ac||_2 \leq \varepsilon \tag{4}$$

The expression $||c||_0|$ represents the l_0 norm of vector h, which corresponds to the count of non-zero elements in the vector. According to [20], it has been proven that when the following condition is met:

$$||c||_0 < \frac{1}{2}spark(A) \tag{5}$$

It is possible to recover only the channel impulse response (c). The parameter $spark(A)$ indicates the smallest count of linearly dependent columns within matrix A. It is clear from analysis that $2 \leq spark(A) \leq rank(A) + 1$. If matrix A represents a partial Fourier matrix with dimensions $P \times C$ and $P < C$, it follows that $||c||_0 < \frac{1}{2}(P + 1)$.

Sparsity in wireless communication refers to a channel impulse response where most energy is concentrated on a few taps, with the rest below the noise threshold. When the channel length (L) is much greater than the number of non-zero taps, leveraging this sparsity enables accurate channel estimation with fewer pilot symbols, enhancing spectrum utilization. Equation (5) helps determine the right pilot overhead level for efficient resource utilization.

$$X = \begin{cases} \frac{P}{2}, P \text{ is even} \\ \frac{P+1}{2}, P \text{ is odd} \end{cases} \tag{6}$$

Based on the analysis, we determine that the channel vector has at most X non-zero taps, with C-X considered noise. This initial estimate of sparsity guides element selection. In cases with higher signal-to-noise ratios, the recovered elements are ordered in descending fashion. We use the difference between neighboring elements to aid selection and gauge sparsity. The support set comprises elements preceding the largest backward difference, as they may contain channel information. When certain conditions are met by the observation matrix (represented by A), the task of recovering sparse signals can be transformed into a convex optimization problem. The parameter δ_x known as the Restricted Isometry Property (RIP) parameter, corresponds to the lowest value of δ that fulfills Formula (7).

$$(1 - \delta)||c||_2 \leq ||Ac||_2 \leq (1 + \delta)||c||_2 \tag{7}$$

Here, c denotes the sparse signal corresponding to x. If the value of δ_x is less than 0.5, the x th order RIP condition is satisfied by the matrix A [21]. Moreover, if the matrix's δ_x parameter is smaller than the value of $\sqrt{2} - 1$, the problem of restoration can be transformed into a minimization problem involving the l_1 norm.

$$\hat{c} = argmin||c||_1, subject to ||r - Ac||_2 \leq \varepsilon \tag{8}$$

Channel Estimation

To address the joint sparsity observed in the channel, a transformed channel vector can be represented as $v = [v_1^T, v_2^T, ..., v_C^T]^T$, where v_i corresponds to the i sub-block of v and contains elements $[c^{(i)}(1), c^{(i)}(2), ..., c^Q(i)]^T$, where $i = 1, 2, ..., C$. The consequence of this transformation is the concentration of non-zero elements within the channel vector. The received pilot signal is similarly transformed to $y = [y_1^T, y_2^T, ...y_{pT},]^T$, where $y_i = [r^{(1)}(i), r^{(2)}(i), ..., y^{(Q)}(i)]^T$, $i = 1, 2, ..., P$. Additionally, the noise is also transformed to $e = [e_1^T, e_2^T, ..., e_P^T]^T$, where e_i corresponds to the i sub-block of e and equals to $[n^{(1)}(i), n^{(2)}(i), ..., n^{(Q)}(i),]^T$, $i = 1, 2, ..., P$. With respect to all sending antennas, the received signal expression may be expressed as follows:

$$y = Mv + e \tag{9}$$

One of the variables is a matrix $M = [M_1, M_2, ..., M_C]$; $M_i = [a^{(1)}(i), a^{(2)}(i), ..., a^{(T)}(i)]$, $i = 1, 2, ..., C$, $a^{(T)}(i)$ is matrix $A^{(T)}$'s i th column. By performing matrix multiplication between Formula (3) and the conjugate transpose of matrix M, denoted as M^H, we can estimate v. This estimation employs compressed sensing techniques in scenarios where the sparsity of the channel is unknown.

$$M^H y = M^H(Mv + e) = v + (M^H M - J) + M^H e \tag{10}$$

Here, J represents the $QC \times QC$ unit matrix. Since the matrix M does not exhibit complete orthogonality, $M^H M - J$ is a matrix with small element values but not zero. The representation of the energy scattering resulting from the observation matrix's non-orthogonality is represented by $e' = (M^H M - J)v + M^H e$. As a result, Eq. (10) can be stated as:

$$M^H y = v + e' \tag{11}$$

Define an $QC \times 1$ vector E during iteration.

$$E = |M^H q| \tag{12}$$

Here, the variable q denotes the iterative residuals, initially set as y. The absolute values of the elements in the conjugate transpose of matrix M multiplied by q are denoted by $|\cdot|$. The elements in vector V' are defined as the summation of the squared values in each set of Q' elements within vector E.

$$V(j) = \sum_{(j-1)\times Q+1}^{j\times Q} |E(i)|^2, \ i = 1, 2, ..., QC; j = 1, 2, ..., C \tag{13}$$

Here, the i th element in vector E is denoted by $E(i)$, and the j th element in vector V is denoted by $V(j)$. To obtain vector Vs, the elements in vector V must be corrected in descending order. The maximum sparsity of the channel is X. After the first iteration, only e' in (11) generates the last C-X elements in Vs. In the wireless communication domain, when considering the subsequent C-X elements, their energy is assigned a specific threshold, denoted as t. In order for a tap energy in the channel to be considered significant, it must exceed this threshold value. Consequently, the support set should only contain vector Vs elements that are greater than the threshold. This algorithm estimates sparsity in two steps: setting an upper limit based on channel characteristics and identifying sparsity by finding the maximum difference within that limit. It surpasses other methods by considering energy dispersion, Gaussian white noise, and using regularization to enhance support set accuracy.

4 Results

The main limitations of the implementation of massive MIMO systems in the 60 GHz frequency band may be high computational complexity, high levels of interference, and high cost, leading to channel estimation errors. Classical algorithms like LS, LMS, and OMP address these issues. LS is simple but neglects noise impact. An innovative LS method [22] aims to reduce noise influence for better OFDM channel estimation. LMS is an adaptive filter with a slow convergence rate, minimizing mean square error to enhance signal quality [23, 24]. OMP is a fast and simple iterative algorithm for signal recovery [25]. In Table. 1 was proposed a list of algorithms with their advantages and disadvantages.

Table 1. List of algorithms for navigating the complexities of 60 GHz 5G wireless communication systems

Refs.	Proposed algorithm	Advantages	Disadvantages
[26]	BSAMP	Efficiently handles channel sparsity with dynamic adjustments. Low computational complexity, adjusts to changing channels	Depends on accurate sparsity levels, relies on precise initial estimation. Limitations in base stations with numerous antennas
[27]	Deep learning with sMPD	Enhances detection performance, reduces computational complexity	Performance depends on channel conditions
[28]	Butterfly optimization algorithm (BOA)	Excels in accuracy and optimal solutions	Parameter selection can be trial-and-error. Randomness may yield varied results
[22]	Novel OFDM channel estimation	Improves estimation precision by reducing noise impact	Assumes AWGN in the channel and relies on known noise variance

(*continued*)

Table 1. (*continued*)

Refs.	Proposed algorithm	Advantages	Disadvantages
[29]	60 GHz wireless systems	Efficiently estimates multiple channel parameters in 60 GHz band	Designed for static indoor settings. May overlook factors in different environments
[30]	Deep learning-based channel estimation	Significantly improves accuracy	Computationally complex, requires a large amount of data for training
[31]	Matching pursuit-based (MP) algortihm	Excels in sparse channels, eliminates zero-tap errors	Effectiveness depends on stopping rule. Requires prior channel probability distribution knowledge
[32]	Forward–backward channel estimation	Copes with tough channel conditions, practical for real-world use	May not work well in different environments. Effectiveness depends on the stopping rule
[33]	OFDM-based channel estimation	Simple lead-frequency interpolation	Neglects channel correlation and noise in complex RF environments
[34]	Channel estimation with cGAN	Improves accuracy, efficient pilot design	Computationally demanding, not effective in noisy or interference-prone settings
[35]	MIMO-OFDM channel estimation	Boosts estimation precision, responsive to changing channels	Increased computational complexity, responsive to training sequence quality. May not be suitable for high-mobility channels
[36]	Leaky least mean square (MLLMS)	Lower computational complexity, better performance in noisy environments	Slightly more computationally complex compared to LMS
[37]	5G high-frequency CSI estimation	RVM (relevance vector machine) excels in accuracy, LS is computationally efficient, OMP closely matches RVM with the right iteration count	LASSO consistently underperforms. OMP is less accurate than RVM
[38]	DNN (deep neural network) and LSTM (long short-term memory) channel estimation	LSTMs handle sequential data and vanishing gradient	LSTMs can be computationally expensive and harder to interpret. Prone to overfitting

5 Discussion

Table. 1 provides a comprehensive comparison of channel estimation algorithms in wireless communication, particularly for MIMO-OFDM and massive MIMO systems. It assesses each algorithm's pros and cons, offering insights into their effectiveness for various scenarios and applications, including V2V communication. Choosing the right channel estimation technique is crucial for accurate and efficient data transfer between vehicles in dynamic contexts. Researchers have explored various algorithms, each with its own advantages and limitations. BSAMP offers low complexity and adaptability but relies on sparsity assumptions and initial accuracy, making it less suitable for large MIMO systems. Deep learning and BOA algorithms show improved performance but may require substantial data and involve randomness. When selecting a V2V communication algorithm, factors to consider include vehicle mobility, computational efficiency, adaptability to different conditions, protocol compatibility, and system-specific requirements. Millimeter-wave (mmWave) technology holds promise for high data rates and low latency in V2V communication. However, in mmWave systems with high-frequency signals susceptible to path loss and atmospheric absorption, accurate channel estimation is crucial. The channel experiences significant variations due to blockages and scattering, impacting reliable communication. In Fig. 1, the three papers focus on channel estimation in wireless communication systems operating at 60 GHz. Their comparison is based on the mean squared error (MSE) of their respective algorithms.

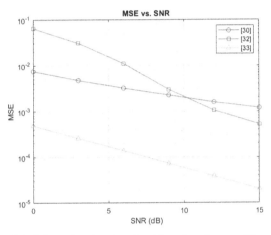

Fig. 1. MSE of channel estimation algorithms based on mmWave(60Ghz)

Comparing three papers' channel estimation approaches yields valuable insights into their performance in different conditions. [29] excels with an efficient algorithm, providing accurate channel estimation even at low SNR levels. [31] introduces the SDCE-NTD method, demonstrating improved MSE performance in both dense and sparse channel scenarios. [32] offers a low-complexity approach with potential for effective channel estimation, though with a slightly higher MSE. In mmWave systems, BOA and MLLMS are well-suited. BOA combines global and local search for complex mmWave channels, while MLLMS, with its simplicity, handles dynamic fluctuations effectively. Selecting the right channel estimation method for mmWave communication systems is critical due to high-frequency challenges like path loss, air absorption, and sensitivity to obstacles. The chosen method should balance complexity, support multiple antennas, real-time processing, and manage data rates and processing demands. It's worth noting that mmWave channels often exhibit sparsity, with a small number of significant paths in the channel impulse response. Figure 2 illustrates the MSE values of three deep learning-based channel estimation algorithms in [30, 33], and [34] at various SNR levels. Each paper focuses on enhancing the accuracy and efficiency of channel estimation through deep learning techniques.

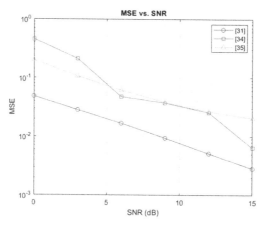

Fig. 2. MSE of channel estimation algorithms based on deep learning

In comparing three algorithms in [30, 33], and [34] across various signal-to-noise ratio (SNR) levels, [30] consistently yields the lowest MSE, demonstrating the effectiveness of the deep learning channel estimation pipeline, even under challenging conditions. [33] performs reasonably well with higher MSE than [30] but lower than [34], indicating its effectiveness in estimation accuracy using inverse convolutional networks and expanded convolutional networks. However, [34] exhibits the highest MSE values but achieves this with fewer pilots, showcasing its ability to optimize pilot design for acceptable channel estimation. For mmWave channel estimation, sparsity-exploiting techniques like LASSO or OMP are well-suited, and adaptability to dynamic channel conditions is crucial, making RLS and MLLMS valuable for continuous updates. Handling interference is essential, with Bayesian learning and deep learning methods being effective in noise mitigation. In massive MIMO mmWave systems, supporting multiple users simultaneously through spatial multiplexing and interference cancellation techniques can significantly enhance system capacity and performance.

6 Conclusion

The paper introduces a channel estimation algorithm for massive MIMO channels that efficiently leverages sparsity within these channels. By utilizing block sparsity and adaptive sparsity level updates, the algorithm achieves low computational complexity while maintaining accurate channel estimation, making it ideal for dynamic channel conditions. Its adaptability allows real-time adjustments, ensuring reliability. This computationally efficient algorithm is practical for large-scale MIMO systems. Additionally, the paper reviews existing literature on mmWave and massive MIMO channel estimation algorithms, highlighting potential areas for further research and improvements.

Acknowledgement. This work was supported by the Ministry of Education and Science of the Republic of Kazakhstan (AP13068587, AP14871109) and Nazarbayev University (grant no. 11022021CRP1507, 021220FD0451).

References

1. He, R., et al.: Propagation channels of 5G millimeter-wave vehicle-to-vehicle communications: recent advances and future challenges. IEEE Veh. Technol. Mag. **15**(1), 16–26 (2020). [Online]. https://doi.org/10.1109/MVT.2019.2928898
2. Jiang, H., Chen, Z., Zhou, J., Dang, J., Wu, L.: A general 3D non-stationary wideband twin-cluster channel model for 5G V2V tunnel communication environments. IEEE Access **7**, 137744–137751 (2019). [Online]. https://doi.org/10.1109/access.2019.2942442
3. Ali, E., Ismail, M., Nordin, R., Abdulah, N.F.: Beamforming techniques for massive MIMO systems in 5G: overview, classification, and trends for future research. Frontiers Inf. Technol. Electron. Eng. **18**(6), 753–772 (2017). [Online]. https://doi.org/10.1631/fitee.1601817
4. Chataut, R., Akl, R.: Massive MIMO Systems for 5G and beyond networks—overview, recent trends, challenges, and future research direction. Sensors **20**(10), 2753 (2020). [Online]. https://doi.org/10.3390/s20102753
5. Vaigandla, K.K., Venu, D.N.: Survey on massive MIMO: technology, challenges, opportunities and benefits. papers.ssrn.com, (2021). [Online]. Available at: https://papers.ssrn.com/sol3/papers.cfm?abstract_id=4232166
6. Borges, D., Montezuma, P., Dinis, R., Beko, M.: Massive MIMO techniques for 5G and beyond—opportunities and challenges. Electronics **10**(14), 1667 (2021). [Online]. https://doi.org/10.3390/electronics10141667
7. Dimce, S., Amjad, M.S., Dressler, F.: mmWave on the road: investigating the weather impact on 60 GHz V2X communication channels. IEEE Xplore 2021. [Online]. https://doi.org/10.23919/WONS51326.2021.9415572
8. Sulyman, A.I., Alwarafy, A., Seleem, H.E., Humadi, K., Alsanie, A.: Path loss channel models for 5G cellular communications in Riyadh city at 60 GHz. IEEE Xplore (2016). [Online]. https://doi.org/10.1109/ICC.2016.7510953
9. Saini, J., Agarwal, S.K.: Design a single band microstrip patch antenna at 60 GHz millimeter wave for 5G application. IEEE Xplore (2017). [Online]. https://doi.org/10.1109/COMPTELIX.2017.8003969
10. Daniels, R.C., Murdock, J.N., Rappaport, T.S., Heath, R.W.: 60 GHz wireless: up close and personal. IEEE Microwave Mag. **11**(7), pp. 44–50 2010. [Online]. https://doi.org/10.1109/mmm.2010.938581
11. Moongilan, D.: 5G wireless communications (60 GHz band) for smart grid—an EMC perspective. IEEE Xplore, 2016. [Online]. https://doi.org/10.1109/ISEMC.2016.7571732
12. Niu, Y., Li, Y., Jin, D., Su, L., Vasilakos, A.V.: A survey of millimeter wave communications (mmWave) for 5G: opportunities and challenges. Wirel. Netw. **21**(8), 2657–2676 2015. [Online]. https://doi.org/10.1007/s11276-015-0942-z
13. Daniels, R.C., Heath, R.W.: 60 GHz wireless communications: Emerging requirements and design recommendations. IEEE Veh. Technol. Mag. **2**(3), 41–50 2007. [Online]. https://doi.org/10.1109/mvt.2008.915320
14. Bosco, B., Emrick, R., Franson, S., Holmes, J., Rockwell, S.: Emerging commercial applications using the 60 GHz unlicensed band: opportunities and challenges. IEEE Xplore 2006. [Online]. https://doi.org/10.1109/WAMICON.2006.351908
15. Marzetta, T.L.: How much training is required for multiuser mimo? IEEE Xplore, 2006. [Online]. https://doi.org/10.1109/ACSSC.2006.354768

16. Ngo, H.Q., Larsson, E.G., Marzetta, T.L.: Energy and spectral efficiency of very large multiuser MIMO systems. IEEE Trans. Commun. **61**(4), 1436–1449 2013. [Online]. https://doi.org/10.1109/tcomm.2013.020413.110848

17. Adnan, N.H.M., Rafiqul, I.Md., Alam, A.H.M.Z.: Massive MIMO for fifth generation (5G): opportunities and challenges. IEEE Xplore 2016. [Online]. https://doi.org/10.1109/ICCCE.2016.23

18. Al-Falahy, N., Alani, O. Y.: Technologies for 5G networks: challenges and opportunities. IT Prof. **19**(1), 12–20 2017. [Online]. https://doi.org/10.1109/mitp.2017.9

19. Huang, C., Liu, L., Yuen, C., Sun, S.: A LSE and sparse message passing-based channel estimation for mmWave MIMO systems. IEEE Xplore 2016. [Online]. https://doi.org/10.1109/GLOCOMW.2016.7848817

20. Donoho, D.L., Elad, M.: Optimally sparse representation in general (nonorthogonal) dictionaries via ℓ 1 minimization. Proc. Nat. Acad. Sci. **100**(5), 2197–2202. 2003. [Online]. https://doi.org/10.1073/pnas.0437847100

21. Candes, E.J., Tao, T.: Decoding by linear programming. IEEE Trans. Inf. Theory **51**(12), 4203–4215 (2005). [Online]. https://doi.org/10.1109/TIT.2005.858979

22. Zheng, Z., Hao, C., Yang, X.: Least squares channel estimation with noise suppression for OFDM systems. Electron. Lett. **52**(1), 37–39 2016. [Online]. https://doi.org/10.1049/el.2015.2678

23. Al-Saggaf, U.M., Moinuddin, M., Arif, M., Zerguine, A.: The q-least mean squares algorithm. Signal Process **111**, 50–60 2015. [Online]. https://doi.org/10.1016/j.sigpro.2014.11.016

24. Hossain, Md. M., Rahman, Md. M., Rana, Md. M.: Least mean square (LMS) for smart antenna. Univ. J. Commun. Netw. **1**(1), 16–21 (2013). [Online]. https://doi.org/10.13189/ujcn.2013.010103

25. Cai, T.T., Wang, L.: Orthogonal matching pursuit for sparse signal recovery with noise. IEEE Trans. Inf. Theory **57**(7), 4680–4688. [Online]. https://doi.org/10.1109/tit.2011.2146090

26. Khan, I., Singh, M., Singh, D.: Compressive sensing-based sparsity adaptive channel estimation for 5G massive MIMO systems. Appl. Sci. **8**(5), 754, 2018. [Online]. https://doi.org/10.3390/app8050754

27. Yan, L., Wang, Y., Zheng, N.: 5G massive MIMO signal detection algorithm based on deep learning. Comput. Intell. Neurosci. 2022, Article ID 9999951, 9 2022. [Online]. https://doi.org/10.1155/2022/9999951

28. Altiraiki, S., Tezel, N.S.: A new approach to pilot contamination in massive MIMO systems for 5G communication networks with butterfly optimization algorithm. J. Polytech. 2020. [Online]. https://doi.org/10.2339/politeknik.726354

29. Fan, D., Zhong, Z., Wang, G., Gao, F.: Channel estimation for 60GHz wireless local area networks with massive receiving antennas. IEEE Xplore [Online]. https://doi.org/10.1109/HMWC.2014.7000215

30. Soltani, M., Pourahmadi, V., Mirzaei, A., Sheikhzadeh, H.: Deep learning-based channel estimation. arXiv:1810.05893 [cs, eess, math, stat], Feb. 2019. [Online]. Available: https://arxiv.org/abs/1810.05893

31. Gao, B., Jin, D., Zeng, L., Xiao, Z., Zhang, C.: Sparse/dense channel estimation with non-zero tap detection for 60-GHz beam training. IET Commun 8(11), 2044–2053, Jul. 2014. [Online]. https://doi.org/10.1049/iet-com.2013.0942

32. Belaoura, W., Ghanem, K., Nedil, M., Bousbia-Salah, H.: Forward–backward processing for efficient underground channel estimation in 60 GHz MISO FBMC systems. Electron. Lett. **55**(2), 92–94 2019. [Online]. https://doi.org/10.1049/el.2018.6406

33. An, X., Zhao, L., Wu, H., Zhang, Q.: Channel estimation algorithm based on attention mechanism. J. Phys.: Conf. Series **2290**, Conf. Ser. 2290 012112 2022. [Online]. https://doi.org/10.1088/1742-6596/2290/1/012112

34. Kang, X.-F., Liu, Z.-H., Yao, M.: Deep learning for joint pilot design and channel estimation in MIMO-OFDM Systems. Sensors **22**(11), 4188 2022. [Online]. https://doi.org/10.3390/s22 114188

35. Sarnin, S.S., Sulong, S.M., Hashim, H.: Channel estimation on the (EW) RLS algorithm model of MIMO OFDM in wireless communication. MATEC Web Conf. **56**, 05014 2016. [Online]. https://doi.org/10.1051/matecconf/20165605014

36. Bhoyar, D.B., Dethe, C.G., Mushrif, M.M.: Modified LLMS algorithm for channel estimation in noisy environment. Univ. J. Commun. Netw. **1**(2), 60–67 2013. [Online]. https://doi.org/10.13189/ujcn.2013.010205

37. Tapio, V., Aminu, M.U., Lehtomäki, J., Juntti, M.: Channel estimation algorithms for hybrid antenna arrays: performance and complexity. IEEE Xplore 2019. [Online]. https://doi.org/10.1109/ISWCS.2019.8877352

38. Mohammed, A.S.M., Taman, A.I.A., Hassan, A.M., Zekry, A.: Deep learning channel estimation for OFDM 5G systems with different channel models. Wirel. Pers. Commun. (2022). [Online]. https://doi.org/10.1007/s11277-022-10077-6

A Survey on Channel Estimation Technique Classifications and Various Algorithms

Koblan Kuanysh[1], Dhananjay Singh[2(✉)], and Ikechi A. Ukaegbu[1(✉)]

[1] Electrical and Computer Engineering Department, School of Engineering and Digital Sciences, Nazarbayev University, Qabanbay Batyr Ave 53Akmola, Astana 010000, Kazakhstan
ikechi.ukaegbu@nu.edu.kz
[2] School of Professional Studies, Saint Louis University, 1N. Grand Blvd. St, Louis, MO 63103, USA
dhananjay.singh@slu.edu

Abstract. The field of wireless communication is experiencing rapid advancements, leading to remarkable achievements in terms of both high data rates and low latency. Notably, the emergence of 5G has significantly influenced the landscape, boasting impressive speeds of up to 20 Gb/s. Furthermore, the utilization of technologies like Orthogonal Frequency Division Multiplexing (OFDM) has played a crucial role in optimizing the efficiency of spectral utilization within communication networks. However, despite these advancements, the progress of wireless communication has introduced challenges in accurately estimating channels. This is primarily due to the presence of varied factors such as distortion, attenuation, fading, scattering, and other interruptions that affect the transmission of radio waves to their intended destinations. To address this issue, researchers have put forward suggestions aimed at enhancing the accuracy of channel estimation and strengthen the signal. In this paper, an extensive and comprehensive review of various channel estimation techniques provided, categorizing them based on their unique characteristics. Additionally, the role of the upcoming technologies, like artificial intelligence in the context of wireless communication will be investigated. Such a review will be invaluable in making informed decisions regarding the selection of channel estimation techniques for emerging wireless communication systems.

Keywords: OFDM · channel estimation algorithm · wireless channel · millimetre wave

1 Introduction

Applying wireless communication not only facilitates data delivery to its destination, but it also plays an important part in accident prevention. The potential of V2V communication to wirelessly communicate data to different nearby automobiles offers great potential in the attempt to avoid accidents and drastically decrease traffic accident mortality. The most important requirements for wireless communication in this regard are fast responsiveness, low latency, and highest reliability. Various approaches are being employed to improve communication in this regard. One such method that used in wireless and mobile communication systems is Orthogonal Frequency-Division Multiplexing due to its exceptional efficiency.

OFDM efficiently transmits data using Fast Fourier Transform (FFT) and Inverse Fast Fourier Transform (IFFT), effectively reducing Inter symbol Interference (ISI) through the insertion of a Cyclic Prefix [1]. This makes it a robust transmission technique, ensuring dependable data transfer even in challenging wireless environments. OFDM is more resistant to signal distortions caused by reflections, diffractions, and other variations in the wireless signal path [2–4].

However, increasing the data rate or decreasing latency of the connection can lead to performance degradation and decreased signal accuracy. To address this issue, proper estimation of channels between antennas becomes crucial. Researchers proposed numerical channel estimation algorithms to tackle this challenge. These methods classified into three practical categories: blind channel estimation, semi-blind channel estimation, and pilot-based channel estimation. Among these techniques, the pilot-aided channel estimation method is the most popular and more practical than other methods due to its reliable and consistent performance [5]. In this approach, the system bandwidth divided into multiple independent subcarrier bandwidths of equal width, with each subcarrier transmitting its modulated signals [6].

However, most existing methods primarily designed for static scenarios, where two connected devices fixed at a single point. Unfortunately, when it comes to dynamic scenarios, where the channels fluctuate rapidly over time, these methods demonstrate deficient performance. In such dynamic scenarios, accurate channel estimations become crucial, particularly in applications like Vehicle-to-Vehicle (V2V) communication, where vehicles need to establish reliable connections while on the move.

The authors in [7] propose a joint channel coding technique to improve channel estimation in pilot-based systems. The technique aims to correct the channel estimation values obtained using traditional DFT-based methods. The [8] proposes a new algorithm for channel estimation in fast fading channels in offset QAM-based filter bank multicarrier (FBMC/OQAM) systems. The paper presents an algorithm that addresses the limitations of existing techniques and significantly improves channel estimation performance in FBMC/OQAM systems operating in fast fading channels. Two specific channel estimation methods proposed in [9], one of them is for slow fading channels and another for fast fading channels, to enhance the performance of OFDM systems in wireless communication. [10] describes the proposal of a location-aided beamforming strategy to achieve ultra-fast initial access in millimetre-wave (mm-wave) communication for the vehicular domain. The paper presents numerical experiments indicating that the availability of location information improves the signal-to-noise ratio (SNR) at the receiver when the distance exceeds eighty meters.

The structure of this paper organized as follows: in Sect. 2, a comprehensive overview of channel estimation classification presented, focusing on three techniques: pilot-aided, semi-blind, and blind, in addition to Decision-directed channel estimation. In Sect. 3, the channel estimation algorithms are illustrated based on the classifications. Furthermore, the paper features a review of channel estimation algorithms in a table, accompanied by concise algorithm explanations and emphasizing their contributions and the comparison plot of the algorithms with each other.

2   Channel Estimation

2.1   Channel Estimation Classification

Pilot-aided Channel Estimation Method. Pilot-based channel estimation technique inserts training patterns into the transmitted signal to predict the channel response. These training sequences are known to both antenna from the transmitter to the receiver and had designed to have specific properties that make them easy to detect and analyse.

As it widely used, enormous investigations have reported in the source related to pilot-based channel estimation. The paper [11] provides a comprehensive examination of pilot-based channel estimation methods, specifically those using block-type and comb-type arrangements. The author explains how channel estimation can achieve using block-type pilots, either with a decision feedback equalizer or without. As well paper [12] investigated the channel estimation based on block-type and comb-type using Least Squares and Linear Minimum Mean Square Error estimators by using the 60 GHz frequency band in contrast with [11]. The paper [13], uses pilot symbols to estimate parameters in OFDM systems, and how they can improve performance in noisy wireless environments. The study finds that using pilot symbols can be more effective than other methods like decision-directed channel estimation and is especially good at handling Doppler.

Despite improvements made by researchers using pilot-assisted channel estimation, the technique still wastes communication bandwidth and relies only on pilot symbols for channel estimation, leading to unavoidable errors during the process.

Semi Blind and Blind Channel Estimation Method. To avoid wasting bandwidth, researchers have investigated blind channel estimation techniques that do not require the use of pilot symbols, which consume valuable channel capacity. Techniques for estimating channels using knowledge about known training symbols and uncertain received signals are known as semi-blind channel estimation techniques.

Various works investigated in semi blind and blind channel estimation. The papers [14] and [15] choose the blind channel estimator approach to estimate the parameters. The paper [14] developed a blind channel estimation method using the noise subspace technique for MIMO-OFDM (Orthogonal frequency-division multiplexing). It can apply to systems no matter the amount of transmitting and receiving antennas, combining, and generalizing existing methods for SISO-OFDM systems. In the article [15], the blind channel estimation for massive MIMO proposed to cope with the pilot contamination, which is a limitation of pilot-based and semi-blind methods. References [16] and [17], they both propose an efficient semi-blind channel estimation approach for the MIMO system. In [16], the authors investigated a semi-blind channel estimation scheme based on an expectation–maximization algorithm and showed the effectiveness of the approach by comparison with known ML (Maximum Likelihood) estimators through the numerical results.

Despite improvements made by researchers using pilot-assisted channel estimation, the technique still wastes communication bandwidth and relies only on pilot symbols for channel estimation, leading to unavoidable errors during the process.

Decision-directed Channel estimation method. The decision-directed technique is often in digital communication systems where the transmitted data is in the form of discrete symbols. In decision-directed channel estimation, the receiver makes use of the received symbols to estimate the channel characteristics. The basic idea behind this technique is to use the received symbols to decide about the transmitted symbols, and then use this decision to estimate the channel characteristics.

The paper [18] investigates decision-directed channel estimation, which is challenging for systems with multiple transmit antennas due to the high computational complexity of inverting a data-dependent matrix. To overcome this, the paper introduces an iterative method that avoids matrix inversion and demonstrates its effectiveness through simulation results. Also, researchers improved the performance of the decision-directed method and decreased the computational complexity. For example, in research [19], the performance of the method enhanced by using demodulator output symbols, DDE applies statistical characteristics of transmitted data to track fast-changing channel transfer functions. While in [20], using the space-alternating generalized expectation–maximization method, pilot overhead reduced without affecting efficiency.

2.2 Channel Estimation Algorithms

Based on the classification of channel estimation, exists numerical channel estimation algorithms. Accurate channel estimation is essential for effective wireless communication and can achieved using various algorithms. The computational complexity and accuracy of the algorithm are the key features of the channel estimation algorithm in wireless communication. The latency, data rate and overall performance of the communication system straightforward depend on the selected algorithm.

We selected a combination of search engines, namely IEEE Xplore, Springer Digital Library, and Google Scholar, recognizing that no single search engine alone provided comprehensive results for our channel estimation research. Our search has conducted using specific keywords, such as "channel estimation," "estimation techniques," "channel estimation classification," "channel estimation AND wireless communication," and "channel estimation AND time varying." We fine-tuned these keywords to suit each search engine's interface, refining them as "channel estimation AND Algorithms," "channel estimation AND performance evaluation," "channel estimation AND V2V communication," and "channel estimation AND OFDM." To strike a balance between broadening the search and obtaining relevant results, we implemented strategies to manage the search process effectively. In certain cases, we included the keyword "mmWave" to account for its associated advantages, such as high data rate transmission, low latency, and improved security. Additionally, the keyword "time varying" incorporated due to the dynamic nature of vehicle mobility and high-speed scenarios on the road. Furthermore, we expanded our exploration by investigating other V2V wireless communication methods. This approach allowed us to gain insights into the challenges faced during the investigation of V2V communication. To ensure up-to-date findings, we applied a time filter to retrieve the most recent papers. Notably, major of the papers we encountered published between 2015 and 2023.

Table 1 summarizes papers with channel estimation algorithms, including their brief explanation and strengths of the proposed algorithm from other conventional algorithms.

Table 1. Comparison of proposed channel estimation algorithms.

Refs.	Proposed algorithm	The contribution of the paper
[21]	Fast LMMSE channel estimation algorithm by the improved most significant taps and calculation by Kumar's fast algorithm	Does not require static knowledge of channel - The computational complexity reduced
[22]	Pilot-aided modified an algorithm for estimating the channel using least squares method in Orthogonal Frequency Division Multiplexing (OFDM) systems	- Noise reduced - Improved channel estimation algorithm
[23]	The method of estimating channels involves utilizing LS and MMSE algorithms and is dependent on a training symbol of the block type that assisted by a pilot	- Lower bit error rate - Decreased symbol error rate - Lower mean square
[24]	Four channel estimation algorithms: Adaptive boosting, least square, best linear Unbiased estimator, Minimum mean square error	- Computational complexity - Useful with high number of carriers - Overall performance
[25]	An algorithm for estimating the channel in which the transfer domain used along with wavelet de-noising and distance decision analysis	- Lower bit error rate - Lower signal-to-noise ratio
[26]	The UAMP-SBL combines unitary approximate message passing and sparse Bayesian learning techniques while incorporating variational inference and unitary transformation methods	- Low complexity - Accurate estimation - Fast sparse Bayesian approach
[27]	The DSACS, Dynamic and Sparsity Adaptive Compressed Sensing algorithm presents a novel solution for AUD and CE in UL grant-free SCMA systems by addressing the challenge of unknown user sparsity	- No need prior knowledge of user sparsity or a potential active user list - Reduced computational complexity
[28]	An improved DCT-based algorithm that uses a simplified denoise filter in the discrete cosine transform domain, which derived from a traditional LMMSE filter	- Low computational complexity - Highly performance
[29]	An algorithm based on a distributed computing technique and channel estimation algorithm inspired by accelerated projection-based consensus (APC). The algorithm designed to process in parallel enhancing its efficiency	- High reliability - Suitable for such scenarios like Massive MIMO systems
[30]	The TOMP algorithm is designed for channel estimation in underwater acoustic OFDM. It streamlines the process by using the orthogonality of specific measurement atoms, and optimizing the selection of measurement atoms	- Improves the accuracy - High speed of signal in reconstruction
[31]	The CPR-Net algorithm enhances signal recovery in time-varying channel OFDM by integrating a deep neural network (FC-DNN) with existing channel estimation methods	- Elevate the accuracy of the signal over time-varying channels using deep learning

(continued)

Table 1. (*continued*)

Refs.	Proposed algorithm	The contribution of the paper
[33]	The pilot-assisted estimating channels in MIMO-single-carrier frequency-domain-equalization over fast time-varying multipath channels. It uses a noise eliminated MMSE and a new polynomial interpolation to obtain accurate estimations	- Improved performance in fast time-varying in MIMO-SCFDE system
[33]	The OMP and MSP algorithms for delay-Doppler (DD) channel estimation in uplink OTFS-MA systems. These algorithms leverage compressed sensing (CS) techniques to handle the sparsity of the DD channel representation	- Normalized MSE and BER performance in rapidly time-varying channels for multi-user
[34]	The RVP-FLMP algorithm to estimate parameters in MIMO systems with a Pop and Beulieu model-based Rayleigh fading. It addresses complexity issues by integrating fractional order calculus into the algorithm	- Faster weight convergence - Improved channel capacity - Lower BER
[35]	The algorithm proposed in the paper is an M-estimator based channel estimation technique for robustification of channel estimation in Rayleigh and Rician fading channels in multi-carrier OFDM 5G wireless communication systems	- Improvements in performance for 5G wireless communication system in Rayleigh and Rician fading channels
[36]	The algorithm for OTFS, which addresses high Doppler effects in dynamic channels. It enhances spectral efficiency in multi-antenna OTFS using pilots with orthogonal matrix-form Frank arrays	- Better NMSE performance - New pilot structure in matrix form
[37]	A novel algorithm for blind iterative channel estimation in OFDM systems has been introduced. This algorithm uses data from an LDPC decoder to iteratively enhance parameter estimation accuracy	- A novel blind iterative channel estimation algorithm with improved Bit Error Rate (BER) performance

Channel estimation algorithms are essential in wireless communication to precisely assess the wireless channel's characteristics. These algorithms vary in complexity and accuracy, with some prioritizing simplicity and others aiming for higher accuracy but increased computational demands. Research papers frequently emphasize improving accuracy, recognizing that advancements in data rate and throughput may come at the expense of estimation precision.

The selection of channel estimation algorithms for a given communication system depends on specific requirements and objectives. For instance, Channel estimation in dynamic wireless channels involves estimating the changing characteristics of the channel to ensure reliable and efficient communication. Numeral parameters impact the estimation process, including the channel impulse response (CIR), channel frequency response (CFR), signal-to-noise ratio (SNR), pilot signal parameters, time and frequency resolution, complexity, and estimation accuracy [38].

Furthermore, Neural networks present a promising avenue for enhancing channel estimation in wireless communication systems. Their capacity to learn intricate relationships between received signals and channel parameters holds potential for improving estimation accuracy [39]. For example, in [40], the authors use the types of the artificial neural networks, called long short-term memory and gated recurrent unit to predict the channel behaviour. In the conventional system, the channel information calculated at the receiver's side and then send to the transmitter side to select appropriate characteristics of the transmission. However, in this system, firstly the channel can be obsolete in a time variant scenario, and requiring complex processing. To cope with this, the authors suggested the machine learning method, to train the flat-fading channel, to predict it further in the context of MIMO communication. The utilization of deep learning techniques, such as deep neural networks and deep reinforcement learning, further bolsters estimation performance. In [41], the authors improved the accuracy of 5G NR channel estimation. In traditional 5G NR, the DMRS pilot, or demodulation reference signal, is used and inserted into the resource grid to further estimate the channel. The accuracy can be enhanced by simply inserting more DMRS signals, which consumes the resource grid. It becomes a challenging task to obtain the channel's behaviour when it comes to reality, as resources are not endless, especially when dealing with non-linear and complex processing. To address this, the authors decided to use neural networks to enhance the performance of channel estimation without consuming the resource grid and adding complexities. [42], just like [41], enhanced the conventional DMRS method, which depletes the resources and degrades performance in high data rate transmission scenarios. The authors utilized LSTM-based neural networks to train the channel correspondence, and they employed ConvLSTM for channel demodulation.

Overall, the utilization of artificial neural networks for channel estimation has opened new avenues and attracted the attention within the research community. Properly using this approach is able to significantly enhance the overall performance of wireless communication systems.

3 Discussions and Future Research

It might be difficult to compare metrics between publications, especially when multiple assessment metrics used to gauge how well a system performs. Additionally, because there are so many distinct channels estimate techniques available, it might be difficult to combine simulation results into a single graph for in-depth comparison because each approach may use a different measure or even a diverse set of metrics. Plots in Fig. 1, like "BER vs SNR," "MSE vs SNR," and "NMSE vs SNR," which serve as graphical representations showing the effectiveness of various channel estimation algorithms under variable signal-to-noise ratio (SNR) situations, are frequently used to illustrate these results. The bit error rate (BER) performance of three channel estimation algorithms in [21, 24], and [31] shown in the Fig. 1a. The BER represented on the y-axis, while the SNR values represented on the x-axis in decibels (dB). The plot clearly demonstrates a trend in which the BER for each of the three algorithms reduces as the SNR rises.

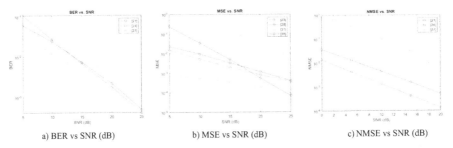

a) BER vs SNR (dB)   b) MSE vs SNR (dB)   c) NMSE vs SNR (dB)

Fig. 1. Comparison plots of channel estimation algorithms.

This suggests that greater SNR levels result in higher-quality transmission and lower error rates. Further-more, we see that the algorithm in the paper [26] beats the other two systems throughout the full range of SNR.

The Fig. 1b demonstrates the Mean Squared Error (MSE) performance of channel estimation algorithms in papers [23, 28, 31], and [35], respectively. Like the previous plot, the x-axis represents the SNR values, and the y-axis represents the MSE. The plot shows that as the SNR increases, the MSE decreases for all systems, indicating better performance. Notably, the channel estimation in [31] achieves significantly lower MSE values compared to the other systems across the entire SNR range. Conversely, the system labelled [35] exhibits higher MSE values, suggesting that it may not perform as well under low SNR conditions.

In these two plots, we can observe that deep learning-aided channel estimation algorithms in [31] outperformed conventional algorithms, such as LS, MMSE, LMMSE, and other traditional methods, in terms of BER and MSE evaluation.

The performance of three algorithms in [21, 26], and [37], as measured by Normalized Mean Squared Error (NMSE), shown in the Fig. 1c. SNR values represented on the x-axis, while NMSE represented on the y-axis. The figure demonstrates that when the SNR rises, all systems' NMSE fall, suggesting increased performance. Throughout the SNR range, the system with the label [21] consistently gets the lowest NMSE values, demonstrating its better performance. The system marked [37] on the other hand displays noticeably larger NMSE values in comparison to the other systems, indicating that it performs poorly in recreating the original signal.

In the plot of NMSE vs SNR, it can be seen that the channel estimation algorithm based on Bayesian learning, aiming to reduce the strength of the noise impulse as presented in [26], demonstrated the best results in terms of NMSE when compared to the blind iterative channel estimation algorithm from [37] and the Fast LMMSE algorithm from [21].

Based on the findings, it is obvious that there are different techniques available to get ideal outcomes in terms of accuracy, complexity, and a variety of other criteria. Choosing conventional and standard channel estimate methods simply for their simplicity, however, may not be the most successful option. Instead, for channel estimation, using the power of artificial intelligence, machine learning, and sophisticated deep learning techniques can be the road to greater outcomes.

Furthermore, it would be also valuable to explore the development of hybrid approaches that combine multiple channel estimation algorithms to leverage their respective strengths. Such hybrid methods may offer improved performance, accuracy, and reliability under varying SNR conditions.

Acknowledgement. This work was supported by the Ministry of Education and Science of the Republic of Kazakhstan (AP13068587, AP14871109) and Nazarbayev University (grant no. 11022021CRP1507, 021220FD0451).

References

1. Waichal, G., Khedkar, A.: Performance analysis of FFT based OFDM system and dwt based OFDM system to reduce inter-carrier interference. In: 2015 international conference on computing communication control and automation, pp. 338–342 (2015). https://doi.org/10.1109/ICCUBEA.2015.71
2. Ye, B., Zhang, Z.: Improved pilot design and channel estimation for 60ghz OFDM based on IEEE 802.11.ad. In: 2013 IEEE wireless communications and networking conference (WCNC), pp. 4129–4133 (2013). https://doi.org/10.1109/WCNC.2013.6555239
3. Guo, D., Fang, Y., Wu, Y.: An improved channel estimation method for V2V OFDM communication system. In: 2018 14th IEEE international conference on signal processing (ICSP), pp. 733–736 (2018). https://doi.org/10.1109/ICSP.2018.8652392
4. Kaur, H., Khosla, M., Sarin, R.K.: Channel estimation in MIMO-OFDM system: a review. In: 2018 second international conference on electronics, communication and aerospace technology (ICECA), pp. 974–980 (2018). https://doi.org/10.1109/ICECA.2018.8474747
5. Qun, W., Xiao, Z., Chengyou, W., Zhiliang, Q.: Channel estimation based on superimposed pilot and weighted averaging. (2022). https://doi.org/10.1038/s41598-022-14482-6.
6. Jiang, T., Chen, D., Ni, C., Qu, D.: Chapter 1—Introduction, Editor(s): Jiang, T., Chen, D., Ni, C., Qu, D, OQAM/FBMC for future wireless communications, Academic Press, 2018, 1–24, ISBN 9780128135570, https://doi.org/10.1016/B978-0-12-813557-0.00001-2
7. Taoliu, T., Zhang, H.: Improved channel estimation method jointing channel coding. In: 2019 3rd international conference on electronic information technology and computer engineering (EITCE), pp. 1624–1628 (2019). https://doi.org/10.1109/EITCE47263.2019.9094926
8. Wu, S., Liu, X., Wei, Y., Bai, X.: Channel estimation for FBMC/OQAM with fast fading channels by Kalman filter. In: 2018 14th international wireless communications mobile computing conference (IWCMC), pp. 987–992 (2018). https://doi.org/10.1109/IWCMC.2018.8450476
9. Samanta, S., Sridha, T.V.: Modified slow fading channel estimation technique and fast fading channel estimation technique for OFDM systems. In: 2018 3rd IEEE international conference on recent trends in electronics, information communication technology (RTEICT), pp. 1638–1643 (2018). https://doi.org/10.1109/RTEICT42901.2018.9012125
10. Garcia, N., Wymeersch, H., Str¨om, E.G., Slock, D.: Location-aided mmWave channel estimation for vehicular communication. In: 2016 IEEE 17th international workshop on signal processing advances in wireless communications (SPAWC), pp. 1–5 (2016). https://doi.org/10.1109/SPAWC.2016.7536855
11. Coleri, S., Ergen, M., Puri, A., Bahai, A.: Channel estimation techniques based on pilot arrangement in OFDM systems. IEEE Trans. Broadcasting **48**(3), 223–229 (2002). https://doi.org/10.1109/TBC.2002.804034
12. El Assaf, A., Kandil, N., Hakem, N., Affes, S., Fortier, P.: A study of 60 GHz channel estimation techniques using pilot carriers in OFDM systems in a confined area, 1–6 (2012). https://doi.org/10.1109/ICWCUCA.2012.6402505

13. Li, Y.: Pilot-symbol-aided channel estimation for OFDM in wireless systems. IEEE Trans. Veh. Technol. **49**(4), 1207–1215 (2000). https://doi.org/10.1109/25.875230
14. Shin, C., Heath, R.W., Powers, E.J.: Blind channel estimation for MIMO-OFDM systems. IEEE Trans. Veh. Technol. **56**(2), 670–685 (2007). https://doi.org/10.1109/TVT.2007.891429
15. Peken, T., Vanhoy, T.B.G.: Blind channel estimation for massive MIMO. Analog Integr. Circuits Signal Process. **91**(2), 257–266 (2017). https://doi.org/10.1007/s10470-017-0943-1
16. Nayebi, E., Rao, B.D.: Semi-blind channel estimation for multiuser massive MIMO systems. IEEE Trans. Signal Process. **66**(2), 540–553 (2018). https://doi.org/10.1109/TSP.2017.277 1725
17. Wan, F., Zhu, W.-P., Swamy, M.N.S.: A semi blind channel estimation approach for MIMO–OFDM systems. IEEE Trans. Signal Process. **56**(7), 2821–2834 (2008). https://doi.org/10. 1109/TSP.2008.917354
18. Deng, X., Haimovich, A.M., Garcia-Frias, J.: Decision directed iterative channel estimation for MMIMO systems 4, 2326–2329 (2003). https://doi.org/10.1109/ICC.2003.1204300
19. Ran, J., Grunheid, R., Rohling, H., Bolinth, E., Kern, R.: Decision- directed channel estimation method for OFDM systems with high velocities. In: The 57th IEEE semi-annual vehicular technology conference, 2003. VTC 2003-Spring., vol. 4, pp. 2358–23614 (2003). https://doi. org/10.1109/VETECS.2003.1208811
20. Ketonen, J., Juntti, M., Ylioinas, J.: Decision directed channel estimation for improving performance in LTE-A. In: 2010 conference record of the forty fourth asilomar conference on signals, systems and computers, pp. 1503–1507 (2010). https://doi.org/10.1109/ACSSC. 2010.5757787
21. Zhou, L.W.: A fast LMMSE channel estimation method for OFDM systems. J Wirel. Com Network (2009). https://doi.org/10.1155/2009/752895
22. Linchao Yang, R.T.: A modified ls channel estimation algorithm for OFDM system in mountain wireless environment. Procedia Eng. **29**, 2732–2736 (2012). https://doi.org/10.1016/j. proeng.2012.01.381
23. Ahmed, A.S., Hamdi, M.M., Abood, M.S., Khaleel, A.M., Fathy, M., Khaleefah, S.H.: Channel estimation using ls and MMSE channel estimation techniques for MIMO-OFDM systems. In: 2022 international congress on human-computer interaction, optimization and robotic applications (HORA), pp. 1–6 (2022). https://doi.org/10.1109/HORA55278.2022.9799887
24. Pradhan, P., Faust, O., Patra, S., Chua, B.: Channel estimation algorithms for OFDM systems **5** (2011). https://doi.org/10.1504/IJSISE.2012.050324
25. Wang, D., Mei, Z., Liang, J., Liu, J.: An improved channel estimation algorithm based on WD-DDA in OFDM system. Mobile Inf. Syst. **2021**, 6540923 (2021). https://doi.org/10.1155/ 2021/6540923
26. Jin, Z., Zhang, X., Tan, S., Xiong, J., Wei, J.: An impulsive noise mitigation and channel estimation algorithm based on unitary approximate message passing. In: 2022 2nd international conference on computer science, electronic information engineering, and intelligent control technology (CEI), pp. 687–690 (2022). https://doi.org/10.1109/CEI57409.2022.9950185
27. Durak, H.M, Ertug, O.: Dynamic and sparsity adaptive compressed sensing based active user detection and channel estimation of uplink grant-free SCMA. Radio Eng. 2021, 576–583 (2021). https://doi.org/10.13164/re.2021.0576
28. Luo, J., Wang, D.: An improved DCT-based channel estimation algorithm with channel PDP modelling. In: 2022 3rd information communication technologies conference (ICTC), pp. 250–254 (2022). https://doi.org/10.1109/ICTC55111.2022.9778367
29. Zuo, C., Deng, H., Zhang, J., Qi, Y.: Distributed channel estimation algorithm for mmWave massive MIMO communication systems. In: 2021 IEEE 94th vehicular technology conference (VTC2021-Fall), pp. 1–6 (2021). https://doi.org/10.1109/VTC2021-Fall52928.2021. 9625267

30. Zhang, S., Xu, L., Yan, S.: A low complexity OMP sparse channel estimation algorithm in OFDM system. In: 2021 IEEE international conference on signal processing, communications and computing (ICSPCC), pp. 1–5 (2021). https://doi.org/10.1109/ICSPCC52875.2021.956 5070

31. Yao, R., Qin, Q., Wang, S., Qi, N., Fan, Y., Zuo, X.: Deep learning assisted channel estimation refinement in uplink OFDM systems under time-varying channels. In: 2021 international wireless communications and mobile computing (IWCMC), pp. 1349–1353 (2021). https://doi.org/10.1109/IWCMC51323.2021.9498717

32. Xie, Z., Chen, X., Liu, Y., Défense, M.: A channel estimation algorithm for MIMO-SCFDE systems over fast time-varying multipath channels. In: 2017 IEEE international conference on signal processing, communications and computing (ICSPCC), pp. 1–6 (2017). https://doi.org/10.1109/ICSPCC.2017.8242580

33. Rasheed, O.K., Surabhi, G.D., Chockalingam, A.: Sparse delay-doppler channel estimation in rapidly time-varying channels for multiuser OTFS on the uplink. In: 2020 IEEE 91st vehicular technology conference (VTC2020-Spring), pp. 1–5 (2020). https://doi.org/10.1109/VTC2020-Spring48590.2020.9128497

34. Sahoo, M., Sahoo, H.K.: Robust variable power fractional LMP based channel estimation for MIMO fading wireless systems. In: 2020 IEEE Calcutta conference (CALCON), pp. 50–53 (2020). https://doi.org/10.1109/CALCON49167.2020.9106502

35. Kumar, T.A., Anjaneyulu, L.: Channel estimation techniques for multicarrier OFDM 5g wireless communication systems. In: 2020 IEEE 10th international conference on system engineering and technology (ICSET), pp. 98–101 (2020). https://doi.org/10.1109/ICSET51301.2020

36. Liang, Y., Wang, Q., Fan, P.: Pilot-aided channel estimation scheme based on frank array for OTFS under rapidly time-varying channels. In: 2022 IEEE 95th vehicular technology conference: (VTC2022-Spring), pp. 1–6 (2022). https://doi.org/10.1109/VTC2022-Spring54318.2022.9860819

37. Li, J., Ding, X., Zhang, Y., Hou, J., Ma, Z., Chen, J., Li, Q.: Code-aided blind iterative channel estimation for OFDM systems. In: 2021 7th international conference on computer and communications (ICCC), pp. 262–266 (2021). https://doi.org/10.1109/ICCC54389.2021.9674615

38. Zhang, Z., Bian, X., Li, M.: Joint channel estimation algorithm based on DFT and DWT. Appl. Sci. 12(15) (2022). https://doi.org/10.3390/app12157894

39. Le, H.A., Van Chien, T., Nguyen, T.H., Choo, H., Nguyen, V.D.: Machine learning-based 5g-and-beyond channel estimation for MIMO-OFDM communication systems. Sensors 21(14) (2021). https://doi.org/10.3390/s21144861

40. Jiang, W., Schotten, H.D.: Recurrent neural networks with long short-term memory for fading channel prediction. In: 2020 IEEE 91st vehicular technology conference (VTC2020-Spring), pp. 1–5 (2020). https://doi.org/10.1109/VTC2020-Spring48590.2020.9128426

41. Dayi, A.B.: Improving 5g nr uplink channel estimation with artificial neural networks: a practical study on NR PUSCH receiver. In: 2022 IEEE international black sea conference on communications and net-working (BlackSeaCom), pp. 129–134 (2022). https://doi.org/10.1109/BlackSeaCom54372.2022.9858290

42. Wang, Y., Chang, J., Lu, Z., Yu, F., Wei, J., Xu, Y.: Channel estimation of 5g OFDM system based on conv-LSTM network. In: 2022 7th international conference on communication, image, and signal processing (CCISP), pp. 62–66 (2022). https://doi.org/10.1109/CCISP5629.2022.9974588

Wearable-Based SLAM with Sensor Fusion in Firefighting Operations

Renjie Wu[1], Boon Giin Lee[1(✉)], Matthew Pike[1], Liang Huang[2], Wan-Young Chung[3], and Gen Xu[4]

[1] School of Computer Science, University of Nottingham Ningbo China, Ningbo 315100, China
{renjie.wu,boon-giin.lee,matthew.pike}@nottingham.edu.cn
[2] Department of Electrical and Electronic Engineering, University of Nottingham Ningbo China, Ningbo 315100, China
liang.huang@nottingham.edu.cn
[3] Department of Artificial Convergence, Pukyong National University, Busan 48513, Korea
wychung@pknu.ac.kr
[4] Ningbo Institute of Materials Technology and Engineering, Chinese Academy of Sciences, Ningbo 315201, China
xugen@nimte.ac.cn

Abstract. In challenging indoor fire rescue scenarios characterized by heavy smoke and dust, conventional cameras struggle to capture high-quality images. Frames with limited visual data fail to provide sufficient information for SLAM (Simultaneous Localization and Mapping) systems to achieve accuracy. This research introduces an innovative solution in the form of a wearable firefighter protective boot integrated with a SLAM system. This system incorporates Pedestrian Dead Reckoning (PDR) and ultrasound sensors to autonomously generate the user's trajectory and an internal structural map. The ultrasound module is strategically positioned on the outer side of the calf, effectively scanning the surrounding boundaries. Additionally, a 9-axis inertial measurement unit, located atop the forefoot, detects walking motions and calculates continuous step positions to determine the trajectory. The Map Point Calculation (MPC) algorithm combines ultrasound range data with the computed trajectory to construct the map model. To validate the system's performance, experiments were conducted within a smoke-filled environment simulated by firefighters at the local fire station. The results unequivocally demonstrate the system's capability to provide highly accurate trajectory estimations and generate precise map points.

Keywords: Simultaneous Localization and Mapping · Pedestrian Dead Reckoning · Wearable Sensing · Sensor Fusion

This work was supported in part by the Zhejiang Provincial Natural Science Foundation of China under Grant LQ21F020024, and in part by the Ningbo Science and Technology (S&T) Bureau through the Major S&T Program under Grant 2021Z037 and 2022Z080.

B. J. Choi et al. (Eds.): IHCI 2023, LNCS 14532, pp. 216–221, 2024.
https://doi.org/10.1007/978-3-031-53830-8_21

1 Introduction

Firefighters operating in environments characterized by dense smoke and raging fires frequently encounter the challenge of losing their orientation due to severely limited visibility. Conventional implementations of Simultaneous Localization and Mapping (SLAM), such as visual SLAM [1], typically relies on calculating geometric relationships between consecutive frames to model trajectories and maps. Unfortunately, the quality of data collected is severely compromised in burning buildings, where smoke and fire severely hamper visibility, consequently undermining the accuracy of SLAM estimates. Alternative SLAM approaches, including Light Detection and Ranging (LIDAR) SLAM [2] and Radio Detection and Ranging (RADAR) SLAM [3], either prove infeasible for firefighting applications or are susceptible to the dynamic nature of smoke and fire scenarios.

Recent advancements in Inertial Measurement Unit (IMU) technology have yielded smaller, more accurate hardware units, making IMUs a more suitable choice for pedestrian dead reckoning (PDR) systems. PDR relies on motion data from IMUs to predict user movement trajectories, encompassing step detection, step-length estimation, and heading estimation [4]. While PDR alone can generate user trajectories, it lacks the capability to reconstruct the surrounding environment [5]. Wang et al., [6] proposed a multi-mode PDR assisted by a map. However, this map is functioned as a prior reference for trajectory optimization regardless of the non map cases. Similar limitation can also be found in [7]. To address this formidable challenge, this article introduces a novel approach, PDR-SLAM, which combines PDR with ultrasound ranging measurements to seamlessly integrate walking trajectories and reconstructed maps. Notably, both motion and ultrasound sensors are sufficiently lightweight to be implemented as wearable devices for firefighting personnel.

2 Method

2.1 System Overview

Figure 1 provides a system overview of PDR-SLAM, illustrating the collection of data through wearable sensors affixed to the firefighter's protective boot. The PDR-SLAM system, responsible for estimating the reconstructed map using the collected sensor data, comprises two core components: the PDR component, which anticipates the user's location, and the Map Points Calculation (MPC) component, which reconstructs a map utilizing ultrasound data. The subsequent sections delve into the details of these two modules.

IMU data features are harnessed to compute the position of the forthcoming step. The acceleration, angular velocity, and orientation data from the IMU are leveraged to determine crucial parameters such as step frequency, step length, and step heading direction [8]. The step and ultrasound map points calculation process, as depicted in Eq. 1, is elucidated below.

$$x_{k+1} = x_k + L_k sin\varphi_k$$
$$y_{k+1} = y_k + L_k cos\varphi_k$$

$$(1)$$

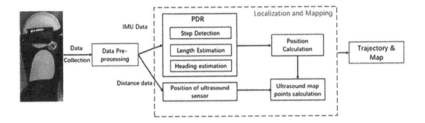

Fig. 1. System overview of the proposed multi-INS.

In Fig. 2, the black marks symbolize the footprints of individual steps, while the arrows denote the direction of movement. Here, i signifies the timestamp of sensor data, k represents the step index computed through step detection, L_k signifies the length of each corresponding step, and φ_k denotes the heading direction for each step, estimated through filtered magnetometer readings. (x_k, y_k) signifies the pedestrian's position coordinates at step k.

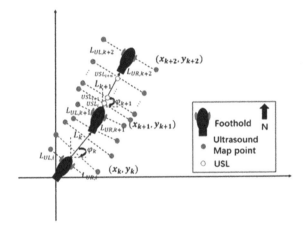

Fig. 2. The computation of step and ultrasound map point coordinates.

2.2   Map Points Calculation (MPC)

In this study, the IMU and ultrasound sensors were seamlessly integrated into a wearable device, ensuring no discernible biases were encountered while walking. Notably, the IMU's x-axis and the ultrasound sensor's ranging direction are arranged in a perpendicular orientation. Employing a geometric relationship approach, the position of the ultrasound sensor on the foot, along with the measured distance, facilitates the calculation of a map point coordinate (refer to Fig. 2).

During movement, the position of the ultrasound sensor on the foot undergoes variations. This dynamic positioning can be defined using its coordinates as follows:

$$x_{USL,i} = x_k + L_k * \frac{T_s(k) - i}{length(k, k + 1)} * sin\varphi_k$$

$$y_{USL,i} = y_k + L_k * \frac{T_s(k) - i}{length(k, k + 1)} * cos\varphi_k$$

(2)

where T_s maps the step index number to the timestamp, $length$ denotes the sample count within the interval $(k, k + 1)$, and $(x_{USL,i}, y_{USL,i})$ signifies the position of the ultrasound sensor. It's worth noting that this assumes a constant swinging velocity for each step, ensuring a uniform distribution of ultrasound sensor locations within the PDR trajectory. The calculation of ultrasound map points' coordinates can be outlined as follows:

$$P_{UL,i} = (x_{USL,i} - L_{UL,i} * sin\varphi_k, y_{USL,i} + L_{UL,i} * cos\varphi_k)$$

$$P_{UR,i} = (x_{USL,i} + L_{UR,i} * cos\varphi_k, y_{USL,i} - L_{UR,i} * sin\varphi_k)$$

(3)

where $P_{UL,i}$ and $P_{UR,i}$ represent the coordinate positions of map points originating from the ultrasound sensors situated on the left and right firefighting protective boots, respectively, as well as $L_{UL,i}$ and $L_{UR,i}$ denote the distances measured by the ultrasound sensors. The coordinate calculation of map points follows the iterative trajectory computing which synchronises localization and mapping.

3   Experimental Setup

As depicted in Fig. 3, the proposed prototype comprises several key components, including a micro-computing unit (MCU) represented by the Seeeduino XIAO, a HC-SR04 ultrasound sensor, a BNO055 IMU, and a firefighter protective boot. For efficient transmission of sensor data to a nearby terminal, the Bluetooth Low Energy (BLE) HM-10 is employed. To ensure the acquisition of high-quality gait inertial data, sensor components are meticulously assembled and affixed to the upper section of the boot. To address potential issues related to sampling rates and data structure inconsistencies, a two-step process is implemented, initially involving data denoising, followed by data synchronization.

The experiment took place within a designated firefighting training facility. This controlled environment was deliberately filled with artificial smoke with visibility limited to less than 1 m, generated using a smoke machine. The layout of the location features a single elongated corridor. The corridor's dimensions were accurately measured using a laser range finder, revealing dimensions of 10 m in length and 1.78 m in width. To assist with data collection, two male firefighters volunteered for the experiment. These firefighters had distinct physical characteristics: one measuring 175 cm in height and weighing 75 kg, while the other stood at 173 cm and weighed 82 kg. They were tasked with navigating the round-trip walking trajectory. Importantly, there were no constraints placed on the participants' walking patterns.

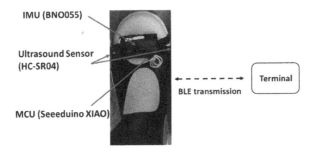

Fig. 3. The configuration of PDR-SLAM components and the data transmission.

4   Results and Discussion

The estimated trajectory and mapping points for the two subjects are presented in Fig. 4. During the localization assessment, the trajectory (red line) closely aligns with the ground truth dashed line. Notably, the estimated trajectories approach a length of approximately 10 m, indicating a high level of accuracy in the system's localization capabilities.

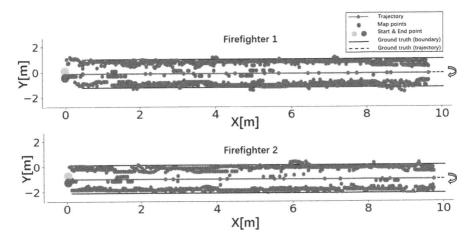

Fig. 4. Outcomes of PDR-SLAM, showcasing both the trajectory (red line) and the mapping results (green points). The ground truth using both black lines and dashed lines. The orange and blue dots indicate both the start and end points of the trajectory. (Color figure online)

Regarding the map evaluation, a majority of the map points are concentrated along the boundaries of the walls, which are represented by two solid lines. However, it's worth mentioning that a few points lie outside the boundary, primarily due to noise stemming from motion artifacts and nearby obstacles. Encouragingly, the results for both subjects exhibit a similar pattern, suggesting that

PDR-SLAM may offer consistent and stable performance. Nonetheless, further experimentation with a larger pool of participants is warranted to validate these findings conclusively.

5   Conclusion

In summary, this study introduces PDR-SLAM, a fusion of PDR and ultrasound ranging for indoor firefighting positioning, seamlessly integrated into a wearable system. The experimental findings within a simulated smoke-filled environment showcase a commendable level of accuracy in both localization and mapping, with trajectory and map points closely aligned with ground truth data. Future endeavors will focus on enhancing computational efficiency through the implementation of a parallel threading mechanism, while also conducting application testing that takes into consideration human factors and user experience.

References

1. Campos, C., Elvira, R., Rodríguez, J.J.G., Montiel, J.M., Tardós, J.D.: ORB-SLAM3: an accurate open-source library for visual, visual-inertial, and multimap slam. IEEE Trans. Rob. **37**(1), 1874–1890 (2021)
2. Hess, W., Kohler, D., Rapp, H., Andor, D.: Real-time loop closure in 2D LIDAR SLAM. In: 2016 IEEE International Conference on Robotics and Automation (ICRA), pp. 1271–1278. IEEE (2016)
3. Holder, M., Hellwig, S., Winner, H.: Real-time pose graph SLAM based on radar. In: 2019 IEEE Intelligent Vehicles Symposium (IV), pp. 1145–1151. IEEE (2019)
4. Hou, X., Bergmann, J.: Pedestrian dead reckoning with wearable sensors: a systematic review. IEEE Sens. J. **21**(1), 143–152 (2020)
5. Zhou, B., et al.: Crowdsourcing-based indoor mapping using smartphones: a survey. ISPRS J. Photogramm. Remote. Sens. **177**(1), 131–146 (2021)
6. Wang, X., Chen, G., Yang, M., Jin, S.: A multi-mode PDR perception and positioning system assisted by map matching and particle filtering. ISPRS Int. J. Geo-Inf. **9**(2), 93 (2020)
7. Liu, F., Wang, J., Zhang, J., Han, H.: An indoor localization method for pedestrians base on combined UWB/PDR/Floor map. Sensors **19**(11), 2578 (2019)
8. Jimenez, A.R., Seco, F., Prieto, C., Guevara, J.: A comparison of pedestrian dead-reckoning algorithms using a low-cost MEMS IMU. In: 2009 IEEE International Symposium on Intelligent Signal Processing, pp. 37–42. IEEE (2009)

The Novel Electrocardiograph Sensor and Algorithm for Arrhythmia Computer Aided Detection

Yeun Woo Jung and Jong-Ha Lee[✉]

Department of Biomedical Engineering, Keimyung University, Daegu, South Korea
segeberg@kmu.ac.kr

Abstract. An electrocardiogram (ECG) system is a system used to analyze signals from the human heart, determine the health status of the heart, or diagnose diseases. Among the ECG signals measured using this system, we would like to propose a Pan-Tomkins algorithm that focuses on P wave and detecting P wave. To compare the performance of the developed algorithm, five actual measurement data of the electrocardiogram device were compared with Holter, and the error rate was 20%. If this is used, it is expected that more patients' cardiovascular diseases can be prevented through wearables. The accuracy was 0.912568, the precision was 0.947368, and the sensitivity was 0.72 in the analysis.

Keywords: ECG · P wave · Artificial Intelligence · Mobile Sensor · Healthcare

1 Introduction

The ECG system measures and analyzes signals from the human heart, aiding in health assessment and heart disease diagnosis [1]. This analysis involves various waveforms (p, q, r, s, t) to determine diseases based on waveform characteristics, with a particular focus on p-wave analysis for heart disease and arrhythmia diagnosis [2]. However, tracking p-waves is challenging due to weak and noisy ECG signals, occasionally abnormal p-wave shapes, making accurate tracing difficult. Many medical professionals recognize p-wave's importance and seek technological solutions. This research uses signal processing algorithms to analyze wearable device data and track p-waves. Digital signal processing, specializing in ECG signal analysis and image processing (noise reduction, object detection), optimizes research through filtering and spectrum analysis. The study compares wearable ECG device data for p-wave detection in cardiovascular diagnoses against Holter data.

2 Material and Methods

2.1 Proposed ECG Sensor

The specifications of the wearable ECG device are as follows. It is designed as a patch-type model using electrodes and is equipped with Bluetooth functionality. It incorporates a secondary derivative algorithm and operates using an ARM module (STM32F030F4P6,

B. J. Choi et al. (Eds.): IHCI 2023, LNCS 14532, pp. 222–227, 2024.
https://doi.org/10.1007/978-3-031-53830-8_22

STMicroelectronics) with a program memory size of 16 kb. The device has a 12-bit ADC resolution and features a 32-bit RISC processor. The sampling rate is set at 300 samples per second (Fig. 1).

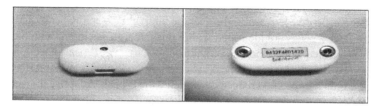

Fig. 1. Proposed ECG device model.

2.2 Pan-Tomkins Algorithm

We utilized the Pan-Tompkins algorithm for comparative analysis, a prominent method for detecting QRS complexes in ECG signals [3]. It involves several steps, including initial Band-pass filtering, differentiation, squaring, moving window integration, and the application of a threshold for QRS complex detection. We conducted experiments using lab-built software [4]. The process began with initial filtering to remove noise and artifacts from the ECG signal, followed by signal normalization. Noise filtering was applied based on the sampling frequency, with low-pass and high-pass filtering used at 200 Hz. For other frequencies, band-pass filtering was employed. The Derivative Filter calculated the signal slope [5]. The squared Derivative-filtered signal underwent nonlinear processing to enhance signal clarity and reduce noise. It was then smoothed using a moving average filter, which calculated the window size to capture QRS complex positions within the signal [6]. To pinpoint QRS complexes accurately, Fiducial Marks (baseline points) were identified. The algorithm scanned the ECG signal, detected peaks that met threshold and minimum peak distance criteria, and used them as baseline points for QRS complex localization. The Thresholding and Decision Rule was applied to set thresholds based on QRS signal characteristics. Detected QRS complexes were classified based on whether they exceeded the threshold, effectively separating them from noise. Following QRS complex identification, the signal's characteristics were analyzed to identify individual signals. This comprehensive process enabled accurate identification and analysis of ECG signals, particularly QRS complexes, which are crucial for various cardiac assessments and diagnoses.

2.3 P-QRS-T Detection

To identify the key points in ECG signals, the Pan-Tompkin algorithm adjusts the R-peak's position and the detection delay, taking the sampling frequency (fs) into account. It considers a windowing time of 0.09 s (90 ms) to pinpoint the R-peak location. Duplicate positions are removed to obtain the actual R-peak positions, selecting the one with the highest signal value. For Q-peak detection, it relies on the detected R-peaks and selects a point 0.1 s (10% of RR interval) before each R-peak with the lowest signal value as the Q-peak. Duplicate points are eliminated to identify the true Q-peaks.

S-peak detection follows a similar approach, starting from the R-peak and looking 0.1 s (10% of RR interval) ahead, selecting the point with the lowest signal value as the S-peak. The main difference between Q-peaks and S-peaks is their relative position within the signal. To detect T-peaks, the average interval between R-peaks and S-peaks is calculated, and T-peaks are identified within this interval. The point with the highest signal value is selected as the T-peak, removing duplicate peaks. P-peak detection uses the interval between already detected R-peaks to estimate the P-peak's position. The highest signal value within the specified range is considered the P-peak. Adaptive thresholding is employed to account for noise or signal fluctuations, and threshold values can be adjusted based on the P-peak's size. This configuration helps distinguish noise or false detections. P-wave height is assumed to be within a certain range, and minimum P-wave amplitude is defined. These techniques are crucial for accurately identifying and analyzing R, Q, S, T, and P peaks in ECG signals, providing valuable information for cardiac assessments and diagnoses (Fig. 2).

$$P(t) = x(t) * w(t) \tag{1}$$

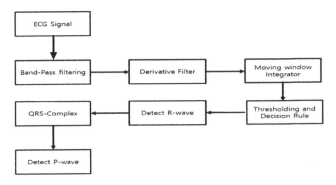

Fig. 2. A schematic diagram of an algorithm

The function that detects P wave is expressed in an equation as follows. w(t) is P wave's it is a window function that models the shape. Use Gaussian functions among window functions did.

$$w(t) = A \cdot e^{-\frac{(t-t0)2}{2a^2}} \tag{2}$$

3 Experimental Results

The RAW data had to be secured first to convert ECG data through the device. After securing RAW data, the value should be set by checking the set sampling rate. The sampling rate represents a continuous interval as a speed to convert an analog signal into a digital form. An analog signal may be digitally converted at a fixed time interval. At this time, sampling rate is very important because the range of the frequency of the signal must be known so that information on the frequency component can be expressed at an appropriate level without losing it. Data were cut and compared for p-wave detection using the Pan-Tomkins algorithm at the same time using the device's sampling rate. The results obtained when comparing the actual ECG data for 5 people are shown in a table (Table 1).

Table 1. Actual ECG Data Comparison Results

| The proposed ECG device | | Holter | |
Table	Number	Table	Number
No. 1	236	No.1	270
No. 2	297	No.2	236
No. 3	198	No.3	247
No. 4	275	No.4	254
No. 5	349	No.5	212

The obtained results when comparing p-wave data from the proposed ECG device and five random Holter data sets through algorithm analysis. When the ECG data of the actual wearable commercialization company was compared and analyzed with Holter, similar results were shown in simple numerical terms. Random sample data from 1 s to 1000 s were compared, and Holter and the error rate showed an average error rate of 21% (Table 2, Table 3, Fig. 3, Fig. 4).

Table 2. Precision Comparison Results

Table	Error rate (%)
No. 1	14
No. 2	20
No. 3	24
No. 4	7
No. 5	39

Table 3. Confusion matrix

	TRUE	FALSE
TRUE	36(TP)	14(FN)
FALSE	2(FP)	131(TN)

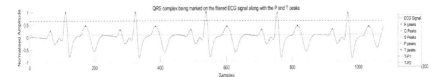

Fig. 3. ECG device graph

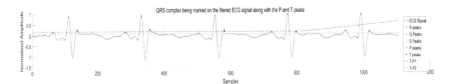

Fig. 4. Holter ECG graph

We numerically labeled the positions of the p-wave on the x-axis, compared them with the x-axis numbers of each dataset, and divided them into true and false based on whether they fell within the error range. As a result, the accuracy was 0.912568, the precision was 0.947368, and the sensitivity was 0.72 in the analysis.

4 Conclusion

The research aimed to detect the p-wave in ECG signals from wearable ECG devices and compare it with Holter monitoring to assess the accuracy of p-wave detection. To achieve this, the Pan-Tomkins algorithm was employed, and the error rate between Holter and p-wave detection was found to be 20%. This demonstrates the potential of signal analysis using wearables. However, the limited amount of data and the fact that the error rate has not yet reached a close proximity to Holter results are acknowledged. Nevertheless, it is anticipated that with ongoing updates to the algorithm and continued comparative analysis with a larger dataset, more reliable results can be obtained in the future.

Acknowledgements. This research is financial supported by "The digital pathology-based AI analysis solution development project" through the Ministry of Health and Welfare, Republic of Korea (HI21C0977) and supported by Basic Science Research Program through the National Research Foundation of Korea (NRF) funded by the Ministry of Education (2022R1I1A3072785).

References

1. Randazzo, V., Ferretti, J., Pasero, E.: ECG WATCH: a real time wireless wearable ECG. In 2019 IEEE International Symposium on Medical Measurements and Applications (MeMeA), pp. 1–6. IEEE (2019)
2. Wasimuddin, M., Elleithy, K., Abuzneid, A.-S., Faezipour, M., Abuzaghleh, O.: Stages-based ECG signal analysis from traditional signal processing to machine learning approaches: a survey. IEEE Access **8**, 177782–177803 (2020)
3. Pingale, S.L.: Using Pan Tompkin 'S Method, ECG signal processing and dignose various diseases in Matlab. In: Proceedings of IRF International Conference, pp. 57–61 (2014)
4. Jindal, B., Devi, R.: MATLAB based GUI for ECG arrhythmia detection using Pan-Tompkin algorithm. In Fifth International Conference on Parallel, Distributed and Grid Computing (PDGC), pp. 754–759. IEEE (2018)
5. Talatov, Y., Mgrupov, T.: Algorithmic and software analysis and processing of ECG signals. In International Multi-Conference on Engineering, Computer and Information Sciences (SIBIRCON), pp. 0403–0406. IEEE (2019)
6. Narayana, K.V.L., Rao, A.B.: Wavelet based QRS detection in ECG using MATLAB. Innov. Syst. Des. Eng. **2**(7), 60–69 (2011)
7. Wang, N., Zhou, J., Dai, G., Huang, J., Xie, Y.: Energy-efficient intelligent ECG monitoring for wearable devices. IEEE Trans. Biomed. Circuits Syst. **13**(5), 1112–1121 (2019)
8. Rekik, S., Ellouze, N.: P-wave detection combining entropic criterion and wavelet transform. J. Signal Inf. Process. **6**(03), 217 (2015)

Optimizing Sensor Subset Selection with Quantum Annealing: A Large-Scale Indoor Temperature Regulation Application

Aurelien Meray$^{(\boxtimes)}$ ⓘ and Nagarajan Prabakar ⓘ

Knight Foundation School of Computing and Information Sciences, Florida International University, Miami, FL, USA
{amera009,prabakar}@fiu.edu

Abstract. Harnessing the potential of quantum computing, we apply D-Wave quantum annealing to tackle the Sensor Subset Selection Optimization (SSSO) problem within the scope of large-scale temperature regulation. Leaning on the principles of superposition and entanglement, the study navigates expansive combinatorial spaces of sensor subsets, showcasing impressive scalability and consistently high-quality solutions. A distinctive hybrid approach, merging classical computation for precomputing Mean Squared Errors (MSEs) and quantum computation to explore vast optimization spaces, is introduced. Approaches like these contribute to the field of Intelligent Human-Computer Interaction by enabling smart environment control, effectively optimizing user interaction with their surroundings and enhancing their overall experience. Results affirm the efficiency of our quantum model across varying complexities, producing solutions that rank within the top 91.43 percentile of potential outcomes. Beyond sensor networks, these methods can influence broader human-computer interaction dynamics. Future research will address real-time MSE calculation and objective function enhancement to increase their robustness, along with experimentation on more comprehensive sensor networks.

Keywords: Quantum Annealing · Sensor Subset Selection Optimization · Hybrid Quantum Implementation · Human-Computer Interaction · Large-Scale Temperature Regulation

1 Introduction

In today's world, human-computer interaction (HCI) plays a crucial role in shaping the quality of life, comfort, and efficiency of various environments [1]. One essential aspect of HCI in large human-inhabited spaces such as convention centers and warehouses, is temperature regulation. Maintaining an optimal environment within these expansive spaces is heavily reliant upon an effective sensor network to monitor and control temperature fluctuations. However, designing and deploying efficient and cost-effective sensor networks that cater to the specific needs of such large spaces presents numerous challenges. One such challenge is the Sensor Subset Selection Optimization (SSSO) problem, which aims to identify the smallest optimal set of sensors from a larger collection that fulfills a specific criterion while minimizing cost and computational resources [2].

B. J. Choi et al. (Eds.): IHCI 2023, LNCS 14532, pp. 228–237, 2024.
https://doi.org/10.1007/978-3-031-53830-8_23

The solution to the SSSO problem can significantly improve human-space interactions by allowing for more accurate and responsive temperature regulation that considers the effects of human activities [3]. In this paper, we explore the potential of applying quantum computing, and more specifically, the D-Wave quantum annealer, to address the SSSO problem in the context of large-scale temperature regulation. This is where our work intersects with Intelligent Human-Computer interaction—quantum computing systems would intelligently compute complex optimization problems that enhance the user's interaction with their environment. We aim to demonstrate how the advancements in quantum computing can lead to enhanced sensor network optimization and subsequently, better human-space interactions.

We begin by outlining the SSSO problem in relation to temperature regulation, setting the stage for our research. We then explain our experimental approach using the D-Wave quantum annealer and highlight its potential to resolve this problem. Finally, we discuss the implications of our findings and future research possibilities, emphasizing quantum computing's transformative potential in addressing complex optimization problems and how it could revolutionize Intelligent Human-Computer interactions by creating efficient, responsive, and user-friendly environments.

2 SSSO Problem in Relation to Temperature Regulation

The Sensor Subset Selection Optimization problem is crucial for effective temperature regulation in large spaces. This problem involves optimally selecting a subset of sensors from a larger collection to fulfill the environmental monitoring needs while minimizing expenditure on resources [4, 5]. Imagine an array of N sensors, numbered uniquely and producing scalar measurements, scattered across a two-dimensional domain. These sensors continuously convey information about the conditions in their environment. An estimation function, $E(\bullet)$, processes the multi-sensor data, generating an estimated field that represents the measured conditions in the domain [6].

The goal is to select a subset of X sensors (S) out of the total N, which most accurately represents the environmental conditions. In our case, we use Gaussian Process Estimation as the estimation function for spatial interpolation, denoted by $E(\bullet)$. This function processes the multi-sensor data, generating an estimated field that represents the measured conditions in the domain.

$\widehat{P}_N$ represents the predicted temperatures using the full sensor array (N sensors), while $\widehat{P}_S$ denotes the predictions of the selected subset (X sensors). The subset S, comprising X sensors, is chosen such that the estimated field $\widehat{P}_S$ derived from S closely approximates the ground truth field or the field $\widehat{P}_N$ estimated from all N sensors.

The objective function F is employed to assess the alignment between the estimated fields derived from the full sensor array $\hat{P}_N$ and the selected subset $\hat{P}_S$. Thus, the SSSO problem aims to find the subset S of X sensors out of N that minimizes this function:

$$\arg_{S \subset 0,1,...,N-1, |S|=X} \min F(\hat{p}_N, \hat{p}_S) \tag{1}$$

For temperature regulation, this means identifying the X sensors from a total of N that most accurately represent the temperature distribution across a vast area, such as a warehouse or convention center. Solving the SSSO problem is challenging due to the enormous number of possible sensor subset combinations.

3 Quantum Motivation

With its complex combinatorial nature, the SSSO presents a unique landscape ripe for the application of quantum computing techniques. Specifically, quantum annealing emerges as a potent tool, suitable for tackling such intricate optimization problems [7]. Quantum annealing can help resolve the SSSO problem by leveraging the principles of superposition and entanglement to sieve through the massive combinatorial space of potential sensor subsets. It provides a way to evolve the system state toward the state corresponding to the minimum of the objective function F. Quantum annealing seeks to find the global minimum of a given function over a defined set, which aligns with our intention of finding the subset of sensors that elicit the smallest value of F [8].

Furthermore, quantum annealing's inherent scalability makes it an appealing option for managing real-world scenarios—where the number of sensor candidates can be significantly large. Compared to classical computational methods, quantum annealing can potentially process larger datasets more efficiently, thereby enabling practical, real-world usage. Finally, while finding the absolute optimal solution may be computationally intensive—if not impossible—for larger sensor networks, quantum annealing techniques have demonstrated the ability to find 'near-optimal' solutions in such situations. This capability could mitigate the computational complexity of the SSSO problem, delivering solutions that, while may not be precisely optimal, are still quite effective for temperature regulation needs. As the field of quantum computing and optimization continues to evolve, the use of quantum annealing and similar techniques present promising avenues for solving optimization problems like SSSO.

4 Materials and Methods

4.1 Large-Scale Indoor Temperature Sensor Dataset

Collecting and managing data from a real-world, large-scale indoor sensor network can be challenging due to issues such as sensor failure, data corruption, and the enormous variety of individual locations and conditions. To tackle these challenges and lend flexibility to our study, we have created a synthetic temperature sensor dataset that caters specifically to the SSSO problem.

Our synthetic sensor dataset simulates the conditions of a large interior space, akin to a conference center or office space, incorporating four rooms and a hallway. This layout creates a myriad of environmental conditions across the space under consideration. Each sensor is placed at a point location on a 2D grid that represents this space, and we have full control over the total number of sensors deployed. Each sensor in our dataset is associated with generated time-series data representing simulated temperature readings. These readings range between 18 and 28 °C. If we generate, for example, 16 sensors, each one collects a time series of simulated temperature data within the mentioned range. The synthetic nature of the data provides the flexibility to experiment with different sensor counts and configurations without the physical constraints of a real-world setting.

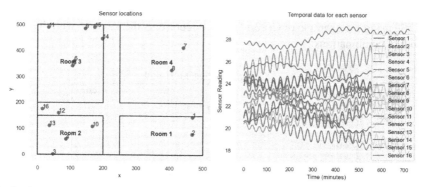

Fig. 1. An example of the synthetic sensor dataset. The left subplot shows the spatial configuration for 16 potential sensor locations in the synthetic indoor environment. The right subplot represents the associated time series of simulated temperature data for each sensor.

Figure 1 presents an instance of our synthetic dataset, illustrating the configuration for 16 potential sensor locations along with their corresponding time series data. This approach affords us considerable flexibility in manipulating our experimental conditions. Moreover, by ensuring the presence of robust and uniform data throughout our study, we are better equipped to perform systematic and controlled evaluations of various solutions to the SSSO problem.

4.2 Our Implementation of the SSSO Problem

In our exploration of the SSSO problem, we opted to use a straightforward objective function to highlight the efficacy of quantum strategies. Our objective function utilizes the Mean Squared Error (MSE) as defined in our previous work [6]. The MSE quantifies the average squared difference between the estimated and actual readings of the sensors, making it an effective measure in the SSSO problem where we aim to minimize this discrepancy. Thus, our implementation of the SSSO problem can be written as:

$$F(\hat{p}_N, \hat{p}_S) = \frac{1}{n} \sum_{i=1}^{n} (\hat{p}_N, \hat{p}_S)^2 \tag{2}$$

where $\hat{P}_N$ denotes the predicted temperatures using the full sensor array and $\hat{P}_S$ denotes the predictions from the selected subset. The n denotes the total number of temperature predictions. Choosing a simple objective function allows us to provide clear evidence of the promise that quantum computation holds in solving complex problems like the SSSO. It provides a clear pathway to understand the potential advantages of quantum annealing and aids in producing interpretable, actionable insights.

4.3 Experimental Setup

In our experiments, we focused on combinations where the number of selected sensors (X) was half of the total available sensors (N). When $X = N/2$, in other words $C(N, N/2)$, this yields the highest number of combinations. For instance, Fig. 2 demonstrates this principle using the example of $C(22,11)$ which represents the scenario with the highest number of combinations.

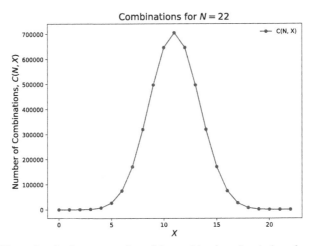

Fig. 2. Chart illustrating the frequency of possible combinations for choice of sensors out of 22.

To maintain simplicity in our setup, we decided to use even values for N. This choice eliminates the issue of having identical combinations for $\lfloor N/2 \rfloor$ and $\lceil N/2 \rceil$ conditions when N is odd. Though it doesn't significantly alter the experimental outcomes, the decision aids in executing easier and more straightforward computations. Every unique N sensors, selecting N/2 sensor subset was run through five iterations, each with a different configuration, to observe variations in the quantum solution. For our experiment, we gradually increased the complexity by starting with smaller N, from six and then building up to 22, conducting a total of 45 experiments.

4.4 Hybrid Quantum Implementation Using D-Wave Quantum Annealer

The D-Wave Quantum Annealer relies on the Binary Quadratic Model (BQM), a data structure that models a system with binary variables interacting pairwise. This BQM describes our problem in a quadratic format optimizing it for quantum annealers and reflects an energy function that the quantum annealer is tasked to minimize [9]. To solve the Sensor Subset Selection Optimization (SSSO) problem, we propose a hybrid implementation bridging classical and quantum computing. First, we use classical computing to precompute the Mean Squared Error (MSE) values, overcoming the limitations of the quantum annealer's inability to handle intricate continuous functions like the MSE. The computed MSE values are then encoded into our BQM, effectively translating the problem into a format suitable for a quantum computer. The D-Wave quantum annealer evaluates the BQM, searching for its lowest energy state that corresponds to the optimal subset of sensors for our application. Our strategy takes advantage of both classical and quantum computing strengths: classical computation handles complex continuous functions while quantum computation tackles vast optimization search spaces.

5 Experimental Results

The validation process was initiated by computing all the potential solutions using a classic brute force approach, where the objective was to determine the entire solution space. This method served a purpose beyond providing a traditional computational contrast. As it painted the widest and most inclusive picture of possible outcomes, it served as a reference field, allowing us to benchmark the solutions generated by the quantum annealer. Table 1 below identifies the percentile ranges in which solutions from our quantum annealing approach positioned themselves in the solution space.

Table 1. Average percentile of all solutions for 5 runs on different initial sensor configurations.

C(N,X)	Number of Combinations	Average Percentile
C(6,3)	20	98.00
C(8,4)	70	93.43
C(10,5)	252	92.14
C(12,6)	924	96.13
C(14,7)	3432	95.57
C(16,8)	12870	92.59
C(18,9)	48620	95.15
C(20,10)	184756	91.43
C(22,11)	705432	94.10

Each percentile value represents the rank of the quantum solution in the sorted list of all possible solutions, i.e., how our quantum solution performs compared to all possible combinations of sensor configurations. As shown in Table 1, the quantum solutions consistently achieved a formidable performance, ranking at or above the 91.43 percentile. This strong percentile score means that the quality of solutions generated by our quantum system is almost at the top among all possible solutions. To aid visual comprehension of these results, Fig. 3, a box plot, depicts the dispersion and central tendency of our quantum solution's performance across varying sensor configurations.

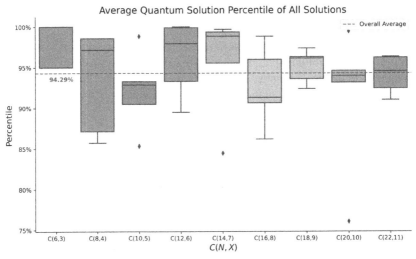

Fig. 3. Box plot showing the distribution of Quantum Solution Percentile across different combinations of sensors, C(N,X), with a red dashed line indicating the overall average.

The overarching average percentile score, represented by the red dashed line, gives an at-a-glance idea of the average solution quality achievable across diverse problem complexities, demonstrating the effectiveness and ability of the quantum approach to produce feasible and optimal solutions.

Another key performance indicator of our analysis lies in the scalability of our quantum solution. The structure of our experiments was designed in such a manner to start with problems of lesser complexity, defined by smaller N values, which would systematically ramp up to higher N values. The intent was to observe how the quantum solution responded to the evolving computational demand induced by an incrementally greater number of sensor candidates.

From our observations, it appeared that the rise in complexity didn't compromise the solution quality significantly. Instead, the quantum solutions consistently averaged around the 94.29 percentile for all experiments, revealing that our quantum model is scalable and maintains high precision even when the solution space widens considerably. To provide a more visually analytical depiction of these results, we created Fig. 4 and Fig. 5. Figure 4 is a distribution plot, specifically for the experimental case of C(22,11), which serves to illustrate the positioning and strength of the quantum solution, denoted by the vertical red line, within the entirety of possible outcomes for this specific scenario.

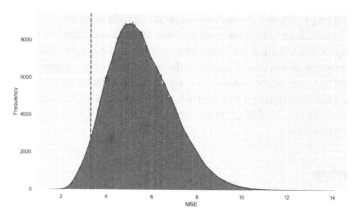

Fig. 4. Distribution plot of solution space for Run 3/5 of C(22,11). Vertical red line indicates where the quantum solution lies among all solutions.

Figure 5 offers a heatmap visualization that allows for an in-depth comparison of three distinct configurations: the ground truth (left), the most optimal 11-subset configuration derived from the brute force approach (middle), and the quantum result (right) for Run 3/5 of the C(22,11) experiment. By presenting the configurations side by side, the heatmap visualization showcases the close resemblance between the quantum and ultimate best solution, emphasizing the effectiveness of the quantum approach in generating comparable solutions for complex combinatorial problems.

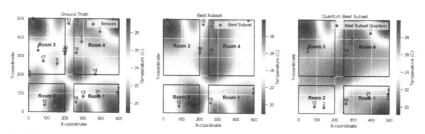

Fig. 5. Heatmap showing the ground truth (left), the most optimal 11-subset configuration (middle), and the quantum result (right) for Run 3/5 of C(22,11) experiment.

Examining the sensor configurations reveals that the brute force solution comprises sensors [1, 2, 6, 7, 9, 10, 12, 15, 16, 18, 19], while the quantum solution consists of sensors [0, 1, 2, 4, 6, 7, 11, 12, 16, 18, 21]. A further comparison between both solution sets reveals the unique elements in each:

- Elements exclusive to the brute force solution: [9, 10, 15, 19]
- Elements exclusive to the quantum solution: [0, 4, 11, 21]

Highlighting these specific sensor differences demonstrates the feasibility and optimality of solutions generated through the quantum annealing process. The resemblance between the quantum and brute force solutions reinforces quantum computing's ability to solve complex optimization problems in a scalable and efficient manner. Our results show that the quantum approach, as currently devised, has been consistently efficient, resilient to scaling, and offers solutions of high quality across a variety of scenarios in the Sensor Subset Selection Optimization problem. This success validates our approach and underscores the potential of quantum computing as a tool for these types of complex optimization problems.

6 Conclusion

Our research has showcased the promise of quantum computing, particularly the D-Wave quantum annealer, in addressing the Sensor Subset Selection Optimization (SSSO) problem within the context of large-scale temperature regulation. By harnessing the principles of superposition and entanglement, quantum annealing efficiently navigates the intricate and vast combinatorial spaces of sensor subsets. This approach, while not without its limitations, has demonstrated remarkable scalability and consistently produced high-quality solutions.

Looking ahead, our work opens doors to broader applications of quantum computing in optimizing human-computer interaction environments. As we continue to explore more extensive sensor networks, refine objective functions, and bridge the gap between classical and quantum computation, we move closer to realizing the potential of quantum techniques in enhancing the comfort, efficiency, and quality of life in large human-inhabited spaces, such as convention centers and warehouses. This research underscores the importance of pushing the boundaries of quantum computing and its relevance in solving complex real-world optimization challenges, even those not traditionally associated with quantum approaches.

Furthermore, our study encourages us to address several areas in future research. Firstly, there is a need to overcome the limitation of precomputing Mean Squared Errors (MSEs) by developing quantum-compatible techniques for real-time calculation. Additionally, future work should include experiments with even more extensive sensor networks, such as the challenging C(100,50) scenario, to assess the scalability and robustness of quantum computing further. Moreover, enhancing objective functions to consider spatial coverage and minimum sensor allocation per area of interest will make our solutions even more applicable to practical scenarios. Lastly, this research reminds us that quantum computing holds untapped potential for solving complex optimization problems in diverse fields, emphasizing the need for continued exploration and innovation in quantum technology.

References

1. Alao, O.D., Joshua, J.V., Akinsola, J.E.T.: Human computer interaction (HCI) and smart home applications. IUP J. Inf. Technol. **15**, 7–21 (2019)
2. Krause, A., Singh, A., Guestrin, C.: Near-optimal sensor placements in Gaussian processes: theory, efficient algorithms and empirical studies. J. Mach. Learn. Res. **9**, 235–284 (2008). jmlr.org
3. Ma, G., Hao, Z., Wu, X., Wang, X.: An optimal electrical impedance tomography drive pattern for human-computer interaction applications. IEEE Trans. Biomed. Circuits Syst. **14**, 402–411 (2020). https://doi.org/10.1109/TBCAS.2020.2967785
4. Onibonoje, M.O., Ojo, A.O., Ejidokun, T.O.: A mathematical modeling approach for optimal trade-offs in a wireless sensor network for a granary monitoring system. Int. J. Technol. **10**, 332–338 (2019). https://doi.org/10.14716/ijtech.v10i2.2099
5. Institute of Electrical and Electronics Engineers. Kerala Section, Institute of Electrical and Electronics Engineers: Spatial correlation based data redundancy elimination for data aggregation in wireless sensor networks. In: International Conference on Innovative Trends in Information Technology (ICITIIT) (2020)
6. Meray, A., Boza, R., Siddiquee, M.R., et al.: Subset sensor selection optimization: a genetic algorithm approach with innovative set encoding methods (2023). TechRxiv. https://doi.org/10.36227/techrxiv.22964987.v1
7. Wang, Y., Wu, S., Zou, J.: Quantum annealing with Markov chain Monte Carlo simulations and D-wave quantum computers. Stat. Sci. (2016). JSTOR
8. Finnila, A.B., Gomez, M.A., Sebenik, C., et al.: Quantum annealing: a new method for minimizing multidimensional functions. Chem. Phys. Lett. **219**, 343–348 (1994). https://doi.org/10.1016/0009-2614(94)00117-0
9. Iturrospe, A.: Optimizing decision making for soccer line-up by a quantum annealer (2021)

Smart IoT-Based Wearable Lower-Limb Rehabilitation Assistance System

Yongfu Wang[1], Boon Giin Lee[1(✉)] [ID], Hualian Pei[2], Xiaoqing Chai[1],
and Wan-Young Chung[3]

[1] Nottingham Ningbo China Beacons of Excellence Research and Innovation Institute, School
of Computer Science, University of Nottingham Ningbo China, Ningbo 315100, China
{yongfu.wang,boon-giin.lee,xiaoqing.chai}@nottingham.edu.cn
[2] The First Affiliated Hospital of Ningbo University, Ningbo 315100, China
[3] Department of Electronic Engineering, Pukyong National University, Busan 48513, Korea
wychung@pknu.ac.kr

Abstract. Venous Thromboembolism (VTE) is a severe medical condition characterized by the development of blood clots within veins, typically occurring
in the deep veins of the legs or as clots that can travel to the lungs. Numerous
previous studies have demonstrated the effectiveness of high-dose lower-limb
physical therapy in aiding patients' functional recovery. However, assessing the
extent of patient compliance with this practice and achieving satisfactory results
have proven challenging, with many issues yet to be resolved before its adoption
in clinical practice. This work proposes a smart unobstructive IoT-based wearable lower-limb rehabilitation assistance system that could trace and evaluate user
lower-limb rehabilitation activities in real-time manner, in the meantime, simultaneous visual guidance cues are presented through visual displays to guide user
completing the rehabilitation tasks, where those tasks are based on the American
Academy of Orthopedic Surgeons (AAOS). The preliminary user study presents
promising benefits of the system contrasting to the typical unguided rehabilitation
training approach in clinical settings.

Keywords: Venous Thromboembolism rehabilitation · Lower-Limb Training ·
Wearable · Internet-of-Thing

1 Introduction

Venous Thromboembolism (VTE) is recognized as a serious medical condition often
characterized by the formation of blood clots within veins, typically in the deep veins of
the legs (commonly recognized as Deep Vein Thrombosis, DVT) or as clots that travel to
the lungs (recognized as Pulmonary Embolism, PE) [1]. The cause of VTE is intricate,

This work was supported in part by the Zhejiang Provincial Natural Science Foundation of China
under Grant LQ21F020024, in part by the Ningbo Science and Technology (S&T) Bureau through
the Major S&T Program under Grant 2021Z037 and in part by the Nottingham Ningbo China
Beacons of Excellence Research and Innovation Institute under Budget Code I01211200008.

usually involves interplay of acquired or inherited tendencies toward thrombosis along with many other risk factors [2]. Each year nearly 10 million people globally are impacted by VTE and around 30% of patients suffer from VTE recurrence within 10 years [2, 3]. In the meantime, VTE disease stands as a significant global contributor to morbidity and mortality [4–6].

Driven by the aforementioned factors, VTE has drawn the large attention from extensive studies, such as ongoing research focused on improving the VTE diagnostic methodology, assessing various Direct Oral Anticoagulants (DOACs) and investigating on safer anticoagulant options [3]. Among which, thromboprophylaxis is one of the popular research trend and has been shown to be an effective method to prevent venous thromboembolism in many studies [3, 7], while the most advocated approach is pharmacological thromboprophylaxis treatment includes injection of aspirin, low-molecular-weight heparin or warfarin. Though it has demonstrated substantial clinical benefits, such as the less bleeding-related adverse effects, fewer work has been focused on mechanical thromboprophylaxis methods, such as Graduated compression stockings (GCS), Intermittent pneumatic compression (IPC). The number of works focused on assisting patients spontaneously conducting mechanical exercises is even fewer, to list a few: [8–10], which motivates our study.

In this paper, we present a smart IoT-based wearable lower-limb rehabilitation assistance system that could trace and evaluate user lower-limb mechanical exercises in real-time manner, in the meantime, simultaneous visual guidance cues are offered to assist the user during the training sections. The preliminary user study result is promising compared to the typical unguided rehabilitation training approach popularly used in clinical settings.

2 System Design

2.1 Hardware Design

The hardware part of the system mainly consists of the Sensing Unit and Client Device. Each individual sensing unit includes: one Seeed Studio XIAO sense controller module, one BNO-055 IMU module, one 100 mAh chargeable lithium battery and one physical switch (see Fig. 1a and Fig. 1b). Specifically, BNO-055 is a 9-DoF Inertial Measurement Unit (IMU) module, which could produce high-precision 3-axis accelerometer and gyroscope information along with Euler angles and quaternion values. XIAO sense module serves as a central controller to communicate with BNO-055 and further relays the processed IMU data to the user client device (such as smartphone, portable laptop) via Bluetooth. The chargeable lithium battery and power switch supplies the power and control the power on-off respectively.

The complete assembled wearable prototype can be seen in Fig. 1c. The user is expected to wear six sets of sensing units in dorsum of foot, lower shin, and shin (left and right) to achieve the optimal trace accuracy regarding diverse lower-limb activities.

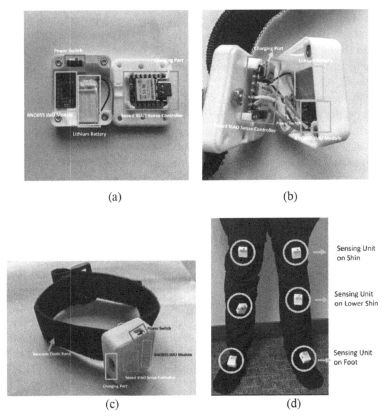

(a)   (b)

(c)   (d)

Fig. 1. Proposed design hardware that include (a) unwired hardware composition of sensing unit, (b) wired hardware composition of sensing unit placed in a 3D printed cover, (c) the overall wearable prototyping of the sensing unit, and (d) the wearable on the lower limb by a participant.

2.2 Data Flow and Application Design

In this pilot work, we have chosen five distinct lower-limb activities from the recommended AAOS list [11, 12], namely "Sit Unsupported Knee Bends", "Ankle Pump", "Ankle Rotation", "Quadriceps Set" and "Straight Leg Raise". The data of sensing units were transmitted to a terminal via Bluetooth connection. Euler angles and quaternion values are extracted from the received data packets and further normalized and re-mapped to project the user lower-limb movements in the simulated scenario modelled using Unity3D tool. A general data flow is shown in Fig. 2.

In the meantime, the rotations of x, y and z angles of the user foot, lower shin and shin are traced and used to identify the current state of the selected activity. The corresponding hints will be presented to the user assisting the user to complete the activity. The overall number of successful training exercises and the training time will be recorded for further analysis purposes.

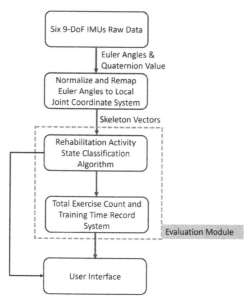

Fig. 2. Flow of Lower-Limb Rehabilitation Assistance System Algorithm

2.3 Lower-Limb Rehabilitation Activity State Classification Algorithm

To develop the state classification algorithm for lower-limb rehabilitation activities, we gathered data from 6 trials involving five selected activities, as outlined in Table 1. Considering the nature of these activities and to prevent overwhelming users with excessive information, our state classification algorithm primarily emphasizes three general states for each activity. These states are (1) approaching the training goal, (2) achieving the training goal, and (3) transitioning away from it. The training goal is defined by the target rotation degrees of the foot, lower shin, and shin, which vary depending on the specific activity.

Table 1. Details Of Five Rehabilitation Activities Collected

Activity ID	Type of Rehabilitation Activity	Trials
F1	Sit Unsupported Knee Bends	6
F2	Ankle Pump	6
F3	Ankle Rotation	6
F4	Quadriceps Set	6
F5	Straight Leg Raise	6

Based on the observation of collected user data, we predefine the corresponding thresholds for each activity to conduct the activity state classification. Let $\theta_{current}$

denote the current angles of foot, lower shin and shin, while using θ_{target} denote the target angles with an offset value β to increase the classification success rate. When $\theta_{current}$ is not within the $\theta_{target} \pm$ offset β range, we classify the current state as approaching the training goal. If $\theta_{current}$ has fits into the θ_{target} range, the current state will switch to the goal attained state. In the meantime, one 10 s timer will be activated. If the user does retreat during the 10 s counting down, the algorithm will guide the user to enter the retreating state and continue with another round of the training. Otherwise, the state will be initialized to the approaching the training goal stage. One iteration of the activity state classification algorithm is listed in Fig. 3.

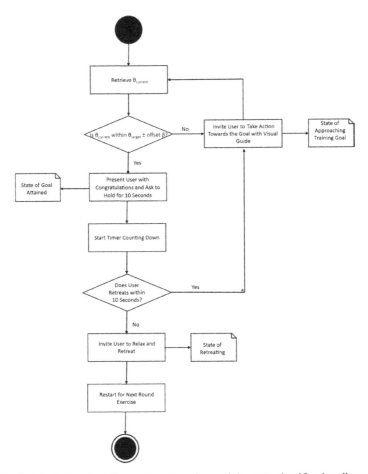

Fig. 3. Illustration depicting an iteration of an activity state classification diagram

3   Conclusion

This work proposes an IoT-based wearable lower-limb rehabilitation assistance system that has two major functionalities. The system is able to trace and evaluate five different lower-limb rehabilitation tasks suggested by AAOS [11, 12] in real-time and the training data will be further stored for subsequent analysis (such as for physical therapists analyzing the patient's recovering progress). The second important feature of the system enables the capability of directly assisting the user to complete the spontaneous training during the training sessions. The preliminary user study result is promising compared to the typical unguided rehabilitation training approach popularly used in clinical settings. In our future work, we will recruit more patients to conduct the field testing and further improve the system.

References

1. CDC: What is Venous Thromboembolism?|CDC. https://www.cdc.gov/ncbddd/dvt/facts. html. Accessed 04 Sept 2023
2. Epidemiology of venous thromboembolism|Nature Reviews Cardiology. https://www.nature. com/articles/nrcardio.2015.83. Accessed 16 Sept 2023
3. Venous thromboembolism - The Lancet. https://www.thelancet.com/article/S0140-673 6(20)32658-1/fulltext. Accessed 16 Sept 2023
4. Goldhaber, S.Z., Visani, L., De Rosa, M.: Acute pulmonary embolism: clinical outcomes in the International Cooperative Pulmonary Embolism Registry (ICOPER). The Lancet **353**, 1386–1389 (1999)
5. Prandoni, P., et al.: The long-term clinical course of acute deep venous thrombosis. Ann. Intern. Med. **125**, 1–7 (1996)
6. Horner, D., et al.: Pharmacological thromboprophylaxis to prevent venous thromboembolism in patients with temporary lower limb immobilization after injury: systematic review and network meta-analysis. J. Thromb. Haemost. **18**, 422–438 (2020). https://doi.org/10.1111/ jth.14666
7. Tian, B., Li, H., Cui, S., Song, C., Li, T., Hu, B.: A novel risk assessment model for venous thromboembolism after major thoracic surgery: a Chinese single-center study. J. Thorac. Dis. **11**, 1903–1910 (2019). https://doi.org/10.21037/jtd.2019.05.11
8. Bijalwan, V., Semwal, V.B., Singh, G., Crespo, R.G.: Heterogeneous computing model for post-injury walking pattern restoration and postural stability rehabilitation exercise recognition. Expert. Syst. **39**, e12706 (2022). https://doi.org/10.1111/exsy.12706
9. Chen, C.-J., Lin, Y.-T., Lin, C.-C., Chen, Y.-C., Lee, Y.-J., Wang, C.-Y.: Rehabilitation system for limbs using IMUs. In: 2020 21st International Symposium on Quality Electronic Design (ISQED), pp. 285–291 (2020). https://doi.org/10.1109/ISQED48828.2020.9137026
10. Chang, Y.-P., Wang, T.-C., Lee, Y.-J., Lin, C.-C., Chen, Y.-C., Wang, C.-Y.: A smart single-sensor device for instantaneously monitoring lower limb exercises. In: 2019 32nd IEEE International System-on-Chip Conference (SOCC), pp. 197–202 (2019). https://doi.org/10.1109/ SOCC46988.2019.1570548017
11. Total Hip Replacement Exercise Guide - OrthoInfo - AAOS. https://www.orthoinfo.org/en/ recovery/total-hip-replacement-exercise-guide/. Accessed 15 Sept 2023
12. Total Knee Replacement Exercise Guide - OrthoInfo - AAOS. https://www.orthoinfo.org/en/ recovery/total-knee-replacement-exercise-guide/. Accessed 15 Sept 2023

Using Machine Learning of Sensor Data to Estimate the Production of Cutter Suction Dredgers

Zahra Zarifianshafiei[1] and Chintan Amrit[1,2](✉) (iD)

[1] Amsterdam Business School, University of Amsterdam, P.O. Box 15953, 1001 Amsterdam,
The Netherlands
zahra.zarifianshafiei@student.uva.nl, c.amrit@uva.nl
[2] LM Thapar School of Management, Thapar Institute of Engineering and Technology,
Dera Bassi Campus, Punjab 140 507, India

Abstract. Production estimation (the excavated soil per time) helps dredging companies to be able to manage the dredging projects efficiently and enables them to predict the time and cost of the project. They can calculate the production with the density and flow sensors which are installed in a dredging ship. However, due to the high price of the density sensor, many companies avoid purchasing vessels with the density sensor and look for a cheaper alternative. In this study, we explore an alternative way to predict density by leveraging data used during production and applying machine learning algorithms on them. In this article we use a dataset that belongs to a Cutter Suction Dredger (CSD) operating in an African country. Our results exceed a prediction accuracy of 80%. However, our models do not predict the density of the dredger operating in a different location. The reason is that the features used to estimate the density classes are influenced by various factors such as ambient conditions, water dynamics, fuel quality, and soil properties and these factors vary based on the region. Therefore, one of the main contributions of our work is checking the generalizability of our trained models and explaining the features that are important for predicting soil density.

Keywords: Dredging · Sensor Data · Machine Learning

1 Introduction

Dredging is the process of excavating and removing underwater sediment, materials, and sand or rock from waterways, seas and oceans to widen, deepen, and clean an excavation area using innovative technology. Dredging productivity estimation is very crucial for construction cost, schedule and resource management. It is also essential to optimize the dredging operation. However, dredging complex environments makes it challenging to estimate productivity accurately [11].

There are some significant reasons that companies seek an innovative way to measure dredging production: First, the different excavation locations have different soil properties. The soil properties have the most significant impacts on a dredging operation.

Consequently, this impacts production and energy consumption. Second, measuring the soil properties in the complex dredging environment is very difficult, and finally, including a density sensor is too expensive for small vessels [3].

Analysing the data gathered using machine learning (ML) offers a different perspective. It helps companies to operate effectively and efficiently in a highly competitive maritime industry. In the shipbuilding industry, a vessel has hundreds of sensors that generate a massive amount of data. Therefore, shipbuilding companies that use ML and information technology can have a competitive advantage compared to their rivals and they could have a leadership position in the shipbuilding market.

This research aims to leverage the machine learning approaches to precisely estimate the cutter suction dredger (CSD) density which is an important factor to calculate the production. This research has been done in collaboration with the R ALPHA[1]. R ALPHA is a shipbuilder company specializing in designing, manufacturing, and providing customers with dredging and offshore equipment and tailor-made solutions and services. It is the global market leader for innovative dredging and mining equipment and vessels. ALPHA Bover[2] CSDs are reliable, fuel-efficient, have low maintenance costs and are highly productive at all dredging depths. In addition, they are equipped with state-of-the-art technology. You can see the R ALPHA CSD Bover in Fig. 1.

Fig. 1. R ALPHA CSD Bover

Our primary research question is:

How can dredging companies leverage machine learning approaches to classify the density of dredged materials to estimate their dredging project's productivity and cost?

We deploy five machine learning algorithms namely Random Forest, Gradient boosting, Support vector machine, Naïve Bayes and Artificial Neural Networks, applied to a dredging dataset to answer the above research question. Our results show that the prediction accuracy of Random Forest and Gradient boosting exceeds 80% which is better than other algorithms used. The main contribution of our work is checking the generalizability of our trained models and explaining the features that are important for predicting soil density.

---

[1] A pseudonym.

[2] Also a pseudonym.

The rest of the paper is structured as follows; Sect. 2 gives an overview of the existing literature. Section 3 describes our research setting and methodology, then in Sect. 4 we provide the results of our analysis, and finally, Sect. 5 concludes the paper.

2 Literature Review

2.1 Dredger Productivity Estimation

Researchers used machine learning approaches to estimate dredging productivity from different perspectives in recent years. In terms of dredger productivity estimation, Yang et al. (2015) [13] proposed an Artificial Neural Networks (ANN) predictor model for predicting the production of cutter suction dredgers. The swing speed (the speed of a cutter head which moves from one side to another side), the velocity of the hydraulic pipeline transportation (flow velocity in the pipeline) and the work pressure of the cutter (the relation between the forces behind the reamer and the amount of excavated soil) are considered as inputs of the model. They also added the Bayesian regularization algorithm to the cost function to help the model reach a more generalized solution. As a result, this model can predict production accurately. However, to have a good performance model, it needs to have more parameters as input of the model.

Bai et al. (2019) [1] proposed a real-time productivity estimation method based on two stages. First, the key features that impact productivity were selected and used as inputs of the four different models, namely: Random Forest, K-Nearest, Naive Bayes, and eXtreme Gradient Enhancement (XGBoost). Next, the authors put four traditional factors that affect productivity as inputs of motioned models in the second stage. These factors are the working voltage of the cutter, the suction vacuum of the submersible pump, the flow rate and the transverse cutter speed (swing speed). Finally, the results of the two stages are compared and evaluated; The result shows XGBoost algorithm predicts productivity more than 90%, which is better than the three other algorithms in both stages.

Wang et al. (2020) [11] proposed a method to estimate the productivity very well, even in the absence of the mixture concentration data. They proposed two ensemble processes, namely, extreme Gradient boosting (XGBoost) and Light Gradient Boosting Machine (LightGBM). In the first model, "Model Average," a weighted average ensemble of mentioned models was built and create a combined model to improve the accuracy. The weights are defined according to the mean square error based on cross-validation results, and the models' predicted values are the output of the model. The second model, "Model stacked generalization," is a hierarchical model combination framework where the combination of the different models creates a meta-model with better or equal performance than each model. The R^2 score of the stacked generalization model for productivity prediction was 0.9281, which was the highest among the algorithms. Wang et al. [11] recommended the XGBoost algorithm in the case of real-time productivity calculation since the XGBoost algorithm can guarantee a certain level of accuracy and low computation time.

Fu et al. (2021) [7] introduced a framework for predicting the productivity of cutter suction dredge. They combined SVR and Elastic Net to remove irrelevant and redundant features. They also added five more features to avoid overfitting based on experts' knowledge. Then they build four different predictive models, SVR (support vector regression), XGBoost, Neural Network, and LSTM (Long-Short Term Memory Network). on a CDS operating in a particular port of China. The result shows that the goodness of fit R^2, was more than 80% in all four models. SVR and Neural Network had the best performance, around 87%. The authors also decided to combine SVR and Neural networks and create a final model. The combined model's determination coefficient R^2 87.61%, which is a too small performance improvement. Believed it needed to apply all five models on optimized construction operation data.

Bai et al. (2021) [2] used an A.I. algorithm to estimate Trailing Suction Hopper Dredger (TSHD). TSHD is a self-propelled and self-loaded dredger with a drag-head system with the advantage of self-navigation and self-load and unloads system compared to other dredgers. Bai and his colleagues have studied the productivity estimation of Cutter suction dredgers in 2019 [1]. They proposed two major modeling stages. First, the ReliefF-Granger algorithm was applied to the real-tome dataset to select the key features that might have the highest impact on productivity. They also proposed the Granger causality test used in economics alongside the ReliefF-Granger algorithm to reach a better feature collection. Second, they propose a stacking strategy combining SVR, Beetle Antennae Search-Back Propagation (BAS-BP) and LSTM algorithms and creating a meta-model. The accuracy of the stacking model was similar to the accuracy of a single model. However, the stacking strategy increased the model performance since it improved the model generalization to predict the productivity of drag heads without time lag which was problematic in their previous study.

2.2 Dredging Productivity Estimation in Similar Areas

In terms of productivity estimation in other similar areas, Schabowicz & Hola (2007) [10] applied a neural network on land excavators' bucket capacity, type, soil properties and transport capacity to estimate the productivity of earthmoving machinery. In comparison with land excavators, CDS operation and working environment are more complex. It is also challenging to know the soil properties in real-time operation. Since the assumptions and operation are totally different, applying this model to the CDS dataset might produce significant errors.

Ebrahimabadi et al. (2015) [5] applied a neural network model to estimate road headers' cutting performance based on the data from a coal mine in Iran. Road headers are mechanical miners who are used in the mining, tunnelling and construction industries. They used geomechanical characteristics such as rock mass properties, rock quality designation and intact rock properties as inputs of the ANN model. As a result, the model was able to predict productivity with 0.97 R2. The result was perfect, but the road header characteristics impact the performance. Therefore, it needs to consider the road header's data as an input of the model.

Zayed & Mahmoud (2014) [16] proposed a Neuro-Fuzzy (N.F.) based predictive model for predicting the productivity of Horizontal Directional Drilling (HDD), which can be applied to different soil types. N.F. model can address non-linear relations, reducing the uncertainty and also model non-numerical data. They considered three types of soil, i.e., clay, rock and sand; the precision of the model was 85% which is a robust prediction. However, it needs to apply this model to other types of soil to validate the result.

YuMin et al. (2011) [15] used a neural network to estimate the production performance of coalbed methane wells in the initial production stages. Due to the heterogeneity of the coalbed, the uniqueness of the production process and the difficulty of recognizing the reservoir's key parameters. It's challenging to predict production with conventional methods. The neural network model can predict the production of gas wells with high accuracy and an average relative error of less than 2% without recognizing reservoirs' key parameters. They recommended applying the method in the early stages of production, in which some reservoir's key parameters are unavailable.

2.3 Soil Classification

In terms of dredger productivity estimation using soil classification, Yue et al. (2015) [14] developed a quantitative classification model for the dredging material under complex conditions in dredging operations to estimate the productivity of CSDs. They used rough-set theory to select important attributes; then, the conditional entropy was adapted to calculate the power of the classification model. According to soil classification, construction technology, and the dredger's performance, a predictive model was built to estimate productivity. The model was able to predict the productivity with the maximum relative error of 11.97% and R2 98%, which is an excellent estimate. However, this model has not been applied to another data point to validate the result. In addition, the dredging performance and productivity are not influenced only by one single material. It needs to evaluate the impact of various materials and the combination of materials on dredging performance and productivity.

Braaksma et al. (2019) [3] propose a novel method to estimate parameters that are difficult to predict, i.e., the average grain size of dredged soil, TSHD's overflow losses, dredging forces and an anchor position. They developed and implemented an overflow loss estimator based on a particle filter. This estimator can inform the operators about when to stop dredging and warn them in the case of overflow.

3   Research Setting and Methodology

Cutter Suction Dredgers (CDS) can dredge sand, clay, slit and even rock sea or riverbed. This research focuses on ALPHA's CSDs which are called Bover[3]. A CSD dredger is generally constructed from the hull, diesel engines, steel pile positioning system, axillary diesel engine-generator system, winch traverse system, reamer lifting system, head cutter, pump system and control system. When the CSD dredger reaches the excavation area, it has positioned itself by lowering the main spud pole at the vessel's bow and placing anchors at port and starboard. The anchor booms are connected to the large arm at the front of a ladder. While the anchors are fixed in the position, the ladder lowers to the seabed. At the end of the ladder, there is a large cutter head that cuts the seabed and caves the material by pulling on one wire and giving slack on the other one. At the end of

---

[3] Also, a pseudonym.

the swing movement, the dredger is pushed forward by a hydraulic cylinder connected to the main spud pole. Inside the CSD dredger, a pipeline vacuums the dredged material (mixture of soil and water) and transforms it into the discharge area, which can be several kilometres away from the dredging location.

Several factors can influence productivity, which are dynamic and interrelated. These correlated factors and the complex dredging environment make it difficult for the operator to understand how much the soil is dredging. At present, most companies are using a semi-empirical formula to build a predictive model to estimate productivity based on historical data from previous dredging construction projects for their dredgers that do not have the density sensor [8]. Mostly they look at any old project in the region to determine soil conditions and use a semi-empirical formula to determine the production. This method is time-consuming and requires input variables manually, and the model's precision is relatively low. The production is calculated based on this formula:

$$P = C \times Q \tag{3.1}$$

where P is productivity (m^3/s), C is mixture volume concentration (%) and Q is the flow (m^3/s). Mixture concentration is calculated from the following formula:

$$C = \frac{\rho_{mixture} - \rho_{water}}{\rho_{soil} - \rho_{water}} \tag{3.2}$$

where mixture subscript means rock or soil plus water, $\rho_{mixture}$ is mixture density (kg/m^3), ρ_{water} is water density(kg/m^3) and ρ_{soil} is soil density (kg/m^3)

In large dredging vessels with density sensors, they can measure mixture volume concentration (C). Otherwise, measuring the C is problematic because of dredging complex environments and unknown soil properties.

As mentioned, density is a necessary variable to calculate production. Researchers have proposed various novel regressive machine learning algorithms to estimate density. In soil mechanics, it is common to discretize soil properties such as grain size, fraction and deformability into different classes since the interpretation of these properties is difficult. The grouping is done also because of the natural distribution that occurs. Therefore, considerations must be made for all members of that group. In this paper, we are looking for a more straightforward way to estimate the density. We also aim to reduce the effect of soil property on the predictive model. Therefore, the density values are discretized and divided into three classes, namely High, Medium and Low. The aim is to predict the class of the density instead of predicting the actual amount of density. According to the density classification model, clients can estimate the density class, and consequently, they can anticipate an average production size in their project according to the density class.

Although the research is predominantly quantitative, we had extensive meetings with the experts in R ALPHA to both understand the dredging process and the data as well as obtain feedback on our analysis.

A dredging vessel contains several sensors which measure and record the dredging operation parameters. In this paper, we built classification models on five operation days data which belong to R ALPHA's CSD operating in Africa. The dredging operation data includes 153056 datasets (rows) with 137 dimensions (columns).

The data analysis process was done using Python. In the first stage of data cleaning, 91 sensors are removed. These sensors were mainly Boolean or had less than two values, and 46 sensors remained for analysis. Fortunately, the dataset did not have any missing values. According to the experts' knowledge, when the discharge pump pressure shows a value less than 1, it means the dredger was not operating. Hence, the rows which had discharge pump pressure values less than 1 were removed.

To eliminate the impact of dimensionality, the features were standardized before processing. It also needed to remove outliers. There are various methods to detect and remove outliers. One of the most common methods to detect outliers is to use the Z-score. In the Z-score method, each instance is subtracted from the mean and divided by the standard deviation [9]. The formula is:

$$Z_i = \frac{X_i - \mu}{\sigma} \tag{4.1}$$

where X_i is the ith instance, μ the sample means of variable, σ: is the standard deviation of the variable. It is common to cut off an instance from the dataset further than the ± 3 Z value (i.e., $|Z| > 3$).

In this study, the target variable is density. The density is a continuous variable, and it changes from 1 to 1.6. The higher density means the dredger excavates more soil. Figure 2 shows the plot box of density.

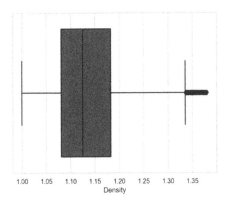

Fig. 2. Density plot box

To apply classification models, it needs to discretize the density. The density values were ordering from low to high and then the three classes based on the data distribution were defined: Low, Medium and High. The density classes are shown in Table 1.

We used two different filter methods to select the best features for modeling parts: Pearson's Correlation and Mutual Information Gain. Alongside these two methods, we use the experts' knowledge to select the features.

In the African dredging dataset, the number of instances in high-density classes is substantially higher than in other classes, and therefore, the dataset is imbalanced. Figure 3 shows the number of instances in each class. We, therefore, first oversampled

Table 1. Density classes

Density Ranges	Classes
1–1.13	Low
1.13–1.26	Medium
1.26>	High

the dataset [4]; since it did not work perfectly on the validation dataset, we decided to apply the under-sampling method [6] on the dredging data.

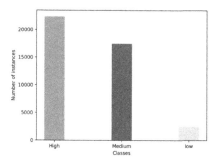

Fig. 3. Unbalanced African dataset

Before starting the modelling process, the dataset was divided into 60% for training and 40% for testing. Then, to validate the results, the model was also applied to the new data point, which came from a different dredging operation location. 10-fold Cross-validation was also applied to the training set to avoid overfitting.

We now discuss the results of applying four traditional classifiers, namely, Random Forest (RF), Gradient Boosting, Support Vector Machine (SVM) and Naive Bays, to predict the density class. We then also use and compare the results with an Artificial Neural Network (ANN).

4   Results and Discussion

In this section, we discuss the results of applying five different ML models to the dredging dataset. We measure the performance of the models and compare them based on the confusion matrix, accuracy, recall, precision F1 score and AUC curve. Here, we only show a part of the results (F1 score and AUC curve), due to space limitations.

According to Fig. 4, ANN has the highest area under the ROC curve (0.9) and the second-best models are RF and Gradient boosting with the 0.83 area under the ROC curve. However, although ANN has the highest AUC, the Accuracy of Random Forest and Gradient boosting are higher than ANN (Table 2). Since ANN works better on a larger dataset and the AUC difference between ANN, RF and Gradient boosting is not significant, we selected RF and Gradient boosting as the best models to predict

the density classes. There are some reasons that the other three models are not working well on the dataset. First, the SVM generally works better on binary classification
rather than multi-classification. Second, ANN has higher performance on larger datasets
rather than smaller ones and finally, Naïve Bayes does not perform well on the dredging dataset because it assumes that each input variable is independent. Another reason
why tree-based models perform better, is that the dredging dataset is a tabular dataset
where tree-based models outperform deep-learning based models. Such an assumption
of independence is unrealistic for real data. Once the density classes are estimated, one
can easily find a threshold for the production. For example, If the models predict that
the density class is medium, the density changes between 1.13 and 1.16. Then, the production can be calculated by multiplying the average flow in this threshold, with the
density. Therefore, a dredging company can find a range for their production based on
the particular density class.

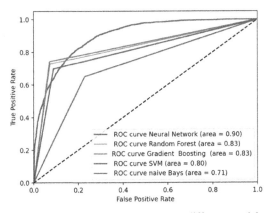

Fig. 4. Using ROC Curve to compare different models.

To validate the results and reach the generalized models, RF and Gradient boosting
were applied to a new dataset containing data from the same dredger operating in a
different location (and therefore a different domain), the Middle East. This dataset contained 17447 rows and after data cleaning, the number of instances reached 9307 and
the dataset was also highly unbalanced. Therefore, we also applied under-sampling to
the new dataset.

Unfortunately, the results of applying RF and Gradient boosting to the validation
dataset were not satisfactory (Accuracy of 47% and 51%, respectively). To find the
reasons why models work on the African dataset perfectly but have weak performance
on the Middle-Eastern validation dataset, it is necessary to know which features have
the highest impact on the models. As demonstrated in Tables 3 and 4, velocity, fuel
consumption, both exhausts left and right, fuel consumption and main engine load are
the most important features to predict the density. Although these features are highly
influenced by soil characteristics, other factors such as fuel quality, weather conditions,
water dynamics and maintenance can impact these features. For instance, humidity
affects the engine burn rate characteristics. An increase in humidity decreases the engine

Table 2. Accuracy of the 5 ML algorithms (Gradient Boosting and Random Forest are our two chosen algorithms – therefore highlighted in red)

Algorithm	Accuracy
Gradient Boosting	0.82
RF	0.79
SVM	0.77
ANN	0.74
Naïve Bays	0.62

burn rate and increases the combustion duration; consequently, more fuel will be needed to continue burning (Wimmer & Schnessl, 2006 [12]). Therefore, humidity might impact the main engine load and fuel consumption. Since the ambient conditions in the two locations are different, it is difficult for a machine-learning algorithm to learn from one dredging dataset and predict another dredger's density which is operating in different conditions. However, the classification models suggested in this study can be used to estimate the density and production of similar CSD dredgers operating in a similar location with the same conditions.

Table 3. RF feature importance

Feature	Rank
Main engine load	0.186911
Main fuel consumption actual	0.177837
Average main exhaust gas right and left	0.157061
Velocity	0.153465
Engine speed	0.091282
Dredge pump vacuum	0.082344
Dredge pump discharge pressure	0.07776
Joined swing winch speed SP	0.038647
Cutter hydro oil pressure	0.034694

Table 4. Gradient boosting feature importance

Feature	Rank
Main engine load	0.207232
velocity	0.158834
Average main exhaust gas right and left	0.157494
Main fuel consumption actual	0.151152
Engine speed	0.087958
Dredge pump vacuum	0.087583
Dredge pump discharge pressure	0.073724
Cutter hydro oil pressure	0.039985
Joined swing winch speed SP	0.036039

5   Conclusions

In this study, we attempted to answer the question of how we could use an ML approach to estimate the dredging production by classifying the density of dredged materials. Estimating the dredging production helps dredging companies calculate the cost and time of dredging projects. We began by analysing the Africa dataset where the density values were divided into three classes high, medium and low and the aim was to estimate the density classes instead of estimating the density values directly. We then applied five different classification algorithms, namely, RF, Gradient boosting, SVM, Naïve Bays and ANN, to the dataset. Among these five models, RF and Gradient boosting estimated the density with an accuracy of 80%, and 82%, respectively. However, to validate the result and to determine the generalizability of the models, the models were applied to a Middle east dredging dataset. Unfortunately, the models did not perform as well in predicting the density classes of the Middle East dredger. We found out that the features that play an important role in density estimation are highly influenced by various factors such as soil characteristics, ambient conditions, fuel quality, water dynamics and maintenance procedures. Although some dredging experts believe that the lack of information about soil mechanics is the main reason that estimating dredged soil density in different locations is a challenging task, we believe the mentioned factors should also be considered in applying the density estimation models in different domains. This understanding of the generalizability of our models and explaining it based on the used features is one of the main contributions of this paper.

References

1. Bai, S., Li, M., Kong, R., Han, S., Li, H., Qin, L.: Data mining approach to construction productivity prediction for cutter suction dredgers. Autom. Constr. **105** (2019). https://doi.org/10.1016/j.autcon.2019.102833

2. Bai, S., Li, M., Song, L., Ren, Q., Qin, L., Fu, J.: Productivity analysis of trailing suction hopper dredgers using stacking strategy. Autom. Constr. **122**, 103470 (2021). https://doi.org/10.1016/j.autcon.2020.103470

3. Braaksma, J., Osnabrugge, J., De Keizer, C.: Estimating the immeasurable: soil properties. Terra et Aqua **117**, 24–32 (2009)

4. Chawla, N.V., Bowyer, K.W., Hall, L.O., Kegelmeyer, W.P.: SMOTE: synthetic minority over-sampling technique. J. Artif. Intell. Res. **16** (2002)

5. Ebrahimabadi, A., Azimipour, M., Bahreini, A.: Prediction of roadheaders' performance using artificial neural network approaches (MLP and KOSFM). J. Rock Mech. Geotech. Eng. **7**(5), 573–583 (2015). https://doi.org/10.1016/j.jrmge.2015.06.008

6. Fernández, A., García, S., Galar, M., Prati, R.C., Krawczyk, B., Herrera, F.: Learning from imbalanced data sets. In: Learning from Imbalanced Data Sets (2018). https://doi.org/10.1007/978-3-319-98074-4

7. Fu, J., Tian, H., Song, L., Li, M., Bai, S., Ren, Q.: Productivity estimation of cutter suction dredger operation through data mining and learning from real-time big data. Eng. Constr. Archit. Manag. **17** (2021). https://doi.org/10.1108/ECAM-05-2020-0357

8. Hardya, T.P.: Analysis of Productivity in Dredging Project A Case Study in Port of Tanjung Perak Surabaya – Indonesia (2016)

9. Mellenbergh, G.J.: Outliers. In: Mellenbergh, G.J. (eds.) Counteracting Methodological Errors in Behavioral Research, pp. 293–308. Springer, Cham (2019). https://doi.org/10.1007/978-3-030-12272-0_17

10. Schabowicz, K., Hola, B.: Mathematical-neural model for assessing productivity of earth-moving machinery. J. Civ. Eng. Manag. **XIII**(1), 47–54 (2007). https://doi.org/10.1080/13923730.2007.9636418

11. Wang, B., Fan, S., Jiang, P., Xing, T., Fang, Z., Wen, Q.: Research on predicting the productivity of cutter suction dredgers based on data mining with model stacked generalization. Ocean Eng. **217**(September), 108001 (2020). https://doi.org/10.1016/j.oceaneng.2020.108001

12. Wimmer, A., Schnessl, E.: Effects of humidity and ambient temperature on engine performance of lean burn natural gas engines, pp. 421–429. American Society of Mechanical Engineers, Internal Combustion Engine Division (Publication) ICE (2006). https://doi.org/10.1115/ICEF2006-1559

13. Yang, J., Ni, F., Wei, C.: A BP neural network model for predicting the production of a cutter suction dredger. In: IC3ME, pp. 1221–1226 (2015). https://doi.org/10.2991/ic3me-15.2015.235

14. Yue, P., Zhong, D., Miao, Z., Yu, J.: Prediction of dredging productivity using a rock and soil classification model. J. Waterw. Port Coast. Ocean Eng. **141**(4), 06015001 (2015). https://doi.org/10.1061/(asce)ww.1943-5460.0000303

15. Lü, Y., Tang, D., Xu, H., Tao, S.: Productivity matching and quantitative prediction of coalbed methane wells based on BP neural network. Sci. China Tech. Sci. **54**(5), 1281–1286 (2011). https://doi.org/10.1007/s11431-011-4348-6

16. Zayed, T., Mahmoud, M.: Neurofuzzy-based productivity prediction model for horizontal directional drilling. J. Pipeline Syst. Eng. Pract. **5**(3), 04014004 (2014). https://doi.org/10.1061/(asce)ps.1949-1204.0000167

17. Grinsztajn, L., Oyallon, E., Varoquaux, G.: Why do tree-based models still outperform deep learning on typical tabular data? Adv. Neural. Inf. Process. Syst. **35**, 507–520 (2022)

Optimizing Microstrip Patch Antenna Array Feeders for Efficient Wi-Fi Energy Harvesting

Minh-Hoang Vu[1], Viet-Thang Tran[2], Yudi April Nando[1], and Wan-Young Chung[1(✉)]

[1] Department of AI Convergence, Pukyong National University, Busan, South Korea
wychung@pknu.ac.kr
[2] Department of Science and Technology, Nguyen Tat Thanh University, Ho Chi Minh City 70000, Vietnam

Abstract. Wi-Fi energy harvesting is one of the most promising solutions for powering devices in daily life. Antennas play a crucial role in this system, influencing energy capture, power transfer efficiency, frequency selectivity, adaptability, and integration. High gain and low return loss are essential criteria. While prior studies explored patch antenna arrays with conventional feeders, this article introduces a novel approach: designing, simulating, and fabricating microstrip patch antenna arrays with 90-degree bend modifications and quarter-wave transformers in the microstrip lines at 2.45 GHz. The primary objective is to minimize reflection and maximize gain. Various prototypes, from single patches to 4 × 8 arrays, are simulated and compared to conventional feeders. Utilizing CST Tools for simulation and Bi-LSTM for optimization, we fabricate a 4 × 4 array of microstrip patch antennas. Measurement results with the N9912A FieldFox HandHeld RF Analyzer and ME1310 Antenna Training Kit (3D) Dream Catcher indicate a return loss of −30.2 dB and a gain of 17.3 dBi. This innovative design enhances Wi-Fi energy harvesting efficiency, especially when integrated with rectifiers, enabling sustainable power for wireless devices.

Keywords: Wi-Fi energy harvesting · microstrip patch antenna array optimization · feeder modifications

1 Introduction

In today's world, billions of devices, from Wi-Fi routers to smartphones, tap into Wi-Fi energy, offering a revolutionary energy source for various applications like the Internet of Things (IoT), sensor networks, wearables, and even building integration [1]. Wi-Fi energy harvesting, unlike conventional sources such as batteries or solar panels, provides a continuous, uninterrupted power supply, drawing energy directly from the airwaves [2]. Recognizing this immense potential, this paper aims to present an efficient energy harvesting system through innovative patch antenna array design and an optimized feed network to minimize losses and maximize power harvesting [3].

The increasing demand for wireless power, coupled with limited ambient Wi-Fi energy, necessitates a novel approach [4]. Antennas are pivotal in energy harvesting,

B. J. Choi et al. (Eds.): IHCI 2023, LNCS 14532, pp. 256–261, 2024.
https://doi.org/10.1007/978-3-031-53830-8_26

impacting energy capture, efficiency, frequency selectivity, adaptability, and integration into wireless applications [5]. Effective antenna performance hinges on high gain and low return loss. Unlike prior research using patch antenna arrays and conventional feeders, this paper pioneers a novel perspective [6]. It focuses on revolutionizing Wi-Fi energy harvesting through microstrip patch antenna arrays with unique 90-degree bend modifications and quarter-wave transformers at 2.45 GHz, aiming to minimize reflection and maximize gain. A comprehensive assessment compares these innovative designs, spanning various prototypes from single patches to expansive 4 × 8 arrays, against conventional feeders.

This research utilizes CST tools for precise simulation and strategically applies Bidirectional Long Short-Term Memory (Bi-LSTM) optimization techniques. Fabricating a 4 × 4 microstrip patch antenna array yields impressive results: a return loss of −30.2 dB and an exceptional gain of 17.3 dBi. These outcomes not only indicate goal achievement but also signify a significant advancement in Wi-Fi energy harvesting through innovative antenna design, primarily by enhancing return loss and gain via feed network modifications.

2 Antenna Design and Methodology

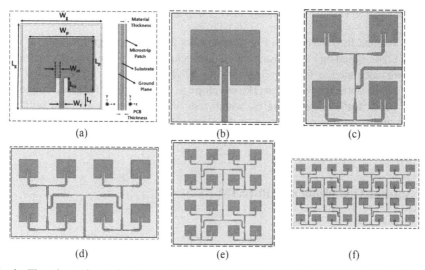

Fig. 1. The microstrip patch antenna, which consists of (a) geometrical structure, (b) single patch, (c) 2 × 2 array, (d) 2 × 4 array, (e) 4 × 4 array, and (f) 4 × 8 array.

The microstrip patch antenna's structure as shown in Fig. 1(a) includes a radiating patch, PCB substrate, and ground plane [7]. An inset feed and feeder with specified dimensions are used. We simulated various configurations, including single antennas and arrays (2 × 2, 2 × 4, 4 × 4, 4 × 8), using CST tools, recording multiple parameters to analyze gain and return loss, as indicated by Fig. 1(b) to Fig. 1(f). This dataset informs subsequent Bi-LSTM optimization. The choice of FR4 substrate (permittivity: 4.3) was based on design simplicity and suitability for Wi-Fi energy harvesting.

$$P_r = \frac{P_t G_t G_r \lambda^2}{(4\pi R)^2} \tag{1}$$

Efficiency in energy harvesting antenna design is paramount and often referenced using Eq. (1) where P_r represents received power and G_r denotes the receiving antenna's gain. Higher gain signifies greater efficiency. Additionally, bandwidth is crucial, especially for Wi-Fi energy harvesting applications covering the 2.40 GHz to 2.48 GHz frequency range.

2.1 Feed Network Optimization

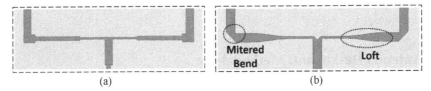

(a)   (b)

Fig. 2. Patch feeder network, which consists of (a) conventional and (b) modified feeders with 90-degree bends and quarter-wave transformers in strip lines.

The antenna feed network in Fig. 2 comprises three microstrip lines (50 Ω, 70 Ω, and 100 Ω) designed on FR4 substrate with the thickness of 1.6 mm. The microstrip line widths are calculated as follows: 3.09 mm for 50 Ω, 1.61 mm for 70 Ω, and 0.72 mm for 100 Ω. To optimize feed network, we propose two methods. First, we implement mitered bends for 90-degree angles, which reduce capacitance associated with such bends, minimizing power reflection to the source. Second, we use a loft to smoothly transition between the microstrip line quarter-wave transformers (50 Ω to 100 Ω), further reducing reflections.

2.2 Antenna Optimization by Bi-LSTM

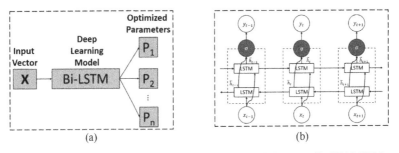

(a)   (b)

Fig. 3. The optimization by Bi-LSTM (a) process and (b) one cell of Bi-LSTM.

Figure 3 illustrates the optimization process of the patch antenna using Bi-LSTM, a machine learning model known for its ability to effectively capture sequential dependencies [8]. Bi-LSTM operates on the principle of recurrent neural networks and can handle various input vectors, including parameters such as return loss and gain. Upon training, this model provides optimized return loss and gain values for each prototype, as presented in Table 1.

Table 1. Comparison of Optimal Simulation Results for Different Patch Antennas by Bi-LSTM

Parameters	Single		2×2		2×4		4×4		4×8	
	Feeder (Normal/N Vs. Modification/M)									
	N	M	N	M	N	M	N	M	N	M
Dimension (mm)	60×60		140×140		240×140		240×260		480×260	
S11 (dB)	−20.3		−30.9	−43.6	−31.3	−38.1	−23.7	−27.9	−24.5	−25.9
Gain (dBi)	5.51		10.6	11.7	12.4	14.4	15.5	17.3	19.5	20.6

2.3 Antenna Array Simulation

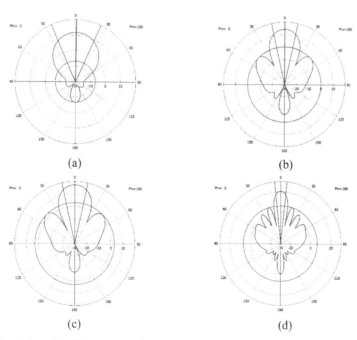

(a)   (b)

(c)   (d)

Fig. 4. Simulation of radiation pattern for (a) 2×2 array, (b) 2×4 array, (c) 4×4 array, and (d) 4×8 array.

In our antenna array design, we opted for four distinct sizes: 2 × 2, 2 × 4, 4 × 4, and 4 × 8, enabling a comprehensive comparison of the optimized antenna and feed network configurations. Figure 4 showcases the radiation patterns for each antenna array. Following the feed network modifications, we achieved gains of 11.7, 14.4, 17.3, and 20.6 dBi, respectively, highlighting the substantial performance enhancements across these array sizes. Meanwhile, the optimized return loss achieves −20.3, −43.6, −38.1, −27.9, −25.9 dB for single, 2 × 2 array, 2 × 4 array, 2 × 4 array, 4 × 4 array, 4 × 8 array, respectively (indicated in Fig. 5(a)).

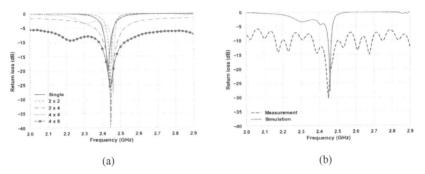

(a)   (b)

Fig. 5. Return loss comparison between (a) simulation result of each antenna array type (b) measurement and simulation of 4 × 4 antenna array.

3 Performance Measurement and Discussion

(a)   (b)

Fig. 6. (a) Fabricated 4 × 4 antenna array and (b) measurement setup.

To validate the performance of our proposed antenna design, we fabricated a 4 × 4 antenna array with the dimension of 260 × 240 × 1.6 mm as illustrated in Fig. 6(a). The measurements were conducted using a FieldFox RF analyzer, and measurement setup is depicted in Fig. 6(b). The comparison between simulation and measurements results, as shown in Fig. 5(b), demonstrates excellent agreement in terms of return loss. In the simulation we achieved a return loss of -27.9 dB, closely matching the measured value of -30.2 dB. Furthermore, we assessed the antenna's power input by placing it 50 cm away from a Wi-Fi router with a monopole antenna, confirming its functionality and efficiency at 5 dBm.

4 Conclusion

In this study, we simulated and fabricated a microstrip patch antenna array with significant feed network modifications, optimizing various antenna prototypes, including single units and arrays of 2×2, 2×4, 4×4, and 4×8 configurations. These modifications, featuring a 90-degree bend and quarter-wave transformer in the microstrip lines, yielded improvements, lowering return loss and enhancing gain compared to traditional feed network designs. We validated our technique's effectiveness with a 4×4 array on an FR4 substrate, achieving -30.2 dB return loss and impressive 17.3 dBi gain. Future research will focus on designing and integrating rectifiers and sensor tags, enabling our Wi-Fi energy harvesting system's use in IoT, wearables, and sensor networks, showcasing its versatile potential.

Acknowledgements. This work was supported by the National Research Foundation of Korea (NRF) grant funded by the Korea government (MSIT) (NRF- 2019R1A2C1089139 and NRF-2019PK1A3A1A05088484).

References

1. Riba, J.R., Moreno-Eguilaz, M., Bogarra, S.: Energy harvesting methods for transmission lines: a comprehensive review. Appl. Sci. **12**(21), 10699 (2022)
2. Pinto, D., et al.: Design and performance evaluation of a Wi-Fi energy harvester for energizing low power devices. In: 2021 IEEE Region 10 Symposium (TENSYMP), 23 August 2021, pp. 1–8. IEEE (2021)
3. Ijemaru, G.K., Ang, K.L., Seng, J.K.: Wireless power transfer and energy harvesting in distributed sensor networks: Survey, opportunities, and challenges. Int. J. Distrib. Sens. Netw. **18**(3), 15501477211067740 (2022)
4. Mujawar, M., Hossen, S.: THz microstrip patch antenna for wearable applications. In: Malik, P.K., Shastry, P.N. (eds.) Internet of Things Enabled Antennas for Biomedical Devices and Systems. STEEE, pp. 173–188. Springer, Singapore (2023). https://doi.org/10.1007/978-981-99-0212-5_14
5. Dhar, A., Pattanayak, P., Kumar, A., Gurjar, D.S., Kumar, B.: Design of a hexagonal slot rectenna for RF energy harvesting in Wi-Fi/WLAN applications. Int. J. RF Microwave Comput. Aided Eng. **32**(12), e23512 (2022)
6. Sarkar, S., et al.: Wi-Fi energy harvester with differential matching analysis for low-power applications. In: 2023 First International Conference on Microwave, Antenna and Communication (MAC), Prayagraj, India, pp. 1–6 (2023). https://doi.org/10.1109/MAC58191.2023.10177093
7. Huang, Y., Boyle, K.: Antennas from Theory to Practice. Wiley (2008)
8. Nguyen, T.H., Phan, Q.B.: Hourly day ahead wind speed forecasting based on a hybrid model of EEMD, CNN-Bi-LSTM embedded with GA optimization. Energy Rep. **8**, 53–60 (2022)

A Resource Efficient Encoding Algorithm for Underwater Wireless Optical Communication Link

Maaz Salman and Wan-Young Chung[✉]

Department of Artificial Intelligence Convergence, Pukyong National University,
Busan 48513, Republic of Korea
chung.wanyoung@gmail.com, wychung@pknu.ac.kr
http://aiotlab.pknu.ac.kr/

Abstract. The underwater communication module demands a substantial amount of power to execute its myriad functions, depleting the energy source at a rapid pace. This research work introduces a resource-optimized encoding algorithm for multi-hop communication, denoted as "resource-efficient communication", which strategically employs an optimal pulse signals for encoding sensor data generated by the underwater node. This significantly mitigates bandwidth usage during transmission, enhances payload security, consequently resulting in reduced power consumption for energy-sensitive sensor nodes. The efficacy of the resource-efficient communication algorithm is assessed by inputting various sensor data over a specific time interval. The evaluation results demonstrate a promising outcome, with a 100% run-time achievement when the sensor data exhibited gradual changes, while it still achieved a commendable 75% run-time in the case of non-deterministic variations in sensor data. The proposed algorithm accomplishes a transmission time of 100 s for steady sensor values and 127 s for fluctuating ones, using a packet size of 10,000 bytes. In contrast, the OOK modulation method requires 160 s for the same task. These results emphasize a significant enhancement in resource utilization efficiency provided by the proposed algorithm compared to conventional communication methods.

Keywords: Resource efficient Encoding · Underwater communication · Multi-hop

1 Introduction

Recent technological breakthroughs in the realm of underwater wireless communication networks (UWCN) have given birth to the emergence of the Internet of Underwater Things (IoUT) [1]. The majority of underwater communication devices rely on battery power, and replacing or recharging these batteries is a

This work was supported by the National Research Foundation of Korea (NRF) Grant funded by the Korea government (MIST) NRF-2019R1A2C1089139.

challenging and costly endeavor due to the harsh underwater environment. Consequently, when designing underwater wireless communication (UWC) systems, it is imperative to factor in the power consumption of underwater nodes to ensure that the network's operational lifespan can be prolonged to a practical duration [2]. Underwater wireless optical communication (UWOC) can be a viable alternative for power efficient communication [3,4]. Compared to non-return-to-zero on-off keying (NRZ-OOK) modulation, the return-to-zero OOK (RZ-OOK) modulation scheme offers the potential for greater energy efficiency in underwater optical communication (UOC). Additionally, the pulse position modulation (PPM) scheme can achieve even more significant power savings in UOC when compared to OOK. However, it's worth noting that PPM may result in lower bandwidth utilization and necessitates more complex hardware [5]. In a recent work by [6], the authors conducted a comprehensive survey on the advancements in UOC, addressing its challenges and future prospects from a layer-by-layer perspective. Their research delves into various energy-efficient routing techniques and energy harvesting methods related to UOC. This study introduces a resource-efficient encoding algorithm designed for transmitting data among multiple nodes in the field of Underwater Optical Communication (UWOC). The proposed algorithm is characterized by its requirement for less intricate hardware and a reduced pulse count for representing sensor values. Consequently, this leads to decreased power consumption and improved overall efficiency. The practical application of the proposed research is depicted in Fig. 1.

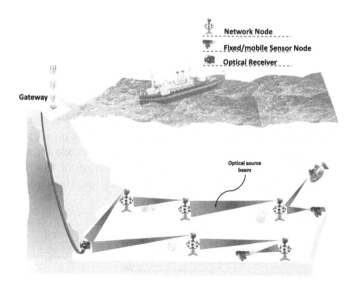

Fig. 1. Practical scenario of muti-hop UOC.

1.1 Secure Resource Efficient Encoding Algorithm

En-Coding. The algorithm commences by retrieving task-specific sensor data. Subsequently, it employs conventional communication, utilizing On-Off Keying (OOK) modulation with 8 bits per packet, for transmitting these sensor readings. Following this initial phase, the algorithm seamlessly transitions to the Channel Optimizer (CoP). The CoP, in turn, assesses the incoming sensor data by comparing it to the previous data. If this difference falls within the range of -4 to 4, the CoP activates the "Resource Efficient Communication" (REC) module within its framework. However, if the difference exceeds this threshold, the CoP continues using conventional communication (ConC) to ensure reliable data transmission.

REC. Within the Resource Efficient Communication (REC) module, the algorithm starts by determining whether the observed difference is positive or negative. A positive difference prompts REC to transmit high pulses, while a negative difference triggers the emission of low pulses. REC employs a sophisticated approach involving the transmission of four distinct combinations of dual pulses, each tailored to correspond to specific values of the observed difference. For example, when the observed difference is either $+3$ or -3, the REC module executes a transmission sequence characterized by the emission of a high pulse immediately followed by a low pulse. The REC utilizes a specific pulse pattern to convey the difference seed. It transmits two high pulses of 5 V to represent a difference seed of $+1$ or -1. Likewise, it sends two low pulses of 0 V to indicate a difference seed of $+2$ or -2. For a difference seed of $+4$ or -4, it uses a sequence of a low pulse followed by a high pulse. These pulse combinations exclusively convey the difference between the current sensor reading and the previous reading, adding an extra layer of security while also minimizing the number of pulses needed to represent the sensor data. Consequently, REC contributes to energy savings within the system and reduces transmission time. The visual representation of these intricate dual pulse combinations is thoughtfully depicted in Fig. 2. The data transmission is intended to occur over a distance of 16 m, with a total of 5 hops involved in the communication process. The pseudo code for the proposed algorithm can be found in Algorithm 1. A representation of this transmission sequence is provided in Fig. 3. In this visual representation, the white box clearly

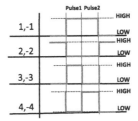

Fig. 2. Depiction of values by using combination of pulses.

signifies the activation of the REC module, whereas the dotted box distinctly marks the utilization of ConC. This innovative technique provides a significant advantage by halving the packet size, achieved through the use of four pulses, each effectively representing one byte of data. This starkly contrasts with conventional communication methods, which demand eight pulses to transmit the same amount of information. This reduction in packet size not only optimizes resource utilization but also streamlines the transmission process, showcasing the algorithm's efficiency and resource-conscious approach.

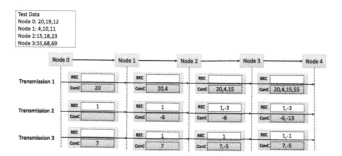

Fig. 3. Sample of transmission sequence.

De-coding. In the receiver module, the process commences with the implementation of a de-mapping algorithm. Initially, this algorithm identifies the sensor data using conventional communication demodulation techniques. Simultaneously, it monitors the communication mode employed during transmission by the Channel Optimizer (CoP) algorithm, distinguishing between ConC and REC modes. If the communication mode is ConC, the algorithm continues the reception process seamlessly by relying on conventional demodulation techniques. However, when the transmission mode is REC, the algorithm activates the REC demodulation technique. This REC demodulation unfolds through a series of meticulous steps. Initially, it determines the integer sign of the difference seed, which can be either $+1$ or -1. Subsequently, it consults a lookup table to ascertain the specific value of the difference seed. In cases with a positive integer sign and a demodulated 2-pulse value equal to 1, the algorithm proceeds to multiply $+1$ by 4. It then subtracts this product from the sensor value obtained through conventional communication. The resulting value represents the actual payload or sensor data. For comprehensive reference, the complete lookup table for other decoding processes is thoughtfully presented in Table 1. To facilitate a comprehensive understanding of the algorithm's inner workings, the pseudo-code is meticulously delineated and detailed in Algorithm 2.

Algorithm 1. Channel Optimizer, CoP

Require: Update SensorData
Ensure:
  NewSensorData ← Update SensorData
  Diff = NewSensorData − Update SensorData
  if $Diff >= -4$ **AND** $Diff <= 4$ **then**
    if $Diff > 0$ **then**
      $FisrtPulse ← HIGH$
      $SecondPulse ← LOW$
    else
      $FirstPulse ← LOW$
      $SecondPulse ← HIGH$
    end if
    if $Diff == 1$ **OR** $Diff == -1$ **then**
      $FisrtPulse ← HIGH$
      $SecondPulse ← HIGH$ **else**
      if $Diff == 2$ **OR** $Diff == -2$ **then**
        $FisrtPulse ← LOW$
        $SecondPulse ← LOW$ **else**
        if $Diff == 3$ **OR** $Diff == -3$ **then**
          $FisrtPulse ← HIGH$
          $SecondPulse ← LOW$ **else**
          if $Diff == 4$ **OR** $Diff == -4$ **then**
            $FisrtPulse ← LOW$
            $SecondPulse ← HIGH$
          end if
        else
          Call **OOK_func**
        end if
      end if
    end if
  end if

Table 1. De-mapping Look-up Table

2-bits	Integer Sign	Sensor Value
1	+1	Sensor Data-4
1	−1	Sensor Data+4
17	+1	Sensor Data-1
17	−1	Sensor Data+1
16	+1	Sensor Data-3
16	−1	Sensor Data+3
0	+1	Sensor Data-2
0	−1	Sensor Data+2

Algorithm 2. De-Coding

Require: Update IntegerSign, LastSenseData(OOK)
Ensure:
  2bits ← Update 2PulsesData
  if *2bits* == 1 **AND** *IntegerSign* == +*ive* **then**
    SensorData ← *LastSenseData* − *IntegerSign* × 4
  else if *2bits* == 1 **AND** *IntegerSign* == −*ive* **then**
    SensorData ← *LastSenseData* − *IntegerSign* × 4
  else if *2bits* == 17 **AND** *IntegerSign* == +*ive* **then**
    SensorData ← *LastSenseData* − *IntegerSign* × 1
  else if *2bits* == 17 **AND** *IntegerSign* == −*ive* **then**
    SensorData ← *LastSenseData* − *IntegerSign* × 1
  else if *2bits* == 16 **AND** *IntegerSign* == +*ive* **then**
    SensorData ← *LastSenseData* − *IntegerSign* × 3
  else if *2bits* == 16 **AND** *IntegerSign* == −*ive* **then**
    SensorData ← *LastSenseData* − *IntegerSign* × 3
  else if *2bits* == 0 **AND** *IntegerSign* == +*ive* **then**
    SensorData ← *LastSenseData* − *IntegerSign* × 2
  else if *2bits* == 0 **AND** *IntegerSign* == −*ive* **then**
    SensorData ← *LastSenseData* − *IntegerSign* × 2

  end if

The flowchart encompassing all the crucial processes, including encoding, decoding, and the step-by-step execution of the REC algorithm, is visually depicted in Fig. 4.

Results and Discussions. The evaluation of the Channel Optimizer (CoP) involved inputting a variety of sensor data, with a specific emphasis on measuring the duration during which these two modules, REC, and ConC, operate within a defined time interval. It is worth noting that a longer duration for which the REC module operates corresponds to a reduced consumption of bandwidth resources by the system. These empirical findings are graphically presented in Fig. 5, which offers a detailed breakdown of the run-time percentages for various sensors under two distinct scenarios: one where the data exhibits consistent variations (St) and another where it fluctuates significantly (Fl). Specifically, the depicted run-time percentages pertain to temperature (T), humidity (H), pressure (Pre), magnetometer (Heading), and proximity (P) sensors.

These percentages are derived from the analysis of 10,000 sensor values processed by the algorithm, with each set of calculations constituting a session that is subsequently repeated in sequence. Analyzing the run-time percentages for all sensors reveals interesting trends. When sensor data shows consistent variations, REC operates at 100% of the time. This high consistency signifies significant bandwidth resource conservation during these transmissions. However, for magnetometer sensor data, which exhibits slower data variation in the form of heading measurements, the pattern is slightly different. In this case,

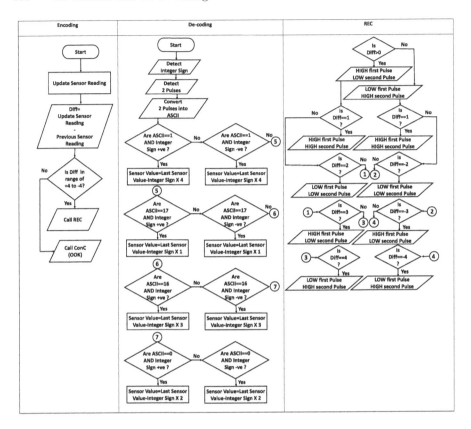

Fig. 4. Flowchart of the encoding, decoding, and the REC algorithm.

REC runs for 80% of the transmission time, with conventional communication (OOK) accounting for the remaining 20%. The evaluation also extends to situations where sensor data undergoes rapid and unpredictable fluctuations over short time intervals. In such dynamic conditions, REC and ConC exhibit varying run-time percentages. For temperature, humidity, pressure, and proximity data, the REC and ConC percentages are (60%, 40%), (60%, 40%), (50%, 50%), and (20%, 80%), respectively. In the case of magnetometer data, the percentages are (80%, 20%). These results demonstrate promising adaptability within the system, even when sensor data displays unpredictable and random fluctuations. In summary, these findings collectively offer compelling evidence of REC's ability to achieve significant bandwidth savings during the transmission process, reaffirming its effectiveness as a resource-efficient communication solution. Additionally, the transmission times for a 10,000-byte sensor payload, specifically for temperature and humidity data, are compared between the proposed algorithm and OOK modulation in Fig. 6. Specifically, for temperature data, the proposed algorithm takes 100 s to transmit a steady packet and 127 s for a fluctuating one, whereas OOK requires 160 s. Similarly, for humidity data, the proposed algorithm takes

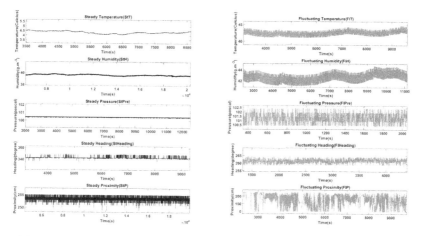

(a) Sensor data with steady and fluctuating variance

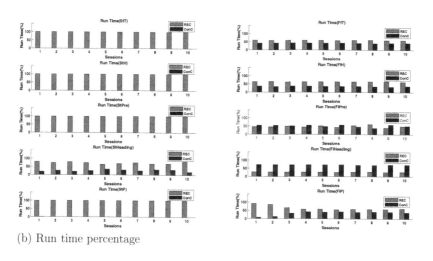

(b) Run time percentage

Fig. 5. Run time percentage of the algorithm with different sensor data.

100 s for a steady packet and 124 s for a fluctuating one, in contrast to the 160 s needed by OOK. In summary, these outcomes collectively furnish compelling evidence of REC's ability to achieve significant bandwidth savings during the transmission process, underscoring its effectiveness as a resource-efficient communication solution. Furthermore, the system is tested in artificially induced real underwater environment. The system's performance was assessed within a water tank, involving a 4-hop communication setup where each hop had the capability to sense various environmental parameters like temperature and pressure. A blue LED (with a wavelength of 470 λ) was employed as the optical source, delivering a power of 102 mW, a luminous intensity of 2500 mcd, and a flux of 1.5 lm. To detect the optical signal at the receiver, a SiPIN photodiode was utilized,

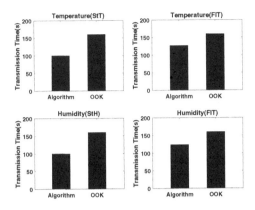

Fig. 6. Transmission time of temperature and humidity data.

featuring a photosensitive area of 10 × 10 mm^2 and a peak sensitivity wavelength of 960 nm. To replicate underwater environmental conditions, including factors like absorption and scattering, water pumps with a water displacement rate of 5 liters per minute were installed at both ends of the water tank. Additionally, small air bubbles were generated using an aerating jet equipped with two outlets, which released air at an airflow rate of 2.5 liters per minute. The ambient light levels in the surroundings were measured to range between 100–150 lux throughout the experiment. To modify the water's salinity and conductivity, and introduce fine suspended particles, the turbidity was adjusted from 0.01 to 50 NTU by introducing a solution of zinc oxide powder into the water. The experimental parameters of the underwater channel is tabulated in Table 2. The evaluation results of this experiment are presented in Fig. 7. The figure illustrates that at a communication link range of 14 m, the system attained a Packet Success Rate (PSR) of 97%, 97%, and 80% at turbidity levels of 0.09, 12, and 45 NTU, respectively. Similarly, at a link range of 16 m, the system achieved a PSR of 95%, 88%, and 73% when operating in turbid water with turbidity levels of 0.09, 12, and 45 NTU, respectively. The performance comparison of REC with other technologies, including OOK, PPM, and DPIM, is depicted in Fig. 8. The required transmission power and bandwidth values for OOK, PPM, and DPIM were sourced from a previous study [7]. The figure illustrates that REC necessitated a transmission power of 0.0005 W and 0.00067 W when the sensor readings were changing gradually (REC-St) and fluctuating rapidly (REC-Fl), respectively. This is in contrast to the required transmission power of 0.001 W, 10.47×10^{-5} watts, and 2×10^{-4} watts for OOK, PPM, and DPIM, respectively. The required bandwidth for REC-St and REC-Fl is 0.5 and 0.75 times less, respectively, compared to OOK, with the bandwidth requirement of OOK normalized to 1. In contrast, other technologies, namely PPM and DPIM, require 6.1 and 3.9 times more bandwidth than OOK, respectively. This illustrates the more efficient utilization of bandwidth by REC in both gradual and rapid sensor data variations.

Table 2. Experimental parameters used in system evaluation

Channel	Aerating Jets	Airflow rate (L/min)	2.5
		No. of outlets	2
	Water Pump (Displacement Rate (L/min))		5
	Lighting Intensity (surroundings) (lux)		100–150
	Turbidity (NTU)		0.01 to 50

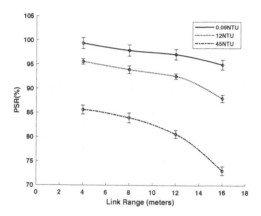

Fig. 7. Packet success rate performance in varying turbid water.

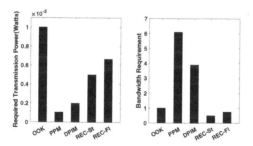

Fig. 8. Required transmission power and bandwidth comparison.

1.2 Future Implications and Applications

The careful design, coding, and evaluation of the algorithm were driven by the growing importance of energy-efficient Underwater Optical Communication (UWOC) modules in diverse settings. The algorithm is purpose-built for small mobile platform applications that demand enhanced mobility, compact design, and minimal energy consumption. It is especially well-suited for stationary network or sensor nodes used in scenarios where the monitored parameters exhibit gradual changes over time. These applications span a wide spectrum, from underwater sensor networks in aquatic research to remote monitoring systems in environmental science. The algorithm's adaptability caters to both high-mobility requirements and the stability needed for continuous data collection and transmission, making it a versatile solution for various contexts. Beyond diving communication devices and offshore fish farms, these modules can be a potential alternative in underwater environmental monitoring, oceanographic research, and marine exploration. Their non-intrusive nature ensures minimal disruption to aquatic ecosystems, and their compact size and energy efficiency make them suitable for extended deployments. Furthermore, the algorithm's optimization holds promise for future underwater communication technologies, fostering advancements in aquatic research, aquaculture management, and environmental conservation efforts.

Conclusion. In our study, we introduce a resource-efficient encoding algorithm meticulously crafted to strategically optimize resource utilization while simultaneously bolstering data frame security. The evaluation of this algorithm yields highly promising results, showcasing a substantial reduction in the number of bits required per data frame compared to conventional techniques when representing a sensor value intended for transmission.

Acknowledgements. This work was supported by the National Research Foundation of Korea (NRF) Grant funded by the Korea government (MIST) NRF-2019R1A2C1089139.

References

1. Islam, K.Y., et al.: A survey on energy efficiency in underwater wireless communications. J. Netw. Comput. Appl. **198**, 103295 (2022)
2. Akyildiz, I.F., et al.: Wireless sensor networks: a survey. Comput. Netw. **38**(4), 393–422 (2002)
3. Hanson, F., Radic, S.: High bandwidth underwater optical communication. Appl. Opt. **47**(2), 277–283 (2008)
4. Salman, M., Bolboli, J., Chung, W.-Y.: Experimental demonstration and evaluation of BCH-coded UWOC link for power-efficient underwater sensor nodes. IEEE Access **10**, 72 211-72 226 (2022)
5. Zeng, Z., et al.: A survey of underwater optical wireless communications. IEEE Commun. Surv. Tutor. **19**(1), 204–238 (2016)
6. Saeed, N., et al.: Underwater optical wireless communications, networking, and localization: a survey. Ad Hoc Netw. **94**, 101935 (2019)
7. Mahdiraji, G.A., Zahedi, E.: Comparison of selected digital modulation schemes (OOK, PPM and DPIM) for wireless optical communications. In: 2006 4th Student Conference on Research and Development, Shah Alam, Malaysia, pp. 5–10 (2006). https://doi.org/10.1109/SCORED.2006.4339297

Containerized Wearable Edge AI Inference Framework in Mobile Health Systems

Lionel Nkenyereye[1] , Boon Giin Lee[2] , and Wan-Young Chung[3]([⊠])

[1] BK21, Education and Research Group of AI Convergence,
Pukyong National University, Busan, Republic of Korea
[2] Nottingham Ningbo China Beacons of Excellence Research and Innovation
Institute, School of Computer Science, University of Nottingham Ningbo China,
Ningbo, Zhejiang, China
[3] Department of Electronic Engineering, Pukyong National University,
Busan, Republic of Korea
wychung@pknu.ac.kr

Abstract. The proliferation of wearable devices and personal smartphones has promoted smart mobile health (MH) technologies. The MH applications and services are extremely responsive to computation latency. Edge computing is a distinguished form of cloud computing that keeps data, applications, and computing power away from a centralized cloud network or data center. In this work, we design a containerized wearable edge AI inference framework. The cloud computing layer includes two cloud-based infrastructures: The Docker hub repository and the storage as service hosted by Amazon web service. The Docker containerized wearable inference is implemented after training a Deep Learning model on open data set from wearable sensors. At the edge layer, the Docker container enables virtual computing resources instantiated to process data collected locally closer to EC infrastructures. It is made up of a number of Docker container instances. The containerized edge inference provides data analysis framework (DAF) targeted to fulfill prerequisites on latency, and the availability of wearable-based edge applications such as MH applications.

Keywords: Wearable sensors · AI inference · Edge intelligence · Activity recognition · Data processing · Deep Learning · Docker Container

1 Introduction

The proliferation of wearable devices and personal smartphones has promoted smart mobile health (MH) technologies. The MH system is the foundation stone of the Healthcare 3.0 [1]. The MH system employs wearable medical sensors, mobile computing, wireless communications, and networking technologies to regularly transmit many sensed data to MH processing platforms. These platforms are commonly located in the cloud to provide anywhere and anytime health-care services, thus improving well-being and quality of life. Furthermore, artificial intelligence (AI) and machine learning (ML) have added smartness to MH applications to analyze MH data efficiently.

The MH applications and services are extremely responsive to computation latency. For instance, the round trip time (RTT) to access MH cloud-based services is very long; thus, it increases latency. This limits the deployment of many time-critical MH applications. Thus, the deployment of computing infrastructures near where data are generated solves the issue of RTT. This computing infrastructure is designated edge computing (EC). EC is a distinguished form of cloud computing that keeps data, applications, and computing power away from a centralized cloud network or data center.

In this work, we design a containerized wearable AI edge inference framework. It is made up of a number of Docker container services instantiated close to end-user wearable sensors. The containerized edge inference provides data analysis framework (DAF) targeted to fulfill prerequisites on latency, and the availability of wearable-based edge applications such as MH applications.

2   Containerized Wearable Edge AI Inference Framework

The MH system exhibits diversity in the devices it incorporates, spanning from wearables (For instance, sensors positioned on the individual's chest, right wrist, left ankle, and cardiac sensors.) to video surveillance cameras and smartphones. Wearable devices are responsible for data collection. Within these devices, a subset is situated on the chest and is capable of providing two-lead electrocardiogram (ECG) measurements, which have the potential for basic heart monitoring. Each sensor can collect patient data, perform simple data processing, and send the information to the gateway (e.g., a smart mobile phone). Just after data collection at the gateway, the mobile phone will send the data to the MH data management platform for storage and further analysis. Healthcare professionals (e.g., doctors, and nurses) observe and read patient data in an emergency. Clinical observation, diagnosis, and medical intervention are carried out. Only healthcare professionals who have authenticated credentials access patient data.

The diagram in Fig. 1 showcases a wearable edge AI inference framework that utilizes containers. This framework streamlines two core components: the pre-trained AI model and the edge AI inference server. The pre-trained model and the inference Docker image are stored in the cloud, while Edge AI inference is a containerized service functioning on edge devices.

The cloud computing layer comprises two cloud-based infrastructures: the Docker Hub repository and Amazon Web Services' storage-as-a-service. The Docker containerized wearable inference is realized following the training of a deep learning model on an open dataset derived from wearable sensors. The MHEALTH dataset [2] encompasses data from 12 physical activities recorded from ten subjects using four different inertial sensors. These sensors capture a total of 23 distinct signal types, which we'll refer to as channels. Most of these

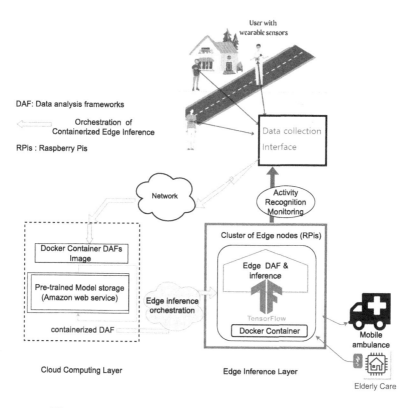

Fig. 1. Containerized Wearable edge inference framework.

channels pertain to body motion, with the exception of two that record electro-diagram signals from the chest. The participants are assigned the responsibility of carrying out 12 different activities, and the sensors on their bodies record data related to acceleration, rotational speed, and magnetic field orientation. This dataset serves as input for edge-based wearable data analytics, focusing on deep learning models for human activity recognition applications. Each log file contains 23 columns for each channel and an additional column for class, representing one of the 12 activities. On average, there are around 100,000 rows for each subject.

We employ the concept of MLOp pipeline [3]. The pipeline steps are detailed as follows: we divide the MHEALTH time series dataset into smaller chunks for classification as proposed in [4]. TensorFlow and Keras [4] are used for training neural networks. The long short-term memory (LSTM) [5] model was tested and showed better accuracy through the epochs. For the model to receive inference requests, with only Python (the programming language) wrapper available, the training process was written in Python using the Flask framework [4]. After ensuring that the model has almost 95 percent precision, it was saved on the cloud-based storage service. In our case, Amazon web service cloud object storage was used to save the pre-trained AI model. The pre-trained model is saved with HDF5. It is loaded while the pre-processing phase starts. Subsequently, the edge AI inference server was encapsulated into a container image and uploaded to our Docker repository account. The Docker repository name for our edge AI inference image is "*nkenye*1982/*mhealth_ai_service*" [6]. We named the AI edge inference server as virtual MH edge service (vMHES).

At the edge layer, the Docker container enables virtual computing resources instantiated to process data collected locally closer to EC infrastructures. Although the computing resources orchestration is isolated, associating a tailored logical edge inference instance would enhance the end-to-end (E2E) performance. The deployment edge node fulfills the role of provisioning and instantiating the wearable sensor data processing. The cluster of worker nodes consists of one server (x64) and two Raspberry Pi 3 (ARM). The server (x64) is used to determine the resource of each wearable sensor service performing data analytics on an x64 worker node. This device runs Ubuntu 18.04 and has an AMD Opteron at 2 GHz and 4 GB RAM. The Raspberry Pi 3 is used to evaluate the provisioning of resources on an ARM device (armhf) to support edge wearable data processing service. This system operates using Ubuntu 20.4 server on a Raspberry Pi with Kernel version 5.4. The hardware configuration consists of 2 GB of RAM and a quad-core Cortex-A72 processor running at 1.5 GHz. All assessments will be conducted on both x64 and armhf platforms. Different container runtimes (e.g., container runtime interface (CRI-O), Docker, and Containerd) were used [7]. To verify this, the orchestrator at the worker nodes is set up using Containerd, Docker, and CRI-O. Additionally, the average loading waiting time of the sensor data processing AI model to load TensorFlow/Keras APIs, dependencies, and Flask starting server was measured.

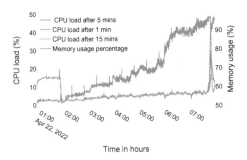

Fig. 2. The average load on the CPU and the usage of memory within the edge cluster system while processing real-time dataset logs for DAF inference requests.

3   Evaluation

The experimental test plan uses Locust [8] to capture the containerized AI requests per second, i.e., the response time. The response time refers to how

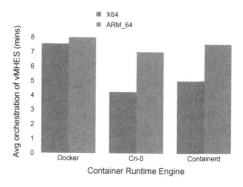

Fig. 3. Allocating resources by retrieving a deep learning image based on mHealth from the Docker registry.

long the AI model (TensorFlow 2's Keras API) orchestrated on the edge node takes to respond to a new prediction's end-user request. The capacity and performance tests are necessary to demonstrate that a containerized DAF instance can successfully process AI models with satisfactory speed when numerous users simultaneously upload MH data.

In Fig. 2, we can observe the computation resource load average within the container instance of the cluster. The CPU load of the active edge container is measured in three observations, reflecting the load over the last 1, 5, and 15 min. It becomes worse as the number of user DAF inference requests rises. The findings indicate that as the volume of data and the number of users' DAF inference requests increase, resource consumption deteriorates. Within the first few minutes, the CPU load average reaches 90% utilization, particularly on the sixth hour of experimentation. This signifies that the container instances became overloaded, leading to a high average load on the computing resources due to the increasing number of DAF inference requests.

As shown in Fig. 3, when orchestrating the scheduling of edge resources by fetching vMHES from the Docker registry, the specific duration varies, typically taking about five minutes on x64 architecture compared to approximately seven minutes on ARM architecture. Interestingly, the CRI-O [7] demonstrates a significantly shorter scheduling time on x64, being twice as fast as on ARM.

The outcomes reveal that Docker and Containerd [9] runtimes are notably less efficient in comparison to CRI-O. This is primarily due to Docker using Containerd as its underlying technology for running containers. The rationale behind CRI-O's superior performance is its lightweight nature, making it an optimal runtime for Kubernetes (K8s). This enables Kubernetes to utilize any runtime that adheres to the Open Container Initiative (OCI) standards as the container runtime.

4   Conclusion

The MH applications and services are extremely responsive to computation latency. For instance, the round trip time (RTT) to access MH cloud-based services is very long. The edge inference layer consists of the edge computing resources and is close to end-users' wearable sensors. The deployment edge node fulfills the role of provisioning and instantiating the wearable sensor data processing. The results show the CPU load average in the first minutes is 90% utilization on the sixth hour of experimentation. This implies that the container instances experienced excessive load, with a high average usage of computing resources as the number of DAF inference requests continued to rise. Furthermore, Docker and Containerd runtimes exhibit significantly lower efficiency compared to CRI-O, primarily because Docker utilizes Containerd for container execution.

Acknowledgment. This work was supported by the National Research Foundation of Korea (NRF) grant funded by the Korea government (MIST) (NRF-2019R1A2C1089139).

References

1. Yang, J., Zhou, J., Tao, G., Alrashoud, M., Mutib, K.N.A., Al-Hammadi, M.: Wearable 3.0: from smart clothing to wearable affective robot. IEEE Netw. **33**(6), 8–14 (2019). https://doi.org/10.1109/MNET.001.1900059
2. Banos, O., et al.: mHealthDroid: a novel framework for agile development of mobile health applications. In: Pecchia, L., Chen, L.L., Nugent, C., Bravo, J. (eds.) IWAAL 2014. LNCS, vol. 8868, pp. 91–98. Springer, Cham (2014). https://doi.org/10.1007/978-3-319-13105-4_14
3. Niranjan, D.R., Mohana: Jenkins pipelines: a novel approach to machine learning operations (MLOps). In: 2022 International Conference on Edge Computing and Applications (ICECAA), Tamilnadu, India, pp. 1292–1297 (2022)
4. bhimmetoglu. A tutorial for datascience for classifying human activity from body motion and vital signs recordings. https://github.com/bhimmetoglu/datasciencecom-mhealth. Accessed 25 Sept 2022
5. Barut, O., Zhou, L., Luo, Y.: Multitask LSTM model for human activity recognition and intensity estimation using wearable sensor data. IEEE Internet Things J. **7**(9), 8760–8768 (2020)
6. DockerHub. Docker image for mobile Health pre-trained models and inference server. https://hub.docker.com/repositories/nkenye1982. Accessed 25 Apr 2023
7. Vaño, R., Lacalle, I., Sowiński, P., S-Julián, R., Palau, C.E.: Cloud-native workload orchestration at the edge: a deployment review and future directions. Sensors **23**, 2215 (2023). https://doi.org/10.3390/s23042215
8. Locust. An open source load testing tool. https://locust.io/. Accessed 25 July 2023
9. Cloud Native Computing Foundation. An Industry-standard container runtime with an emphasis on simplicity, robusteness and portability. https://containerd.io/. Accessed 25 July 2023

Social Computing and Interactive Elements

Leveraging Explainable AI to Analyze Researchers' Aspect-Based Sentiment About ChatGPT

Shilpa Lakhanpal[1(✉)], Ajay Gupta[2], and Rajeev Agrawal[3]

[1] California State University Fullerton, Fullerton, CA 92831, USA
shlakhanpal@fullerton.edu
[2] Western Michigan University, Kalamazoo, MI 49008, USA
[3] Office of the Deputy Assistant Secretary of the Army (Procurement), Crystal City, VA, USA

Abstract. The groundbreaking invention of ChatGPT has triggered enormous discussion among users across all fields and domains. Among celebration around its various advantages, questions have been raised with regards to its correctness and ethics of its use. Efforts are already underway towards capturing user sentiments around it. But it begs the question as to how the research community is analyzing ChatGPT with regards to various aspects of its usage. It is this sentiment of the researchers that we analyze in our work. Since Aspect-Based Sentiment Analysis has usually only been applied on a few datasets, it gives limited success and that too only on short text data. We propose a methodology that uses Explainable AI to facilitate such analysis on research data. Our technique presents valuable insights into extending the state of the art of Aspect-Based Sentiment Analysis on newer datasets, where such analysis is not hampered by the length of the text data.

Keywords: ChatGPT · ABSA · Explainable AI

1 Introduction

ChatGPT [1] is a generative artificial intelligence (AI) chatbot which has revolutionized the landscape of Natural Language Processing (NLP). Up until the year 2022, NLP had made strides in various tasks such as text classification, question answering, summarization, sentiment analysis, named entity recognition, etc. But ChatGPT is the first of its kind to be released into the public domain and perform a variety of these NLP tasks with effortless ease and be accessible to an ever-increasing user base. It is the fastest-growing consumer application in history, reaching a 100 million monthly active users in January 2023 within just 2 months of its release [2]. It is accessible as a question answering interface, where it provides intelligent, coherent and pertinent human-like answers to questions posed by the users, surpassing any other AI chatbot of its generation in popularity [3].

Ever since its release on November 30, 2022, ChatGPT has dominated public discourse with people taking to the social media and writing about their experience with ChatGPT. There are millions of tweets on Twitter. The public discourse ranges from sentiment of marvel to skepticism to apprehension as to the various levels of usefulness, applicability, and possibilities that ChatGPT is poised to bring in the near future [4–6]. The opinion paper [6] assimilates 43 contributions from experts in the fields of computer science, marketing, information systems, education, policy, hospitality and tourism, management, publishing, and nursing. These experts appreciate and recognize ChatGPT's potential to offer significant advantages to these fields benefitting their overall productivity. They also consider its limitations, possible disruptions it may cause to accepted practices, threats it may pose to privacy and security, and the consequences of biases, misuse, and misinformation arising from its use. Naturally, the euphoria surrounding ChatGPT has spurred the research community to investigate ChatGPT from various points of view including but not limited to how people across various sections of society view it and perceive its utility as, what ethical questions it raises, how the technology behind it can be improved etc. In fact, it is this sentiment about ChatGPT conveyed by the researchers in their papers that we investigate in our work. We look at how researchers across the world perceive ChatGPT. In doing so we investigate whether and how well some of the widely used sentiment analysis language models are able to capture the research community's exploration thus far. Particularly, since the focus of their research papers is ChatGPT, are the sentiment analysis language models able to capture their sentiment towards ChatGPT with aspect to various issues, fields and domains?

2   Sentiment Analysis in Research Articles

Sentiment analysis is an active field in NLP, where the goal is to identify the sentiment expressed in the text and classify it as a fixed polarity value such as positive, negative, or neutral. Sentiment analysis is used to extract sentiment from a wide arena of user bases such as social media, networking posts, customer reviews about products, state of current affairs, restaurants etc. Such analysis is important as for example, businesses can gain insight into customers' opinions about their products and accordingly adapt their marketing strategies. While majority of efforts are focused on the classification and analysis of sentiment in customer reviews [7], and social media posts [8], lesser effort is dedicated to extracting and interpreting sentiment from research articles. With the success and now prolific use of transformers [9] and transformer-based models in various NLP tasks [10]; the task of sentiment analysis is also benefitted by these models. Even so, transformer-based models such as Bidirectional Encoder Representations from Transformers (BERT) [11] are still largely being used for analyzing social media content [12] or product reviews [13]. The use of research article data figures heavily in the field of biomedical research [14], where various tasks of NLP (not including sentiment analysis) are being performed by transformer-based models. Sentiment analysis lags behind in the field of biomedical research, because of lack of domain-specific sentiment vocabulary [15]. To the best of our knowledge, text data from Computer Science research articles has not been used for sentiment analysis, and in our work, we use such data.

3   Aspect-Based Sentiment Analysis (ABSA)

Largely sentiment analysis is performed either at the document-level or the sentence-level, where a single polarity value of either positive, negative or neutral is identified as the entire sentiment of the document or the sentence respectively. A single polarity value may not be an accurate sentiment representation of the entire document or the sentence as fine-grained sentiments may be need to be extracted towards several aspects in the document or sentence [16]. It is this Aspect-Based Sentiment Analysis (ABSA) that is gaining traction in the recent years. To understand what ABSA entails, let us look at an example laptop review: *"I feel the latest laptop from Mac is really good overall. The resolution is amazing. The laptop is sleek, and light, making it easy to carry around. However, it's a bit pricey. Honestly, for this price, I expected a better battery life."* At the aspect-level, we get information about the aspects such as *resolution, design, price* and *battery*. While the review is positive with respect to the former two aspects, it is negative towards the latter two. Sometimes, the aspects are not explicitly mentioned, and can only be inferred. Such as the aspect *design* can be inferred by *"sleek and light"*.

ABSA research involves sentiment elements such as aspect category, aspect term, opinion term and sentiment polarity [16]. Aspect categories need to be pre-defined such as food, service for the restaurant domain. Aspect term denotes the actual item mentioned or referred to, in the text such as ice-cream or pizza belonging to a food category. Opinion term is term used to convey feeling or sentiment about the aspect term. And sentiment polarity denotes whether the opinion is positive or negative, etc. In the restaurant domain, for an example review: *"The ice-cream is heavenly"*, the *aspect category* is *food, aspect term* is *ice-cream, opinion term* is *heavenly and sentiment polarity* is *positive.*

Many advancements have been done in aspect term extraction, aspect category detection, opinion term extraction, and aspect sentiment classification or some combinations thereof [16]. But only so many datasets have been extensively used. So far, the SemEval datasets [17–19] made public as part of shared work held during the International Workshop on Semantic Evaluation, held annually from 2014 to 2016, have been the most widely used for ABSA. Despite their popularity, most sentences in the dataset only include one or more aspects with the same sentiment polarity, effectively reducing the ABSA task to sentence-level sentiment analysis [20]. Also, these SemEval datasets contain reviews from only one domain, either restaurant or laptop. In [21], SentiHood dataset based on neighborhood domain is introduced, where work has been done to identify sentiments towards aspects of one or more entities of the same domain. With the use of transformer-based models, successful work [22] has been done to jointly detect all sentiment elements from a given opinionated sentence, for one or more aspects, even when aspects may have contrasting sentiments. The problem still remains that datasets used commonly have unique data structure and ABSA tasks successful on one dataset cannot be easily translated to another dataset.

A major step required in sentiment analysis is annotating the dataset for training the classifier. This step is expensive and often impossible in ABSA tasks as they require aspect-level annotations. To address this problem, transfer learning can be applied where we can leverage knowledge learned from one domain and apply on another. This involves taking a pre-trained language model such as BERT, which has been trained on a large dataset, and fine-tuning it for downstream tasks such as sentiment analysis. Fine-tuning requires labeled dataset which is specific to the task. Fine-tuned models of BERT exist for sentiment analysis such as:

- nlptown/bert-base-multilingual-uncased-sentiment [13], which we will refer to as the nlptown model

  Fine-tuned models of BERT exist for ABSA such as:

- yangheng/deberta-v3-base-absa-v1.1 [23], which we will refer to as the yangheng model

These models can be further fine-tuned for ABSA tasks on target aspect-level labeled dataset. Which brings us back to the original problem of unavailability of labeled datasets. In the next section we describe our dataset.

4 Dataset

We collected data from arXiv [24] using the arXiv API [25]. The data comprises of metadata of research papers documenting research focused on ChatGPT, it's applications and implications. Specifically, we collected data from 868 papers submitted between December 8, 2022 and July 24, 2023 that contain the term chatgpt in either the title or the abstract or both.

The items from the metadata that we used for analysis are the titles and abstracts. For each paper, we added the title as a sentence before the abstract and refer to the resulting text as the abstract.

4.1 Challenges in Analyzing Sentiment in This Dataset

For aspect terms to be extracted, much work is done in supervised learning, which requires labeled data. But research is still lacking in unsupervised learning [26]. Our dataset is unlabeled; hence it becomes difficult to extract aspects.

Most datasets [18, 19, 21] used in ABSA tasks are from the restaurant, laptop or neighborhood domain. Majority of the reviews in these contain 1–3 sentences each. All papers in our dataset are about ChatGPT, so we can qualify our domain as the ChatGPT domain. Each abstract in our dataset contains on an average 8 sentences, with the largest containing 19. Although models using transformers can capture some long-term dependencies, they may still struggle with long documents [27], where important context or sentiment clues are spread far apart. The model's ability to maintain relevant context over extended lines of text can impact its overall performance. In long text, sentiments towards aspects might change over time, leading to aspect-level sentiment shifts. Handling these dynamic shifts is still an ongoing challenge.

5 Using Explainable AI to Analyze Aspect-Based Sentiment

In Fig. 1, we present a flowchart summarizing our methodology.

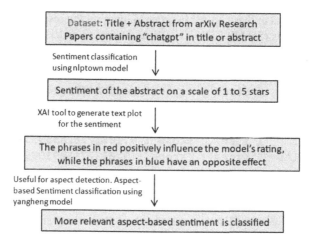

Fig. 1. Flowchart summarizing our methodology

5.1 Example 1

In Fig. 2, we present an abstract [28] from our dataset.

Why Does ChatGPT Fall Short in Providing Truthful Answers? Recent advancements in Large Language Models, such as ChatGPT, have demonstrated significant potential to impact various aspects of human life. However, ChatGPT still faces challenges in aspects like truthfulness, e.g. providing accurate and reliable outputs. Therefore, in this paper, we seek to understand why ChatGPT falls short in providing truthful answers. For this purpose, we first analyze the failures of ChatGPT in complex open-domain question answering and identifies the abilities under the failures. Specifically, we categorize ChatGPT's failures into four types: comprehension, factualness, specificity, and inference. We further pinpoint three critical abilities associated with QA failures: knowledge memorization, knowledge recall, and knowledge reasoning. Additionally, we conduct experiments centered on these abilities and propose potential approaches to enhance truthfulness. The results indicate that furnishing the model with fine-grained external knowledge, hints for knowledge recall, and guidance for reasoning can empower the model to answer questions more truthfully.

Fig. 2. An abstract [28] from our dataset

This abstract indicates that this paper investigates why ChatGPT falls short in providing truthful answers, categorizing its failures into comprehension, factualness, specificity, and inference. It proposes enhancing truthfulness by furnishing the model with fine-grained external knowledge, knowledge recall hints, and reasoning guidance. We create a sentiment analysis pipeline for the nlptown model using the transformers library from HuggingFace [10]. We use it to classify the overall sentiment of the abstract. The result is depicted in Fig. 3. nlptown model rates sentiment on a scale of 1 to 5 stars, with 1 indicating most negative, and 5 indicating most positive. The model gives this abstract a 1 star rating with 10.22% probability, a 2 star rating with 32.27% probability and a 3 star rating with 37.04% probability indicating a very negative-neutral range sentiment with a cumulative probability of 79.53%. We turn to Explainable AI (XAI) to understand how the model arrives at these scores in order to interpret the results. We use SHapley Additive exPlanations (SHAP) [29] to visualize the text to understand which words or phrases influenced the model's decision.

```
[[{'label': '3 stars', 'score': 0.37044963240623474},
  {'label': '2 stars', 'score': 0.32270216941833496},
  {'label': '4 stars', 'score': 0.17089851200580597},
  {'label': '1 star', 'score': 0.10217782855033875},
  {'label': '5 stars', 'score': 0.033771809190511703}]]
```

Fig. 3. Sentiment classification for the abstract in Fig. 2 using nlptown model

In Fig. 4, we present the SHAP text plot for the 3 stars rating, since it has the highest probability. The words and phrases highlighted in red positively influence the model's prediction towards this rating, while the words and phrases in blue have an opposite effect. The darker colors indicate a stronger influence. In Fig. 4, the text indicating that ChatGPT has demonstrated significant potential to impact human life is highlighted in blue, which means it contributes negatively towards the rating. The text indicating ChatGPT's failure to provide truthful answers or lack of truthfulness is highlighted in red in ALL instances. The text proposing approaches to enhance ChatGPT's truthfulness is also highlighted in red. The text in red pushes the model towards the 3 stars rating. **The resulting overall sentiment is neutral (3).**

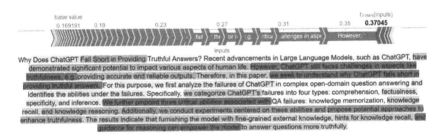

Fig. 4. SHAP text plot for 3 stars (neutral) rating for the abstract in Fig. 2

We can use Explainable AI to infer aspect terms, which is otherwise impossible for datasets such as ours. And then we can perform ABSA using the yangheng model. As a case in point, we apply the yangheng model to extract the sentiment of this abstract towards the aspect *truthfulness*. The results are depicted in Fig. 5.

```
[[{'label': 'Negative', 'score': 0.6775800585746765},
  {'label': 'Positive', 'score': 0.24405549466609955},
  {'label': 'Neutral', 'score': 0.07836440950632095}]]
```

Fig. 5. Sentiment of the abstract in Fig. 2 towards the aspect truthfulness

It labels the text conveying a negative sentiment towards the aspect truthfulness with 67.76% probability. Which is accurate since the text conveys that as it stands currently, ChatGPT lacks truthfulness. Hence, together both models can help understand the sentiment at a finer level.

5.2 Example 2

We present another abstract [30] from our dataset in Fig. 6.

A New Era of Artificial Intelligence in Education: A Multifaceted Revolution. The recent high performance of ChatGPT on several standardized academic test has thrust the topic of artificial intelligence (AI) into the mainstream conversation about the future of education. The objective of the present study is to investigate the effect of AI on education by examining its applications, advantages, and challenges. Our report focuses on the use of artificial intelligence in collaborative teacher-student learning, intelligent tutoring systems, automated assessment, and personalized learning. We also look into potential negative aspects, ethical issues, and possible future routes for AI implementation in education. Ultimately, we find that the only way forward is to accept and embrace the new technology, while implementing guardrails to prevent its abuse.

Fig. 6. Another abstract [30] from our dataset

This abstract discusses the impact of AI in education and learning, focusing on its applications, benefits, and challenges. It highlights ChatGPT's success in academic tests and explores its role in collaborative teacher-student learning, intelligent tutoring systems, automated assessment, and personalized learning. The abstract acknowledges potential negative aspects and ethical concerns, but ultimately suggests accepting and embracing AI while implementing safeguards to prevent misuse.

We use the nlptown model to classify the overall sentiment of the abstract. The result is depicted in Fig. 7.

```
[[{'label': '4 stars', 'score': 0.5352276563644409},
  {'label': '5 stars', 'score': 0.35541731119155884},
  {'label': '3 stars', 'score': 0.07598904520273209},
  {'label': '2 stars', 'score': 0.023732537403702736},
  {'label': '1 star', 'score': 0.009633398614823818}]]
```

Fig. 7. Sentiment classification for the abstract in Fig. 6 using nlptown model

It indicates a positive – very positive range sentiment with a cumulative probability of 89.06%. In Fig. 8, we present the SHAP text plot for the 4 stars (positive) rating.

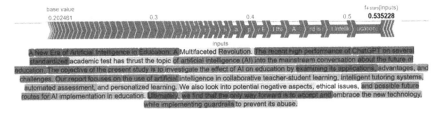

Fig. 8. SHAP text plot for 4 stars (positive) rating for the abstract in Fig. 6

The text has major portions recognizing the advantages and applications of ChatGPT in education and learning and these are highlighted in red, indicating that these portions contributed positively towards the 4 stars rating. The text about potential negative aspects is highlighted in blue. Hence **the overall rating is positive (4)**.

We find that the yangheng model **labels the sentiment of this abstract towards aspects such as education and learning as positive**, as depicted respectively in Fig. 9 and Fig. 10.

```
[[{'label': 'Positive', 'score': 0.5271556973457336},
  {'label': 'Neutral', 'score': 0.4476832151412964},
  {'label': 'Negative', 'score': 0.025161078199744225}]]
```

Fig. 9. Sentiment of the abstract in Fig. 6 towards the aspect education

```
[[{'label': 'Positive', 'score': 0.7254205346107483},
  {'label': 'Neutral', 'score': 0.2606089413166046},
  {'label': 'Negative', 'score': 0.013970516622066498}]]
```

Fig. 10. Sentiment of the abstract in Fig. 6 towards the aspect learning

6 Conclusion

We observe in Example 5.2, that the overall sentiment and the sentiments towards the aspects under consideration are positive. But in Example 5.1, the overall sentiment is neutral but the sentiment towards the aspect under consideration is negative. We have presented a methodology to analyze finer aspect-based sentiment from text in research papers. Our method also provides insights into the detection of aspects from text using the XAI framework, SHAP. In this work, we have presented a novel contribution.

References

1. OpenAI ChatGPT. https://openai.com/chatgpt. Accessed 08 Aug 2023
2. ChatGPT sets record for fastest-growing user base - analyst note. https://www.reuters.com/technology/chatgpt-sets-record-fastest-growing-user-base-analyst-note-2023-02-01. Accessed 08 Aug 2023

3. ChatGPT Grew Another 55.8% in March, Overtaking Bing and DuckDuckGo. https://www.similarweb.com/blog/insights/ai-news/chatgpt-bing-duckduckgo. Accessed 08 Aug 2023
4. Haque, M.U., Dharmadasa, I., Sworna, Z.T., et al.: "I think this is the most disruptive technology": Exploring Sentiments of ChatGPT Early Adopters using Twitter Data. arXiv preprint arXiv:2212.05856 (2022)
5. Tlili, A., Shehata, B., Adarkwah, M.A., et al.: What if the devil is my guardian angel: ChatGPT as a case study of using chatbots in education. Smart Learn. Environ. **10**, 15 (2023)
6. Dwivedi, Y.K., Kshetri, N., Hughes, L., et al.: Opinion Paper: "So what if ChatGPT wrote it?" Multidisciplinary perspectives on opportunities, challenges and implications of generative conversational AI for research, practice and policy. Int. J. Inf. Manag. **71** (2023)
7. Fang, X., Zahn, J.: Sentiment analysis using product review data. J. Big Data **2**, 5 (2015)
8. Pak, A., Paroubek, P.: Twitter as a corpus for sentiment analysis and opinion mining. In: Proceedings of the Seventh International Conference on Language Resources and Evaluation (LREC'10). Valletta, Malta (2023)
9. Vaswani, A., Shazeer, N., Parmar, N., et al.: Attention is all you need. In: Proceedings of the 31st Conference on Neural Information Processing Systems (NIPS 2017), Long Beach, CA (2017)
10. Wolf, T., Debut, L., Sanh, V., et al.: Transformers: state-of-the-art natural language processing. In: Proceedings of the 2020 EMNLP (Systems Demonstrations), pp. 38–45. ACL (2020)
11. Devlin, J., Chang, M., Lee, K., et al.: BERT: pre-training of deep bidirectional transformers for language understanding. In: Proceedings of the NAACL-HLT 2019, pp. 4171–4186. ACL, Minneapolis (2019)
12. Muller, M., Salathe, M., Kummervold, P.E.: COVID-Twitter-BERT: a natural language processing model to analyse COVID-19 content on Twitter. Front. Artif. Intell. **6** (2023)
13. nlptown/bert-base-multilingual-uncased-sentiment. https://huggingface.co/nlptown/bert-base-multilingual-uncased-sentiment. Accessed 12 Aug 2023
14. Lee, J., Yoon, W., Kim, S., et al.: BioBERT: a pre-trained biomedical language representation model for biomedical text mining. Bioinformatics **36**(4), 1234–1240 (2020)
15. Zunic, A., Corcoran, P., Spasic, I.: Sentiment analysis in health and well-being: systematic review. JMIR Med Inform. **8**(1) (2020)
16. Zhang, W., Li, X., Deng, Y., et al.: A Survey on Aspect-Based Sentiment Analysis: Tasks, Methods, and Challenges. arXiv preprint arXiv:2203.01054 (2022)
17. Pontiki, M., Galanis, D., Pavlopoulos, J., et al.: SemEval-2014 task 4: aspect based sentiment analysis. In: Proceedings of the 8th International Workshop on Semantic Evaluation (SemEval 2014), pp. 27–35. ACL (2014)
18. Pontiki, M., Galanis, D., Papageorgiou, H., et al.: SemEval-2015 task 12: aspect based sentiment analysis. In: Proceedings of the 9th International Workshop on Semantic Evaluation (SemEval 2015), pp. 486–495. ACL, Denver (2015)
19. Pontiki, M., Galanis, D., Papageorgiou, H., et al.: SemEval-2016 task 5: aspect based sentiment analysis. In: Proceedings of the 10th International Workshop on Semantic Evaluation (SemEval-2016), pp. 19–30. ACL, San Diego (2016)
20. Chebolu, S.U., Dernoncourt, F., Lipka, N., et al.: Survey of Aspect-based Sentiment Analysis Datasets. arXiv preprint arXiv:2204.05232 (2022)
21. Saeidi, M., Bouchard, G., Liakata, M., et al.: SentiHood: targeted aspect based sentiment analysis dataset for urban neighbourhoods. In: Proceedings of COLING 2016, the 26th International Conference on Computational Linguistics: Technical Papers, pp. 1546–1556. ACL, Osaka (2016)
22. Zhang, W., Deng, Y., Li, X., et al.: Aspect Sentiment Quad Prediction as Paraphrase Generation. arXiv preprint arXiv:2110.00796 (2021)
23. Yang, H., Li, K.: PyABSA: a modularized framework for reproducible aspect-based sentiment analysis. arXiv preprint arXiv:2208.01368 (2022)

24. arXiv. https://arxiv.org. Accessed 08 Sept 2023
25. arXiv API Access. https://info.arxiv.org/help/api/index.html#arxiv-api-access. 12 Aug 2023
26. Trusca, M.M., Frasincar, F.: Survey on aspect detection for aspect-based sentiment analysis. Artif. Intell. Rev. **56**(5), 3797–3846 (2022)
27. Park, H.H., Vyas, Y., Shah, K.: Efficient Classification of Long Documents Using Transformers. arXiv preprint arXiv:2203.11258 (2022)
28. Zheng, S., Huang, J., Chang, K.C.: Why Does ChatGPT Fall Short in Providing Truthful Answers? arXiv preprint arXiv:2304.10513 (2023)
29. Lundberg, S.M., Lee, S.: A unified approach to interpreting model predictions. In: Proceedings of the 31st Conference on Neural Information Processing Systems (NIPS 2017), Long Beach, CA (2017)
30. Kamalov, F., Gurrib, I.: A New Era of Artificial Intelligence in Education: A Multifaceted Revolution. arXiv preprint arXiv:2305.18303 (2023)

Wi-SafeHome: WiFi Sensing Based Suspicious Activity Detection for Safe Home Environment

Gayathri Gorrepati, Ajit Kumar Sahoo, and Siba K. Udgata[✉]

WiSeCom Lab, School of Computer and Information Sciences,
University of Hyderabad, Hyderabad, India
{ajit,udgata}@uohyd.ac.in

Abstract. Recently, WiFi signals are being used for sensing task based applications in addition to standard communication activities. Specifically, the Channel State Information (CSI) extracted from WiFi signals through channel estimation at the receiver end provides unique information about environmental dynamics. This CSI data is used for various tasks including motion and human presence detection, localization, environmental monitoring, and a few other sensing applications. By analyzing both the amplitude and phase of the CSI data, we can gain intricate insights into how signal transmission paths are affected by physical and environmental changes. In this study, we focus on leveraging low-cost WiFi-enabled ESP32 micro-controller devices to monitor suspicious-related activities within indoor environments. We conducted exhaustive experiments involving four distinct suspicious-related human activities leaving a room, entering a room, sneaking into a room without formal entry, and engaging in suspicious activities within a room. To enhance the detection rate for these activities, we employ feature engineering techniques on the received CSI data. Additionally, we applied a low-pass filter to eliminate noise from the received signal effectively. To achieve accurate suspicious activity classification, we harnessed various lightweight machine learning (ML) algorithms, which include Support Vector Machine (SVM), Random Forest (RF), Gradient Boosting, Extreme Gradient Boosting (XG Boost), and K-Nearest Neighbor (KNN). Our results reveal that KNN outperformed the other ML models, achieving an accuracy rate of 99.1% and F1-Score of 0.99. This suggests that KNN is a robust choice for effectively classifying suspicious activities based on WiFi CSI data.

Keywords: WiFi sensing · Channel State Information · Smart Home · Suspicious Activity Detection · ESP32 · Machine Learning

1 Introduction

In contemporary times, the applications of WiFi signals have expanded beyond conventional data transmission. WiFi technology is now harnessed for diverse

© The Author(s), under exclusive license to Springer Nature Switzerland AG 2024
B. J. Choi et al. (Eds.): IHCI 2023, LNCS 14532, pp. 291–302, 2024.
https://doi.org/10.1007/978-3-031-53830-8_30

sensing and activity detection purposes. Any disruption in the signal's trajectory from the transmitter to the receiver leads to alterations in the signal's characteristics. These alterations in the signal serve as valuable indicators for detecting a wide range of activities [1]. One of the important applications of WiFi sensing is Human Activity Recognition (HAR), which has a huge potential for surveillance-related applications because of its non-invasive and non-line-of-sight (NLoS) nature. Fall detection for elderly persons, human movement detection, gesture recognition, and activity monitoring are essential applications of HAR [2]. Wearable sensor devices lack universal convenience and aren't suitable for continuous usage. Likewise, camera-based models are reliant on lighting conditions and necessitate a clear line of sight. Furthermore, privacy issues can restrict the applicability of this model in certain situations [3]. Suspicious activity is defined as unauthorized entry into a building or an indoor area and performing some activities in and around the facility [4]. Existing detection systems monitor suspicious activities using acoustic, seismic, ultrasonic, passive infrared, microwave, or camera-based sensors. Real-time activity detection can be possible by embedding lightweight machine-learning algorithms in low-cost WiFi devices [5]. WiFi sensing can be a potential solution for suspicious activity detection systems. The advantages of WiFi sensing are it is ubiquitous and can sense through the walls, with no need for line-of-sight and lighting conditions. The WiFi sensing model reuses the infrastructure of wireless communication, so it is low-cost and easy to deploy [6,7]. Disturbances within the region of interest can be gleaned from specific characteristics present in the received WiFi signal. These characteristics, such as the Received Signal Strength Indicator (RSSI) and Channel State Information (CSI), offer valuable insights. RSSI offers a single value that amalgamates various signals resulting from multi-path effects. It provides a broad overview of the received signal and communicates information pertaining to the data-link layer of the communication model. The WiFi channel, segmented into multiple sub-carriers through Orthogonal Frequency Division Multiplexing (OFDM), permits the receiver to estimate the CSI using the received signal. Channel State Information offers a comprehensive and precise breakdown of each sub-carrier.

Suspicious activity detection is essential for indoor safety applications. Most of the existing works have used devices that have high deployment costs and high power consumption. In this work, we used a low-cost, low-power, and portable WiFi-enabled ESP32 device for the experiments. Since each sub-carrier contains essential information, we extracted different statistical features from correlated sub-carriers and used them to detect suspicious activities. Here, we consider suspicious activities as any kind of activity that is performed by trespassers in a home environment or surroundings when the residents are not around.

After an exhaustive literature survey and critical analysis of existing works, we noted the following gaps or limitations.

- Most of the reported work uses wearable devices, which are not very convenient and require high deployment costs.
- Vision-based activity detection requires high computational cost, line of sight, and light condition, and also raises privacy concerns.

- There is almost no study that focuses on suspicious activities in a home environment.
- Most of the studies use a single transmitter to transmit the signals, which is not enough for the entire home.

The main contributions of this research work are as follows:

- We formulated the problem of suspicious activity detection in a home environment using Wi-Fi signal-based CSI parameter
- We propose a lightweight machine-learning model for various suspicious activities using WiFi-enabled low-cost ESP32 modules
- In this work, we designed the model architecture for the deployment of the transmitter, repeater, and receiver, and conducted exhaustive experimentation to collect data as there is no benchmark dataset available for the same.
- We used one transmitter, one receiver, and multiple repeaters deployed in different rooms in a home for better coverage and monitoring of suspicious activities throughout the home
- Performed feature engineering on amplitude data of the CSI values and selected important and easy-to-extract statistical features for analyzing different activities.
- Experiments are conducted in different environments including dark environments to validate the classification accuracy of the proposed model.
- Tested various lightweight machine learning algorithms to select the best.

2   Background

In this section, we will discuss the deployment of a transmitter, repeater, and receiver in the home environment using the Fresnel zone concept. We will also discuss the principle for CSI value extraction.

2.1   Fresnel Zone Based Transmitter, Repeater and Receiver Deployment

In a free-space wireless communication scenario, the Fresnel zone takes the form of concentric ellipses with the transmitter (Tx) and receiver (Rx) as their focal points. Among these ellipses, the innermost one, known as the First Fresnel Zone (FFZ), holds particular significance. Approximately 70% or more of the signal energy is conveyed from the transmitter to the receiver through this zone, making it the primary conduit for maintaining a strong and reliable wireless link [8]. As the distance between the Tx and Rx increases, additional Fresnel zones emerge, with their sizes growing accordingly. While these additional zones contribute to signal propagation, their impact on signal strength diminishes with distance. Nevertheless, the First FZ remains of paramount importance for the majority of wireless communication links, particularly in line-of-sight scenarios. Practical wireless communication scenarios can be influenced by various factors, such as obstacles, terrain, and environmental conditions, which can alter the characteristics of the Fresnel zones and affect signal propagation. For instance, obstructions within the Fresnel zone may lead to signal blockage or diffraction, resulting in signal degradation and reduced link quality. Therefore, comprehending the Fresnel zones' characteristics and their effects on signal propagation is crucial for designing and optimizing wireless communication systems to ensure reliable and efficient data transmission. In this work, fresnel zones are considered for placing the transmitter and receiver at certain points, such that the scattered data is minimal. Dan Wu *et al.* proposed in [9] that utilizing the Fresnel zone model, the WiDir model is designed to infer the walking direction of a user or object by analyzing phase change dynamics from multiple WiFi subcarriers. This approach enables non-intrusive direction detection using existing WiFi signals without the need for extra hardware or sensors. The Fresnel zones provide a theoretical concept for the analysis and design of WiFi sensing systems [10]. In this work, we use the Fresnel zone concept to decide the deployment location of the transmitter (one), repeater (many), and receiver (one) for efficient CSI data collection at the receiver end.

2.2   Channel State Information

In indoor scenarios, wireless signals traveling from the transmitter to the receiver are subject to environmental changes induced by physical obstacles. These alterations within the transmission medium can be effectively assessed through the utilization of CSI data at the receiver's end. The amplitude and phase variations within the CSI serve as valuable parameters for the analysis of various activities. These characteristics are derived from the calculation of the signal's channel matrix, denoted as H, through the mathematical equation: This equation is central to understanding and utilizing CSI data for diverse applications.

$$y = Hx + n$$

The above equation defines the received signal (y) as the outcome of multiplying the transmitted signal (x) by the channel matrix (H), and adding the noise (n). Here, y represents the received signal, x is the signal transmitted, and n stands for the noise component. The channel matrix (H) comprises 64 subcarrier values, each expressed as a complex number, encompassing both amplitude and phase information [11]. Certainly, you can represent each element H_{ij} in the channel matrix as:

$$H_{ij} = |H_{ij}| \cdot e^{i\theta_{ij}}$$

Here, $|H_{ij}|$ represents the magnitude or amplitude of the complex number H_{ij} and $e^{i\theta_{ij}}$ represents the phase of the complex number H_{ij}. This representation separates the magnitude and phase components, making it clearer to understand how the channel affects the transmitted signals in terms of amplitude and phase. In wireless communication systems, these amplitude and phase values are essential for signal processing, beamforming, and other operations to optimize the quality of communication channels.

The IEEE 802.11n standard employs the Orthogonal Frequency Division Multiplexing (OFDM) modulation scheme, utilizing a total of 64 sub-carriers within the signal. However, only 52 of these sub-carriers are operational, while the remaining 12 are designated as NULL sub-carriers. Among the 52 active sub-carriers, four are specifically allocated for pilot signals, serving the purpose of synchronization between the transmitter and receiver. The amplitude and phase values of each of these sub-carriers can be determined using the following formulas [12],

$$Amplitude = \sqrt{imaginary^2 + real^2}$$

$$Phase = \text{atan2}(imaginary, real)$$

We considered amplitude data of 52 sub-carriers for further analysis.

3   Methodology

The WiFi sensing-based suspicious activity detection model uses variations across multiple sub-carriers to detect activities. At first, the raw CSI data is parsed to extract the amplitude and phase. Then we pre-process the amplitude data to remove noise and outliers. The essential features are extracted from the amplitude data for classification using machine learning models. The overview of the proposed suspicious activity detection workflow is shown in Fig. 1.

3.1   Feature Extraction

In the realm of machine learning, features play a pivotal role. To streamline the dimensionality of our training input data, we engage in feature engineering. This process involves extracting crucial statistical features that effectively capture the variations occurring over specific time intervals. Our emphasis is on lightweight algorithms, ensuring that these features are suitable for real-time applications.

We implement a sliding window mechanism to pinpoint windows that are particularly sensitive and encompass specific activities. Among the sub-carriers, higher variance denotes a greater degree of variation in the dataset over a defined period, making it more susceptible to disturbances compared to other sub-carriers. Conversely, lower variance signifies fewer variations, implying a lack of activity. Five significant statistical features, specifically the mean, variance, mean absolute deviation, amplitude range, and Interquartile range (IQR), are extracted from the data [13]. The feature matrix (W) has dimensions equal to the *number of samples* multiplied by 5.

The matrix presented below is a snapshot of data within a specific timeframe. This window has dimensions of $m \times n$, where m signifies the number of packets and n denotes the number of sub-carriers. In this matrix, $x_{i,j}$ symbolizes the amplitude of the ith packet within the jth sub-carrier.

The matrix, $W_{(m,n)}$, encapsulates information about the sub-carriers within a window of size m.

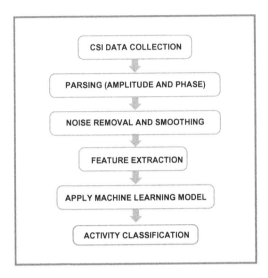

Fig. 1. Proposed architecture for WiFi-based suspicious activity detection system

$$W_{(m,n)} = \begin{bmatrix} x_{1,1} & x_{1,2} & \cdots & x_{1,n} \\ x_{2,1} & x_{2,2} & \cdots & x_{2,n} \\ \vdots & \vdots & \vdots & \vdots \\ x_{m,1} & x_{m,2} & \cdots & x_{m,n} \end{bmatrix}$$

We employ a row-wise sliding technique, with a chosen step size, to traverse the data. With each window's shift, we measure statistical features, which are then compared to those of subsequent windows. This process allows us to identify the specific activity windows necessary for analysis.

Upon the selection of a particular window, we extract a feature matrix, which encompasses the features related to each activity. In this context, f_{s_k} signifies the feature vector of the kth sub-carrier. Each feature vector comprises five distinct values, each corresponding to an extracted feature attribute.

$$Feature(W) = \left[f_{s_1}, f_{s_2}, f_{s_3} \cdots f_{s_k} \cdots f_{s_n} \right]$$

3.2 Machine Learning Models

The central objective of this research is classifying the detection of suspicious activities through the application of machine learning models. In pursuit of this goal, we have opted for low-complexity machine learning models, specifically the Support Vector Machine, K-Nearest Neighbor, and Random Forest, for the purpose of classification. The training data and test data ratio is 90:10. After a series of exhaistie experiments for the optimal value of hyperparameters for these machine learning models, the hyperparameters selected for these machine learning models are presented in Table 1.

4   Experiment Setup and Data Collection

In this section, we discuss the details of the experimental setup, the devices used, and the data collection methods.

4.1   Experimental Setup

For the experiment, we have used the ESP32 microcontroller, which is a cost-effective WiFi-enabled device, as a receiver (Rx). The experiment is conducted in an ideal home environment. The transmitter (Tx) is placed outside the room, which is a commodity WiFi router in the corridor fitted to the ceiling, and the receiver is placed inside the room. The receiver is connected to the computer system for real-time activity analysis and visualization as shown in Fig. 2. The distance between the transmitter and receiver is 8 m. Each room in the house has a commodity Wi-Fi router configured as a repeater (RTx). Each RTx is connected to Rx, and the Fresnel zones FZ1, FZ2, and FZ3 are between the receiver (Rx) and transmitter (Tx), receiver (Rx), and repeater (RTx) respectively.

4.2   Data Collection

Each activity was performed for 10 s and repeated 60 times with six different subjects (persons) over one week. Each day we conduct the experiments repeatedly by allowing different persons to perform different types of suspicious activities. We performed four different types of activities of suspicious nature in the region of interest. Activities such as leaving the room, entering the room, moving around in front of the door, and sneaking into the room but not entering are performed. The data is collected at the receiver end at a sampling rate of 100 packets per second and transmitted to the computing system for analysis.

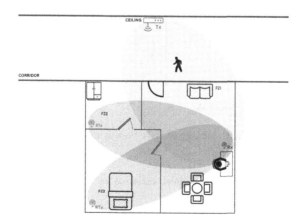

Fig. 2. A scenario of an experimental setup for suspicious activity detection

5   Results and Discussions

Amplitude serves as a primary choice for activity detection, primarily owing to its capacity to capture variations induced by obstructions in the transmission path between the transmitter and receiver. In contrast, within a static environment, the amplitude information exhibits minimal fluctuations. Initially, our focus revolved around statistical parameters, encompassing measures such as variance, mean, minimum, maximum, average, standard deviation, range, interquartile range (IQR), skewness, and kurtosis. Then we selected five essential features for the suspicious activity analysis. Initially, we performed sub-carrier and frame-wise analysis on the amplitude part of the CSI data. The phase component is not used for activity recognition because the phase values of WiFi signal are affected by Channel Frequency Off-set (CFO) and Sampling Frequency Off-set (SFO) [14].

5.1   Sub-carrier Wise Analysis

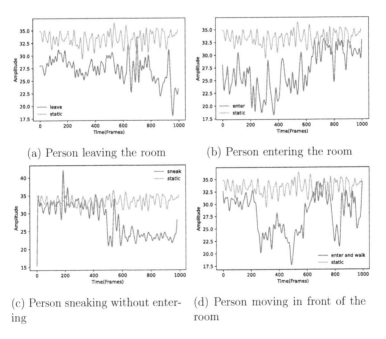

(a) Person leaving the room   (b) Person entering the room

(c) Person sneaking without enter-   (d) Person moving in front of the
ing   room

Fig. 3. Graphs for sub-carrier with highest variance for an activity vs. the corresponding sub-carrier of static state

Within each frame, there is a total of 52 sub-carriers, and the variance of the data for each sub-carrier is a key indicator. A higher variance within the data suggests greater variations among the values across the frames. For each distinct activity, we computed the variance of each sub-carrier during the activity's timeframe and identified the sub-carrier exhibiting the highest variance. This sub-carrier with the maximum variance signifies heightened sensitivity to environmental alterations compared to the others. Graphs were generated to illustrate the relationship between the amplitude of the maximum variance sub-carrier for each activity and the corresponding sub-carrier for the static state. In Fig. 3(a), which pertains to *leaving the room* variations are most prominent within the timeframe of 600 to 1000. Figure 3(b) depicts *entering the room* characterized by heightened variation between the timeframes of 200 to 700. Figure 3(c) showcases *sneaking without entering the room* demonstrating activity primarily within the timeframe of 400 to 600. Finally, Fig. 3(d) illustrates *moving around in front of the door* with the maximum variance observed in the timeframe spanning 200 to 1000.

5.2   Frequency Domain Analysis

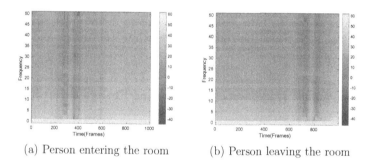

(a) Person entering the room   (b) Person leaving the room

Fig. 4. Frequency domain graphs

We conducted a Fast Fourier Transform (FFT) on the filtered amplitude data to shift from the time domain to the frequency domain. By utilizing this frequency domain data, we generated spectrograms, which unveil discernible differences among activities, particularly during human entries and exits. Figure 4(a) portrays the frequency domain data associated with a person entering a room, while Fig. 4(b) illustrates the same individual leaving the room. These graphs shed light on the way intensity (frequency/amplitude) diminishes as frequency increases. Notably, for the human entering the room, there's a specific timeframe where density decreases with an increase in frequency, rendering it notably less dense compared to surrounding timeframes. Conversely, when a person is leaving the room, the intensity decreases over a timeframe, but this graph is the inverse of the entering activity. During entry, CSI attenuation is more pronounced at the activity's outset, whereas during exit, attenuation escalates after a certain duration, pushing the lower-intensity plot to the opposite extreme. Through frequency domain analysis, we've effectively distinguished between a person entering and leaving a room.

5.3   Machine Learning-Based Activity Classification

A range of machine learning algorithms, including Support Vector Machine, Random Forest, Gradient Boosting, Extreme Gradient Boosting, and K-Nearest Neighbor, have been employed to classify the activities. The performance of all five machine learning models used in this study is summarized in Table 1, showcasing their accuracy and F1-score values. Notably, the K-Nearest Neighbor (KNN) model has outperformed the others, achieving an accuracy of 99.1% and an impressive F1-score of 0.99. The confusion matrix illustrated in Fig. 5 provides a comprehensive overview of the top 4 classification model's performance on the test dataset. They serve as a visual representation of the model's accuracy, showcasing both correct and incorrect predictions for each class. Furthermore, it highlights the specific classes that the model finds challenging to differentiate. In this context, classes 0, 1, 2, and 3 correspond to distinct activities, namely, entering the room, leaving the room, sneaking into the room, and moving in front of the room respectively.

Table 1. Comparison of classification Accuracy and F1-score for machine learning models

ML Models	Hyperparameters	Accuracy	F1-score
Random Forest	$n\_estimators = 10$	83.3%	0.81
XG Boosting	default	93%	0.93
Gradient Boosting	$n\_estimators = 3, max\_features = 4$	83.3%	0.81
SVM	kernel = 'linear', C = 0.1, degree = 2	97.5%	0.97
KNN	K = 3	**99.1%**	**0.99**

(a) Gradient Boosting   (b) Extreme Gradient Boosting

(c) Support Vector Machine   (d) K-Nearest Neighbor

Fig. 5. Confusion matrices of four top-performing machine learning models

6   Conclusions and Future Work

This research introduces an innovative device-free WiFi-based model for detecting suspicious activities in a smart home environment. By leveraging the channel variations induced by reflection, scattering, and diffraction, we have created an efficient detection system capable of monitoring potentially suspicious human behaviors. The experiments conducted simulate activities such as room entry, exit, sneaking, and movement within a secure facility. Key to this model's success are the use of low-cost, low-power, easily deployable, and portable WiFi-enabled ESP-32 devices. For classification, five distinct models such as SVM, RF, GB, XGB, and KNN have been employed, each capitalizing on statistical features extracted via a sliding window mechanism. Among these models, KNN stands out with an outstanding classification accuracy of 99% and an impressive F1-score of 0.99.

One of the model's current challenges lies in effectively tracking and detecting multiple individuals entering or leaving a room simultaneously. To address this challenge, future enhancements may involve advanced technologies, such as Multiple Input, Multiple Output (MIMO) WiFi devices. Additionally, the potential extension of this model to other commodity WiFi devices is a promising avenue for further research and development.

References

1. Ma, Y., Zhou, G., Wang, S.: WiFi sensing with channel state information: a survey. ACM Comput. Surv. (CSUR) **52**(3), 1–36 (2019)
2. Sahoo, A.K., Akhil, K., Udgata, S.K.: Wi-fi signal-based through-wall sensing for human presence and fall detection using esp32 module. Intell. Syst. **431**, 1–6 (2022)
3. Wang, T., Yang, D., Zhang, S., Wu, Y., Xu, S.: Wi-Alarm: low-cost passive intrusion detection using WiFi. Sensors **19**(10) (2019)
4. Tripathi, R.K., Jalal, A.S., Agrawal, S.C.: Suspicious human activity recognition: a review. Artif. Intell. Rev. **50**, 283–339 (2018)
5. Sahoo, A.K., Kompally, V., Udgata, S.K.: Wi-fi sensing based real-time activity detection in smart home environment. In: 2023 IEEE Applied Sensing Conference (APSCON), pp. 1–3 (2023)
6. Jiang, H., Cai, C., Ma, X., Yang, Y., Liu, J.: Smart home based on WiFi sensing: a survey. IEEE Access **6**, 13317–13325 (2018)
7. Sruthi, P., Udgata, S.K.: An improved Wi-Fi sensing-based human activity recognition using multi-stage deep learning model. Soft Comput. **26**(9), 4509–4518 (2022)
8. Zhang, D., Zeng, Y., Zhang, F., Xiong, J.: Chapter 11 - WIFI CSI-based vital signs monitoring. In: Wang, W., Wang, X. (eds.) Contactless Vital Signs Monitoring, pp. 231–255. Academic Press (2022)
9. Wu, D., Zhang, D., Xu, C., Wang, Y., Wang, H.: WiDir: walking direction estimation using wireless signals. In: Proceedings of the 2016 ACM International Joint Conference on Pervasive and Ubiquitous Computing, pp. 351–362 (2016)
10. Zhang, D., Zhang, F., Wu, D., Xiong, J., Kai, N.: Fresnel Zone Based Theories for Contactless Sensing, pp. 145–164, March 2021
11. Atif, M., Muralidharan, S., Ko, H., Yoo, B.: Wi-ESP-A tool for CSI-based device-free Wi-Fi sensing (DFWS). J. Comput. Des. Eng. **7**(5), 644–656 (2020)
12. Hernandez, S.M., Bulut, E.: Lightweight and standalone IoT based WIFI sensing for active repositioning and mobility. In: 2020 IEEE 21st International Symposium on A World of Wireless, Mobile and Multimedia Networks (WoWMoM), pp. 277–286. IEEE (2020)
13. Natarajan, A., Krishnasamy, V., Singh, M.: A machine learning approach to passive human motion detection using WiFi measurements from commodity IoT devices. IEEE Trans. Instrum. Meas. 1 (2023)
14. Yousefi, S., Narui, H., Dayal, S., Ermon, S., Valaee, S.: A survey on behavior recognition using WiFi channel state information. IEEE Commun. Mag. **55**(10), 98–104 (2017)

Comparative Analysis of Machine Learning Models for Menu Recommendation System

Aleksandr Kim, Ji-Yun Seo, and Sang-Joong Jung[✉]

Dongseo University, 47 Jurye-ro, Sasang-gu, Busan 47011, Korea
sjjung@dongseo.ac.kr

Abstract. In an era characterized by an unprecedented abundance of food-related content and a growing diversity of user preferences, personalized recommendation systems play a vital role in enhancing user experiences across various online platforms. This research paper studies multiple single models to be combined within an Ensemble Model in order to provide more accurate and diversified recommendations in the future works. The primary objective of this study is to evaluate and compare the performance of four distinct models – SVM, Random Forest, LSTM RNN, and Collaborative Filtering, considering precision and recall as key evaluation metrics, in order to identify the most effective approach for enhancing user experience and increasing customer satisfaction. Through rigorous tests and performance evaluation, we analyze the strengths and weaknesses of each model in terms of recommendation accuracy, scalability, and real-world applicability. Furthermore, this study creates a foundation for upcoming work by proposing the combination of the most efficient models within an Ensemble Model. By harnessing the collective capabilities of these diverse models, we are planning to build a powerful recommendation system and improve recommendation quality.

Keywords: SVM · LSTM · Collaborative Filtering · Random Forest

1 Introduction

In recent years, the food industry has witnessed a drastic change, primarily due to the widespread popularity of food delivery and online ordering platforms [1]. The convenience and accessibility offered by these platforms provided consumers with an unprecedented range of culinary choices. As a result, the sheer volume of food-related content and the diversity of user preferences have reached its highest levels. This paradigm shift in food consumption has not only provided new opportunities but has also created significant challenges in effectively matching users with the most relevant and appealing meal options. Several studies and reports have highlighted in popularity of food delivery and online ordering services. For instance, a report by Statista [2] noted that the online food delivery worldwide revenue of US$923.1 billion in 2023 is expected to increase to US$1,465.6 billion by 2027. This growth emphasizes the increasing reliance of consumers on digital platforms to fulfill their culinary desires. In parallel with the growth of the food delivery industry, personalized recommendation systems became a critical component in enhancing user experiences on these platforms. Such systems leverage advanced algorithms to analyze user behaviors, preferences, and item features to provide tailored customized meal recommendations.

2 Related Works

This section explores some of the key approaches and methods that have been investigated, with a specific focus on the four models under evaluation in this study: Collaborative Filtering (CF), Long Short-Term Memory Recurrent Neural Network (LSTM RNN), Random Forest (RF), and Support Vector Machine (SVM). CF [3] has been a base technique refined for years in the field of recommendation systems. CF is based on the idea that users who have interacted with similar items in the past are likely to have similar preferences in the future. User-based CF identifies users with similar behavioral patterns [4] while item-based CF computes the similarity between items [5]. RF is an ensemble learning method that combines multiple decision tree classifiers to make predictions [6]. In the context of recommendation systems, RFs can be used to model user preferences and item features, providing personalized recommendations. SVMs are a supervised learning method that aims to find an optimal hyperplane in a high-dimensional space to maximize the margin between different classes [7]. SVMs can be applied to linearly separable data using a linear kernel [8], but they can also handle non-linearly separable data by transforming it into a higher-dimensional feature space using kernel functions such as polynomial, radial basis function (RBF), or sigmoid kernels [9]. LSTM RNNs have become an architectural innovation in the field of deep learning for sequential data analysis [10]. LSTMs use a complex gating mechanism that enables them to capture long-range dependencies and store information over extended time intervals [11]. With their superior capacity to model sequential data, LSTMs have significantly improved the state-of-the-art in various applications [12]. Table 1 displays the performance of the mentioned models.

Table 1. Selected models' performance in the related works

Model		Dataset	Task	Metric	Score
CF	Matrix Factorization [13]	MovieLens	Recommendation	Recall	72.94
	Neural CF [14]	Pinterest	Recommendation	Recall	87.77
RF	Random Forest [15]	Segment	Classification	Precision/Recall	97.7/97.7
	Extra Trees [16]	HCV	Classification	Precision/Recall	99.8/94
SVM	Linear SVM [17]	CALTECH	Classification	Recall	63.2
	HFM SVM [18]	Aberdeen 2D face	Classification	Accuracy	98.33
LSTM RNN	LSTM RNN [19]	Honeycomb-structures dataset	Classification	Recall	96
	L Mixed [20]	IMDB Movie Reviews	Sentiment Analysis	Accuracy	95.68

3   Methodology

This study was conducted to estimate the performance of single models listed above. In later studies selected models would be performing as a part of an Ensemble Model which will address the drawbacks of each of the models and increase overall performance in the future studies. This section explains the data prerequisites and chosen architecture for each model, as well as the evaluation process.

3.1   Data Acquisition and Preprocessing

The utilized real-world dataset for this study is the commercial dataset provided by a local company. The data was collected from more than 8,000 stores and 220,000 users all over Busan City, South Korea. The raw dataset contained three main cuisine section sheets that we combined to create a single dataset (Table 2-a). Subsequent data analysis concluded the list of columns we were able to utilize in the study, while the rest were dropped or preprocessed. The cuisine category was added to the dataset as a separate column. The descriptive address column was simplified to the district name. Order datetime turned into month and hour columns. Selected meals in the raw dataset, presented by the "○" sign, were replaced by 1 if purchased and 0 otherwise. Eventually, categorical data was label encoded to properly train ML models (Table 2-b).

Table 2. Example of raw data provided for the study.

Model	Accuracy	Precision	Recall
LSTM RNN	0.9272	0.9272	0.8959
Collaborative filtering	0.9132	0.9281	0.8975
Random Forest	0.8908	0.9462	0.8579
SVM	0.7148	0.7148	0.7148

The second step of the data preprocessing varied depending on the model we evaluated. Data for the Random Forest model was fed without any additional changes. Unlike Random Forest, the LSTM RNN model requires feature standardization. Therefore, StandardScaler was applied to the dataset. In addition, the input data was reshaped to fit the LSTM model format. In the case of the SVM Classifier, we had to transform our multi-label classification task into multi-class classification. SVMs are primarily designed for binary and multiclass classification tasks. Hence, we took 10 columns of target variables consisting of 0s and 1s and turned them into binary numbers. Then converted them into decimal format numbers and fed them to SVM as output. Finally, for our Collaborative Filtering model, we merged the District, Month, and Time columns to create a pivot table and dropped the Cuisine column. Originally, we were planning to build a user-based CF model using user numbers as IDs, but due to data insufficiency decided to categorize users by District, Month, and Time.

3.2 Model Architecture

That's the current architecture for each of our models. Random Forest model consists of 300 decision trees, each with a maximum depth of 25 levels. During the construction of each tree, only 50% of the available features are considered at each split (max_features = 0.5), which adds randomness to the mod el and reduces overfitting. This ensemble of decision trees works together to make predictions, and the final prediction is often based on a majority vote (for classification tasks) or an average (for regression tasks) from all the individual trees. The architecture of the SVM classifier is defined by the choice of a linear kernel, the regularization strength of 1.0 (C), the 'ovo' decision function shape for multiclass classification, and the random seed 42 for reproducibility. The classifier will learn a decision boundary that best separates the classes in the training data based on the specified parameters. The RNN model consists of an LSTM layer with 50 units and ReLU activation [21], followed by a Dense layer with 10 units resulting in 11,510 parameters in total. The model is trained using the Adam optimizer [22] and the Mean Squared Error loss function for 50 epochs. For the CF model, we calculated pairwise Cosine Similarities between all samples in our pivot table.

4 Experiments and Results

After defining the model architectures, we split the dataset and trained them on the training dataset. For most of the models, we used precision, recall, and accuracy as evaluating metrics during the test phase. The initial results achieved by each model are displayed in Table 3. The LSTM RNN model showed the highest results because of its efficiency in working with implicit datasets compared to traditional methods. In addition, the sparse dataset affected the performance of RF and SVM whereas CF managed to achieve better results. The hyperparameter tuning, however, allowed us to drastically increase the performance of RF. Eventually, due to inefficiency and poor results for the current dataset, the SVM model was decided to be omitted in the subsequent research (Fig. 1).

Table 3. Performance of implemented models on testing data

Model	Accuracy	Precision	Recall
LSTM RNN	0.9272	0.9272	0.8959
Collaborative filtering	0.9132	0.9281	0.8975
Random Forest	0.8908	0.9462	0.8579
SVM	0.7148	0.7148	0.7148

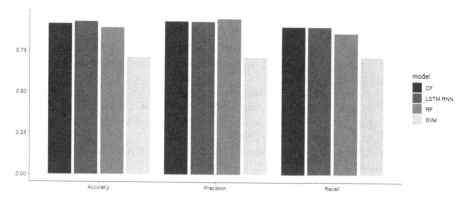

Fig. 1. Visualized performance comparison of implemented models

5   Conclusion

In this study, we evaluate single models on the commercial menu dataset. Our primary aim is to establish a solid foundation for future research, particularly the development of a robust meal recommendation system. The future system will be constructed through an Ensemble Model, carefully built by combining the best-performing models selected in this study. Moreover, we are planning to implement the Meta-learner to improve the predictive performance, address overfitting, and add flexibility to our recommendation system. The final objective is to create a Novel Meal Recommendation System with an Ensemble Model that provides accurate and personalized rec- recommendations. We anticipate that our findings and methodologies will significantly contribute to the development of recommendation systems and data-driven decision-making in the culinary domain.

Acknowledgement. This work was supported by the Technology development Program (RS-2023-00223289) funded by the Ministry of SMEs and Startups (MSS, Korea).

References

1. Chakraborty, D., et al.: Consumers' usage of food delivery app: a theory of consumption values. J. Hosp. Mark. Manag. **31**(5), 601–619 (2022)
2. Statista Research Department: Online Food Delivery – market data analysis & forecast. Statista (2023). https://www.statista.com/study/40457/food-delivery/
3. Koren, Y., Rendle, S., Bell, R.: Advances in collaborative filtering. In: Ricci, F., Rokach, L., Shapira, B. (eds.) Recommender Systems Handbook, pp. 91–142. Springer, New York (2021). https://doi.org/10.1007/978-1-0716-2197-4_3
4. Roy, R., Dutta, M.: A systematic review and research perspective on recommender systems. J. Big Data **9**(1), 59 (2022)
5. Sarwar, B., et al.: Item-based collaborative filtering recommendation algorithms. In: Proceedings of the 10th International Conference on World Wide Web (2001)
6. Rigatti, S.J.: Random forest. J. Insur. Med. **47**(1), 31–39 (2017)

7. Meyer, D., Wien, F.T.: Support vector machines. In: The Interface to LIBSVM in Package e1071, 28(20), 597 (2015)
8. Fletcher, T.: Support vector machines explained. Tutorial paper, 1–19 (2009)
9. Brereton, R.G., Lloyd, G.R.: Support vector machines for classification and regression. Analyst **135**(2), 230–267 (2010)
10. Sherstinsky, A.: Fundamentals of recurrent neural network (RNN) and long short-term memory (LSTM) network. Physica D: Nonlinear Phenomen. **404**, 132306 (2020)
11. Graves, A., Graves, A.: Long short-term memory. In: Supervised Sequence Labelling with Recurrent Neural Networks, pp. 37–45 (2012)
12. Greff, K., et al.: LSTM: a search space odyssey. IEEE Trans. Neural Netw. Learn. Syst. **28**(10), 2222–2232 (2016)
13. Rendle, S., et al.: Neural collaborative filtering vs. matrix factorization revisited. In: Proceedings of the 14th ACM Conference on Recommender Systems (2020)
14. He, X., et al.: Neural collaborative filtering. In: Proceedings of the 26th International Conference on World Wide Web (2017)
15. Ali, J., et al.: Random forests and decision trees. Int. J. Comput. Sci. Issues (IJCSI) **9**(5), 272 (2012)
16. Alotaibi, A., et al.: Explainable ensemble-based machine learning models for detecting the presence of cirrhosis in Hepatitis C patients. Computation **11**(6), 104 (2023)
17. Ladicky, L., Torr, P.: Locally linear support vector machines. In: Proceedings of the 28th International Conference on Machine Learning (ICML-11) (2011)
18. Yang, C., et al.: Hybrid fuzzy multiple SVM classifier through feature fusion based on convolution neural networks and its practical applications. Expert Syst. Appl. **202**, 117392 (2022)
19. Hu, C., et al.: LSTM-RNN-based defect classification in honeycomb structures using infrared thermography. Infrared Phys. Technol. **102**, 103032 (2019)
20. Sachan, D.S., Zaheer, M., Salakhutdinov, R.: Revisiting LSTM networks for semi-supervised text classification via mixed objective function. In: Proceedings of the AAAI Conference on Artificial Intelligence, vol. 33, no. 01 (2019)
21. Nair, V., Hinton, G.E.: Rectified linear units improve restricted Boltzmann machines. In: Proceedings of the 27th International Conference on Machine Learning (ICML-10), pp. 807–814 (2010)
22. Kingma, D.P., Ba, J.: Adam: a method for stochastic optimization. arXiv preprint arXiv:1412.6980 (2014)

A Novel Similarity-Based Method for Link Prediction in Complex Networks

Abhay Kumar Rai[1]([⊠]), Rahul Kumar Yadav[2], Shashi Prakash Tripathi[3], Pawan Singh[1], and Apurva Sharma[4]

[1] Department of Computer Science, Central University of Rajasthan, Ajmer 305817, Rajasthan, India
abhay.jk87@gmail.com
[2] Analytics and Insights Unit, Tata Consultancy Services, Noida 201309, India
[3] Analytics and Insights Unit, Tata Consultancy Services, Pune 411057, Maharashtra, India
[4] Department of Computer Science, Banasthali Vidyapith, Rajasthan 304022, India

Abstract. In complex systems with interactive elements, link prediction plays an important role. It forecasts future or missing associations among entities of a complex system using the current network information. Predicting future or missing links has a wide variety of application areas in several domains like social, criminal, biological, and academic networks. This paper presents a novel method for finding missing or future links that uses the concepts of proximity between the vertices of a network and the number of associations of the common neighbors. We test the performance of our method on four real networks of varying sizes. We tested it against six state-of-the-art similarity-based algorithmss. The outcomes of the experimental evaluation demonstrate that the proposed strategy outperforms others. It remarkably improves the prediction accuracy in considerable computing time.

Keywords: Link prediction · network features · social network analysis · similarity-based methods · similarity scores

1 Introduction

The utility of network theory-based techniques has increased in solving real-world problems in recent years. We can model any complex system as a network or graph and extract usable information from it using these techniques. Many physical, social, academic, and biological systems can be modeled as networks. The persons or entities of such systems are designated as nodes, and their associations are modeled as the edges between the nodes. Finding future associations is one of the usable information in the majority of real-world systems, and the problem of finding such associations in a complex system is called the link prediction problem. Forecasting probable friends in a social network [1], finding invisible relationships between the members of a criminal or terrorist network

[2], suggesting similar products on a product-selling website [3], and recommending potential experts in an academic network [4] are a few applications of link prediction.

The popularity of finding probable associations in complex systems is steadily expanding due to its vast range of applications. As a result, we require a method for finding probable links that give a sufficient level of accuracy in a reasonable computing time. Our research proposes a new similarity formula for forecasting future associations based on two basic network features. It gives better link forecasting accuracy than other methods within a considerable time.

1.1 Motivation

While designing a similarity formula for link prediction, similarity-based approaches utilize a few network features. Zhou et al. [1] predicted the relationship between any two vertices of a graph by measuring the amount of information flow via their common neighboring nodes. The Katz index [5] considers all possible paths of different lengths for computing the similarity scores among unconnected vertices of a graph. Tong et al. [6] consider a specified count of random walks between different node pairs of a network for finding missing links. Mishra et al. [7] designed a method based on edge relevance and merged node concepts for predicting missing links. SimRank [8] is a link forecasting technique that estimates how shortly two random walkers meet who have started from two different nodes in a network. Local path index [9] has derived from the Katz index. It is based on the fact that path lengths having a higher value than a specified length l do not contribute to the similarity scores. So, it only considers finite length paths for estimating alikeness among the unconnected nodes.

The above techniques are suitable for dealing with link prediction problems but they have some limitations. Global feature-based methods like Katz take more time to execute, while local similarity-based approaches have low prediction accuracy. Hybrid methods like MNERLP-MUL [7] give adequate accuracy within a considerable time, but they are not appropriate for all categories of datasets.

The observation from the existing literature on similarity-based link prediction techniques motivates us to design a novel similarity-based method that gives better accuracy in considerable time and also works well for all kinds of networks. Our proposed formula for calculating alikeness among unconnected node pairs in a graph has two components. It utilizes two simple structural features of a network. The first component considers information flow between any two vertices via their common neighbors. And the second one uses the closeness concept between the nodes for finding probable links. In an input network, we compute the alikeness scores between the pairs of vertices not currently connected using the proposed formula. We predict the missing links using these scores.

1.2 Contributions

The main findings of our work are given below:

- We have developed a novel similarity-based technique that forecasts future or missing links using two simple features in a network.

- We have conducted the experiments on four datasets of different sizes and characteristics.
- The proposed method has been compared with local, global, and hybrid feature-based approaches.

A summary of the remainder work is presented here. Section 2 gives a brief description of works related to link prediction tasks. Section 3 proposes a new similarity formula for finding probable associations. Section 4 explains the characteristics of the datasets in brief. Section 5 tells about the evaluation strategy, and Sect. 6 summarizes the principal findings of this study.

2   Related Work

The available link forecasting methods are classified into similarity and learning-based approaches. Based on the type of network features used, the similarity-based techniques may belong to local, global, or hybrid categories. Learning-based strategies are categorized based on the models used for embeddings. They generate either node or edge embeddings for the supplied input graph.

Similarity-based algorithms presume that entities get associated with other similar objects in any complex system [10–14]. These strategies are based on the fact that there will be a similarity between the pair of vertices of a graph if they are associated with alike vertices or if they have proximity to each other in a network. Mostly, we measure the vicinity between an individual pair of vertices in a network by utilizing some distance function. These algorithms use a similarity formula for determining the similarity scores between every disconnected node pair in an input graph. They use neighborhood or path-based information for connection prediction. They are easier to implement than learning-based strategies. This work focuses only on similarity-based approaches.

Similarity-based approaches that belong to local category methods use confined graph structural features to measure the sameness between the two individuals in a network. In general, they are more time-efficient than non-local approaches. Additionally, these methods facilitate the users to efficiently tackle the problem of finding missing or future links in changing and dynamic networks like online social networks. Preferential Attachment [15], Resource Allocation Index [1], and the Adamic/Adar Index [16] are a couple of examples of local similarity-based approaches. The main problem with these methods is that they utilize only adjacent node information for link prediction. These approaches restrict the forecasting of probable links between the set of unconnected nodes whose path length is more than two. It creates a severe problem in real-world complex networks, as in such networks various associations exist at distances greater than two or more, especially in large-size networks.

Unlike local approaches, global techniques employ full topological information of the networks to score each link. The Matrix Forest index [17], Katz index [5], and RWR (Random Walk with Restart) [6] are a couple of examples of techniques based on global similarity. Most of them are path-based approaches. Their computational cost might render them impracticable for large networks.

Hybrid approaches, which strike a balance between local and global approaches, have lately evolved. These methods are widely used because they provide more accurate link predictions in considerable time. These methods are faster to run than global methods and give greater accuracy than other latest global and local algorithms. Local Path Index [9], ComSim [18], FriendREC [19], MNERLP-MUL [7], ALP [21], and CCPA [20] are a couple of examples of hybrid approaches. The newest among them is MNERLP-MUL.

Based on the study of existing link prediction approaches, we identified two simple features that are most prominent in forecasting the future of missing links. Additionally, they require less time for execution if we design the similarity formula using these features. The suggested method differs from others in the following ways: (i) Our method requires lesser execution time than global approaches; (ii) it captures two structural features in contrast to the existing local similarity-based algorithms; and (iii) contrary to other hybrid methods, which use complex features, our method only needs two simple features to perform the link prediction task. (iv) We contrasted our approach with the current similarity-based link prediction methods, and (v) it performs better than others on all datasets.

3   Defining Proposed Similarity Formula

A simple undirected graph G(V, E) without edge weights and self-loops has been considered for defining the proposed similarity formula. The graph must not comprise more than one connected component. We have designed the proposed formula while keeping two network features in mind. The formation of future associations between any two unconnected vertices in a network depends on two prominent characteristics of the nodes: the amount of information that these nodes can pass coming from each of their adjacent nodes decreases as their degree increases, resulting in a decrease in their influence on the neighbors; and their closeness value. The closeness between any two nodes depends on the shortest path length between these vertices. A lower shortest path length represents more closeness between the nodes, while a longer shortest path length represents less closeness between the nodes.

The resource allocation index served as the source for our first feature [1]. According to this index, a higher possibility of future connection development is inferred by the large amount of information flow between any two vertices via their shared neighbors. The second feature is inferred from the closeness-centrality [22]. It depends on the shortest path length between an individual pair of vertices. It implies that if two objects in a complex system are similar in some way, the shortest distance between them will be shorter.

With the aid of these ideas, we explicitly establish the following new similarity measure for determining similarity scores between any two vertices:

$$RSim_{pq} = \mu.RA + (1 - \mu)\frac{N}{sh_{pq}}$$

Where RA is determined by the formula $\sum_{x \in \tau(p) \cap \tau(q)} \frac{1}{|\tau(x)|}$. p and q are the vertices of the input network, $\tau(p)$ is a set of nodes adjacent to p, N is the counts of nodes in a network, and sh_{pq} is the length of the shortest path between the vertices p and q. Parameter $\mu \in [0, 1]$ is a variable to be set by the user and is used to regulate the relative weight and influence of the two components of the similarity formula.

4 Datasets

All methods, including the suggested approach, have been implemented using Python programming language. The effectiveness of all the algorithms has been evaluated using four real-world datasets.

Table 1. A summary of the datasets used for comparison

Name	Type	Nodes	Edges	Description
Dolphin	Undirected	62	159	A dataset of associations among dolphins
Football	Undirected	115	613	A dataset of football games played in USA
fb-pages-food	Undirected	620	2102	A dataset of food related pages
arenas-email	Undirected	1133	5451	A dataset of email communications

Dolphin: This dataset [23] contains a public network of bottlenose dolphins inhabiting New Zealand's Doubtful Sound. The behavioral information on 62 dolphins has been captured from the years 1991 to 2004. In this network, a single node represents a dolphin, and links represent frequent connections between dolphins.

Football: This network [24] was created by compiling data on games played throughout the regular season between US college football teams. Links between the teams through which they are connected symbolize games, while nodes in the network stand in for teams (designated by university name).

fb-pages-food: This network [25] was created by gathering data on associations between the 621 blue-tick official food-related pages from the Facebook website. The connections between the food pages are shown by the links in this network.

arenas-email: The associations established between the users due to email communications between them at a university in southern Catalonia, Spain, are represented as a dataset [26]. An individual user is represented by a node in the network, and an edge denotes that an email was sent from one node to the other. The frequency of email communications between any two individuals and the direction of communications is not recorded.

5   Performance Evaluation

This part of our work compares the effectiveness of the suggested approach to other link prediction techniques. We have calculated the AUC and accuracy values for all algorithms on four real-world datasets to check the effectiveness of each method.

5.1   Evaluation Strategy

We took the same evaluation strategy for all the algorithms. Self-loops and multi-edges have been eliminated after reading the dataset as a graph. For positive sample generation, we arbitrarily pruned 30% of the graph's links. Similarly, we have taken 30% of the input graph's disconnected node pairs to generate negative samples.

We calculated the similarity scores between the given network's disconnected node pairs using all the algorithms used for comparison on the deleted graph. The existence or non-existence of an association between a specific pair of vertices is determined by the mean similarity score. An association with a similarity score higher than the mean value between the two individuals is forecasted to be the future link between these nodes.

5.2   Performance Evaluation of the Suggested Strategy in Comparison to Others

As evaluation metrics, we employed AUC and accuracy. We took the Resource Allocation Index as a local approach; and Katz, RWR, and SimRank as global approaches. In the hybrid category, we have taken LPI and MNERLP-MUL. The MNERLP-MUL is the newest among all the methods. While comparing the suggested technique, we employed the mono-layer version of the MNERLP-MUL. For checking the effect of μ used in the proposed method, we have tried all possible values of this parameter to get its optimal value. The suggested approach performs optimally when μ is set at 0.8. We also used the best value for each parameter supplied by other techniques as well.

Table 2. AUC values for all the methods

Dataset/Algorithm	Dolphin	Football	fb-pages-food	arenas-email	Average
RA	0.7139	0.8254	0.8513	0.7847	0.7938
Katz	0.7748	0.8255	0.9265	0.8588	0.8464
RWR	0.5953	0.5225	0.5668	0.5454	0.5575
MNERLP-MUL	0.7171	**0.8631**	0.8228	0.7984	0.8004
SimRank	0.4839	0.5156	0.4627	0.5272	0.4974
LPI	0.7121	0.8250	0.8353	0.7785	0.7877
RSim	**0.8167**	0.8303	**0.9504**	**0.8746**	**0.8680**

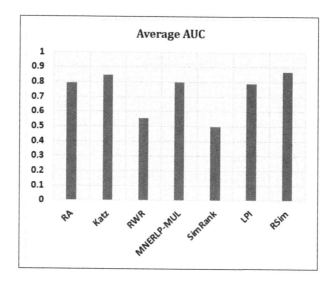

Fig. 1. Average AUC values of all the methods

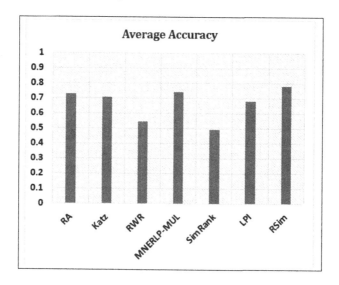

Fig. 2. Average accuracy values of all the methods

We have listed the AUC values for each approach on four datasets in Table 2. If we consider global similarity-based methods, Katz gives the best AUC values compared to others on all the datasets. However, it takes a very high execution time compared to other global approaches. Global approaches like RWR and SimRank take very little time to execute compared to Katz, but their AUC results are quite low on all the datasets compared to Katz. Even their perfor-

Table 3. Accuracy results for all the methods

Dataset/Algorithm	Dolphin	Football	fb-pages-food	arenas-email	Average
RA	0.7021	0.7650	0.7159	0.7220	0.7263
Katz	0.6809	0.7568	0.6850	0.7021	0.7062
RWR	0.5745	0.5301	0.5375	0.5336	0.5439
MNERLP-MUL	0.7021	**0.7842**	0.7384	0.7352	0.7400
SimRank	0.5000	0.5137	0.4697	0.4920	0.4939
LPI	0.6596	0.7568	0.6451	0.6581	0.6799
RSim	**0.7021**	0.7541	**0.8955**	**0.7709**	**0.7807**

mance is lower than other local and hybrid similarity-based approaches. In the hybrid category, MNERLP-MUL performs better on the Football dataset. The designed technique produces better outcomes on the rest of the three datasets. Overall, the suggested strategy outperforms all other methods on three networks and is the second-best performer on the Football dataset. The MNERLP-MUL method gives the best result only for the Football dataset, but it does not perform well on the rest of the datasets. Hence, MNERLP-MUL is suitable for only a few datasets. Even Katz and RA perform better than MNERLP-MUL on the rest of the three datasets.

Here, it is to be noted that the majority of the existing similarity-based methods used for link forecasting perform better on some datasets, while the same approaches do not perform well on other datasets. However, the proposed method performs well on average for all the datasets in terms of AUC. The last column of Table 2 contains average AUC values for each approach. Figure 1 shows a histogram of the average AUC results of all the strategies presented in the last column of Table 2. These AUC results indicate the proposed method as the best-performing technique.

Table 3 comprises the accuracy values of all the methods on four datasets. Among global similarity-based approaches, the Katz method outperforms SimRank and RWR on all the datasets. In the hybrid similarity-based category, MNERLP-MUL produces the best accuracy value on the Football network, but for the other three datasets, our strategy gives the best performance. Altogether, the proposed method gives the best accuracy values on Dolphin, fb-pages-food, and arenas-email networks. For the Football dataset, MNERLP-MUL gives the best accuracy value.

We have recorded the average accuracy values of each approach in the last column of Table 3. Figure 2 shows a histogram of the average accuracy values for all the approaches presented in the last column of Table 3. The average accuracy results indicate our strategy is the best-performing technique.

The discussion above shows how effective the suggested strategy is. On average, it outperforms all other similarity-based approaches on AUC and accuracy metrics. Another important conclusion about our method is that it works well with larger datasets. So, it is appropriate for real-world networks because they tend to be vast in size.

6   Conclusion

In this paper, we have presented a new similarity formula for link prediction. The suggested approach outperforms previous cutting-edge similarity-based link prediction approaches on AUC and accuracy metrics. We drew comparison algorithms from all three categories of similarity-based methods. All the algorithms have been evaluated on four different sizes of real datasets. The experimental outcomes reveal that the suggested strategy surpasses other existing approaches on AUC and accuracy across all datasets. Therefore, the proposed method is suitable for forecasting missing or probable associations in large-size real networks.

References

1. Zhou, T., Lü, L., Zhang, Y.C.: Predicting missing links via local information. Eur. Phys. J. B **71**, 623–630 (2009)
2. Rai, A.K., Kumar, S.: Identifying the leaders and main conspirators of the attacks in terrorist networks. ETRI J. **44**(6), 977–990 (2022)
3. Dhelim, S., Ning, H., Aung, N., Huang, R., Ma, J.: Personality-aware product recommendation system based on user interests mining and metapath discovery. IEEE Trans. Comput. Soc. Syst. **8**(1), 86–98 (2020)
4. Kong, X., Shi, Y., Yu, S., Liu, J., Xia, F.: Academic social networks: modeling, analysis, mining and applications. J. Netw. Comput. Appl. **132**, 86–103 (2019)
5. Katz, L.: A new status index derived from sociometric analysis. Psychometrika **18**(1), 39–43 (1953)
6. Tong, H., Faloutsos, C., Pan, J. Y.: Fast random walk with restart and its applications. In: Sixth International Conference on Data Mining (ICDM'06), pp. 613–622. IEEE (2006)
7. Mishra, S., Singh, S.S., Kumar, A., Biswas, B.: MNERLP-MUL: merged node and edge relevance based link prediction in multiplex networks. J. Comput. Sci. **60**, 101606 (2022)
8. Jeh, G., Widom, J.: Simrank: a measure of structural-context similarity. In: Proceedings of the Eighth ACM SIGKDD International Conference on Knowledge Discovery and Data Mining, pp. 538–543 (2002)
9. Lü, L., Jin, C.H., Zhou, T.: Similarity index based on local paths for link prediction of complex networks. Phys. Rev. E **80**(4), 046122 (2009)
10. Li, S., Huang, J., Liu, J., Huang, T., Chen, H.: Relative-path-based algorithm for link prediction on complex networks using a basic similarity factor. Chaos Interdiscip. J. Nonlinear Sci. **30**(1), 013104 (2020)
11. Son, L.H., Pritam, N., Khari, M., Kumar, R., Phuong, P.T.M., Thong, P.H.: Empirical study of software defect prediction: a systematic mapping. Symmetry **11**(2), 212 (2019)
12. Wang, J., Rong, L.: Similarity index based on the information of neighbor nodes for link prediction of complex network. Mod. Phys. Lett. B **27**(06), 1350039 (2013)

13. Zhou, M.Y., Liao, H., Xiong, W.M., Wu, X.Y., Wei, Z.W.: Connecting patterns inspire link prediction in complex networks. Complexity **2017**, 8581365 (2017)

14. Rai, A.K., Tripathi, S.P., Yadav, R.K.: A novel similarity-based parameterized method for link prediction. Chaos Solitons Fractals **175**, 114046 (2023)

15. Mitzenmacher, M.: A brief history of generative models for power law and lognormal distributions. Internet Math. **1**(2), 226–251 (2004)

16. Adamic, L.A., Adar, E.: Friends and neighbors on the web. Soc. Netw. **25**(3), 211–230 (2003)

17. Chebotarev, P.Y.E., Shamis, E.V.: A matrix-forest theorem and measuring relations in small social group. Avtomatika i Telemekhanika **9**, 125–137 (1997)

18. Yadav, R.K., Rai, A.K.: Incorporating communities' structures in predictions of missing links. J. Intell. Inf. Syst. **55**, 183–205 (2020)

19. Yadav, R.K., Tripathi, S.P., Rai, A.K., Tewari, R.R.: Hybrid feature-based approach for recommending friends in social networking systems. Int. J. Web Communities **16**(1), 51–71 (2020)

20. Ahmad, I., Akhtar, M.U., Noor, S., Shahnaz, A.: Missing link prediction using common neighbor and centrality based parameterized algorithm. Sci. Rep. **10**(1), 364 (2020)

21. Ayoub, J., Lotfi, D., El Marraki, M., Hammouch, A.: Accurate link prediction method based on path length between a pair of unlinked nodes and their degree. Soc. Netw. Anal. Min. **10**, 1–13 (2020)

22. Wasserman, S., Faust, K.: Social network analysis: methods and applications (1994)

23. Lusseau, D., Schneider, K., Boisseau, O.J., Haase, P., Slooten, E., Dawson, S. M.: The bottlenose dolphin community of doubtful sound features a large proportion of long-lasting associations: can geographic isolation explain this unique trait?. Behav. Ecol. Sociobiol. **54**, 396–405 (2003). http://www-personal.umich. edu/~mejn/netdata/

24. Girvan, M., Newman, M. E.: Community structure in social and biological networks. Proc. Natl. Acad. Sci. **99**(12), 7821–7826 (2002). http://www-personal. umich.edu/~mejn/netdata/

25. Rossi, R., Ahmed, N.: The network data repository with interactive graph analytics and visualization. In: Proceedings of the AAAI Conference on Artificial Intelligence, vol. 29, no. 1 (2015). https://networkrepository.com/fb-pages-food.php

26. Guimera, R., Danon, L., Diaz-Guilera, A., Giralt, F., Arenas, A.: Self-similar community structure in a network of human interactions. Phys. Rev. E **68**(6), 065103 (2003). http://konect.cc/networks/arenas-email/

Semantic Segmentation Based Image Signature Generation for CBIR

Suneel Kumar[1], Mona Singh[1], Ruchilekha[2], and Manoj Kumar Singh[1(✉)]

[1] Department of Computer Science, Institute of Science, Banaras Hindu University,
Varanasi, India
{suneel.kumar17,mona.singh10,manoj.dstcims}@bhu.ac.in
[2] DST Centre for Interdisciplinary Mathematical Sciences,
Banaras Hindu University, Varanasi, India
ruchilekha.cims@bhu.ac.in

Abstract. Content-Based Image Retrieval (CBIR) leveraging semantic segmentation integrates semantic understanding with image retrieval, enabling users to search for images based on specific objects or regions within them. This paper presents a methodology for constructing image signatures, a pivotal element in enhancing image representation within a CBIR system. The efficiency and effectiveness of a CBIR system significantly hinge on the quality of the image signature, which serves as a compact and informative representation of raw image data. Our proposed methodology begins with emphasizing clear object or region boundaries through pixel-level semantic segmentation masks. A pretrained semantic segmentation model, such as DeepLab v3+, is employed to generate pixel-wise object class predictions, yielding the necessary segmentation masks. Subsequently, each image is segmented into meaningful regions based on these masks, and relevant features are extracted from each segmented region using a pre-trained Deep Convolutional Neural Network (DCNN) models AlexNet, VGG16 and ResNet-18. During the retrieval phase, when a user queries the system with an image, the query image is segmented using the pre-trained semantic segmentation model, and features are extracted from the segmented regions of the query image. These query features are utilized to search the database for the most similar regions or images. Similarity scores, calculated using Euclidean distance, are used to rank the database entries based on their similarity to the query, allowing for efficient retrieval of the top-k most similar regions or images. We found that for some classes semantic segmented based retrieval better performance in comparison to image based.

Keywords: CBIR · Semantic Segmentation · DeepLabV3+ · Euclidean distance

1 Introduction

In today's digital age, the vast growth of images has revolutionized the way we communicate, learn, and share information. With the rise of social media,

B. J. Choi et al. (Eds.): IHCI 2023, LNCS 14532, pp. 319–329, 2024.
https://doi.org/10.1007/978-3-031-53830-8_33

e-commerce platforms, and the internet at large, billions of images [2,8] are uploaded, shared, and stored daily. This exponential growth in visual data has led to an increasing need for efficient and effective methods to retrieve and organize these images. This is where the field of image retrieval [2,5,24] comes into play.

Image retrieval [18] is a fundamental area of computer vision and information retrieval that deals with the task of searching for and retrieving relevant images from a large collection based on user-defined queries or similarity criteria. In essence, it seeks to bridge the semantic gap [22] between the high-level concepts that humans associate with images and the low-level pixel data that computers can process.

The primary goal of image retrieval is to enable users to find images that match their information needs, whether for personal use, professional research, or commercial applications. This can include finding similar images to a given query image, searching for images containing specific objects or scenes, or even identifying images with particular aesthetic qualities.

To achieve these goals, image retrieval systems employ a wide range of techniques and methodologies, often combining computer vision, machine learning, and information retrieval principles. These systems must address the challenges of scalability, efficiency, and accuracy, given the immense volume of visual data available.

Image retrieval indeed encompasses various approaches, with two prominent categories being tag-based image retrieval (TBIR) and content-based image retrieval (CBIR). These two methods utilize different aspects of images and metadata to retrieve relevant images based on user queries. Let's delve into each of these approaches: First, TBIR relies on metadata or textual annotations, often in the form of keywords or tags, associated with images. These annotations are typically added by users or generated through automated processes like image tagging algorithms or crowd-sourcing. The primary idea is to use these tags to index and retrieve images. Tag-based retrieval leverages human-generated descriptions, making it intuitive for users to find images using familiar keywords. In TBIR mainly two challenges: Ambiguity and Tagging Effort. In ambiguity, different users may use different tags for the same image, leading to ambiguity and potential retrieval issues. While in tagging effort, manually tagging images can be time-consuming and may result in incomplete or inaccurate annotations. So, CBIR become to address the challenges of ambiguity and tagging effort in image retrieval, offering a more efficient and accurate way to search and retrieve images from large collections.

CBIR focuses on the visual content of images, such as colors, textures, shapes, and other low-level features. Instead of relying on textual annotations, CBIR systems analyze the pixel data within images to determine similarity and retrieve relevant images. CBIR systems are not dependent on human-generated tags, making them more objective and potentially more accurate in retrieving visually similar images. But this also have two major challenges: semantic gap and feature extraction. The "semantic gap" refers to the disparity between low-level image features and high-level human perception. This gap arises because computers typically process images based on numerical features like color histograms, texture, shape, and spatial relationships, while humans understand images in terms of objects, scenes, emotions, and context. Bridging this gap remains a significant challenge in CBIR. In feature extraction, extracting meaningful and discriminative features from images requires sophisticated algorithms and techniques.

2   Related Work

CBIR has been a subject of extensive research and development in the field of computer vision and information retrieval. Over the years, various techniques and approaches have been proposed to address the challenges and improve the effectiveness of CBIR systems. Here is an overview of some key research areas and contributions related to CBIR.

In early CBIR systems often used color histograms [11,21] to represent image content. These histograms capture the distribution of colors in an image. In texture analysis methods [9], such as Gabor filters and local binary patterns [16], have been employed to capture textural information in images, and shape descriptors [4] techniques like shape context and contour-based representations help in identifying objects based on their shapes. IBM introduced the initial commercial CBIR system, known as the "Query by Image Content system" [7]. This system utilizes color percentage, color layout (including RGB and YIQ), and Tamura texture features for retrieving images. MIT's Media Laboratory [17] was introduced first academic CBIR system, "Photobook". It incorporates color, texture, and shape features for image retrieval. It uses word components for texture features and shape boundaries for shape features. Bag of Visual Words(BoVW) [25] models have been employed to bridge the semantic gap by clustering and quantizing local features such as Scale-Invariant Feature Transform (SIFT) and Speeded Up Robust Features (SURF) and building histograms of visual words were used and it was further customised with unique characteristics.

Deep Learning have revolutionized CBIR by learning hierarchical features directly from image data. Models like AlexNet, VGG, and ResNet have been adapted for image retrieval tasks. A novel image retrieval method was demonstrated, utilizing neural codes [3] extracted from a pre-trained Convolutional Neural Network (CNN) model, which was subsequently fine-tuned on diverse datasets. Principal Component Analysis (PCA) was uses to compress neural codes, and the compression procedure produced encouraging outcomes. To enrich feature representations, CNN features from various layers were concatenated to create new feature vectors, as proposed in a prior study [14]. To improve retrieval performance, this method further included a retraining mechanism using significant feedback [23]. In order to provide an efficient image description for both the classification problem and image retrieval system, [19] suggested combining machine learning and deep learning methods. This approach integrated the LBP, HOG, and enhanced AlexNet (CNN) descriptors. Using the PCA technique, [13] suggested modifying the VGG16 model for better feature representation and dimension reduction. Distributed frameworks like Hadoop and Spark have been applied to process and retrieve images from massive datasets [15].

The field of CBIR continues to evolve, driven by advances in deep learning, the availability of large-scale image datasets, and the increasing demand for efficient and effective image retrieval solutions in various domains, including e-commerce, healthcare, and multimedia management. Researchers are continually exploring new techniques and methodologies to further enhance the performance and usability of CBIR systems. But bridging the semantic gap remains a fundamental challenge in CBIR.

3   Background

3.1   Deep Convolutional Neural Network

A Deep Convolutional Neural Network (CNN) [12], often referred to simply as a ConvNet, is a specialized type of artificial neural network designed for processing and analyzing visual data, particularly images and videos. Deep CNNs have revolutionized computer vision tasks, such as image classification, object detection, and image segmentation, by significantly improving accuracy and reducing the need for manual feature engineering. CNNs consist of multiple convolutional layers. Each layer applies convolutional operations to the input image to extract features. Convolutional filters or kernels slide over the input image, performing element-wise multiplications and summing the results to create feature maps. These layers can capture patterns like edges, textures, and shapes. Activation functions like ReLU (Rectified Linear Unit) are applied element-wise to the feature maps after convolutional layers. ReLU introduces non-linearity and helps the network learn complex representations. After activation function, pooling layers are often used to reduce the spatial dimensions of feature maps, making the network more computationally efficient. Common pooling operations include max-pooling, which retains the maximum value in a local region, and average-pooling, which calculates the average. Convolutional layers are usually followed by one or more fully connected layers. These layers perform traditional neural network operations and are responsible for making high-level decisions, such as classifying objects in an image.

Deep CNNs have shown to perform very well in a range of computer vision applications, including image classification, object detection, facial recognition, medical image analysis, and more. They have become a cornerstone technology in the field of artificial intelligence and have enabled breakthroughs in visual understanding and analysis. DCNN based CBIR approaches have shown good results in recent research and have grown more famous than handcrafted feature extraction techniques.

3.2   Semantic Segmentation

Semantic segmentation [6] is a computer vision task that involves classifying each pixel in an image into a specific object class or category. Unlike image classification, which assigns a single label to the entire image, semantic segmentation provides a fine-grained understanding of the image by segmenting it into regions and assigning a class label to each pixel within those regions. This technique is crucial for various applications, including autonomous driving, medical image analysis, and scene understanding in robotics.

The primary objective of semantic segmentation is to perform pixel-wise classification. Each pixel in the input image is assigned a label corresponding to the object or category it belongs to. Common classes include "car," "tree," "road," "building," and more. Deep neural networks, particularly Convolutional Neural Networks (CNNs), are commonly used for semantic segmentation tasks. These networks are designed to learn hierarchical features and spatial relationships within the image. Many semantic segmentation models adopt an encoder-decoder architecture. The encoder extracts features from the input image using convolutional layers and downsampling operations (e.g., max-pooling). The decoder then upscales the feature maps to the original image resolution using techniques like transposed convolutions or bilinear interpolation. This process helps in capturing both local and global contextual information. Skip connections or skip connections are often employed to preserve fine-grained details during the upsampling process. These connections allow the network to combine low-level features (from the encoder) with high-level features (from the decoder) to improve segmentation accuracy.

DeepLab v3+ [6] is a state-of-the-art deep learning model for semantic segmentation that builds upon the DeepLab family of models. It is known for its high accuracy and has been widely used in various computer vision tasks, including object segmentation, autonomous driving, and scene understanding. DeepLab v3+ utilizes a convolutional neural network (CNN) architecture. Typically, a pretrained CNN like ResNet, MobileNet, or Xception is used as the backbone network. ASPP is a module that captures multi-scale contextual information from different regions of the image. The decoder module upscales the feature maps to the original image resolution, and skip connections are used to fuse information from different scales. The final layer produces per-pixel class predictions. DeepLab v3+ is a powerful tool for semantic segmentation tasks, but it may require significant computational resources for training. Utilizing pretrained models and transfer learning can help reduce the training time and resource requirements while still achieving excellent segmentation results.

4   Methodology

Content-Based Image Retrieval (CBIR) using semantic segmentation combines the power of semantic understanding with image retrieval to enable users to search for images based on the objects or regions within those images. In this approach, each image is segmented into meaningful regions, and these regions are used to index and retrieve images.

In this section, we will discuss the proposed method for image signature construction, which plays a crucial role in improving image representation within a CBIR system. The efficiency and effectiveness of a CBIR system heavily rely on the quality of the image signature. The signature construction technique is a methodology used to convert raw image data into a compact and informative representation that can be efficiently used for retrieval and comparison in a CBIR system. Image signature is crucial due to following reason:

- **Fast Retrieval**: A well-constructed image signature allows for quick and efficient retrieval of similar images from a large database.
- **Compact Representation**: The signature is typically much smaller in size compared to the raw image data, making it more manageable for storage and processing.
- **Discriminative Power**: A good image signature retains essential information about the image content, ensuring that similar images are accurately matched during retrieval.
- **User Experience**: The quality of the image signature directly impacts the user experience by providing relevant and accurate search results.

The image signature construction is a critical component of a CBIR system, as it transforms raw image data into a compact, informative, and efficient representation that plays a fundamental role in the system's efficiency and effectiveness. Properly designed image signatures are key to achieving accurate and fast image retrieval in CBIR systems.

In the proposed methodology, firstly gather a dataset of images with associated pixel-level semantic segmentation masks. Ensure that each image contains clear object or region boundaries with corresponding class labels. We use a pre-trained semantic segmentation model (e.g., DeepLab v3+) to segment objects or regions within the images. This model should provide pixel-wise object class predictions, resulting in semantic segmentation masks. For each image, segment the objects or regions of interest based on the semantic segmentation masks. After that, extract relevant features from each segmented region. Here we also use pre-trained DCNN model to extract high-level features specific to each region.

These extracted feature organize with image filename in which each segmented region should be linked to its corresponding image and class label. When a user queries the system with an image then segment the query image using the pre-trained semantic segmentation model. Extract features from the segmented regions of the query image. Use these query features to search the database for the most similar regions or images. For extracting the most similar images, we calculate the similarity scores (Euclidean distance) between the query features and the features of regions or images in the database. Then rank the database based on their similarity scores to query. Retrieve the top-k most similar regions or images. Figure 1, illustrate the proposed methodology in which firstly make the feature dataset using semantic segmentation of each images. Then perform retrieval task with respect to the query image. Algorithm 1, illustrate the feature extraction and retrieval process.

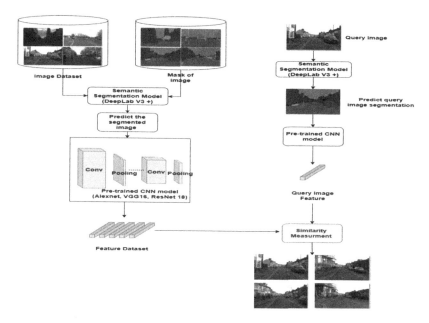

Fig. 1. Proposed methodology

5   Experimental Result and Discussions

In this study, we use CamVid [1] dataset which is a popular computer vision dataset for semantic segmentation tasks, it consists of 701 images, each with corresponding masks for semantic segmentation. The CamVid dataset contains 4 classes. These classes typically represent common objects or regions of interest that you need to segment within the images. All images in the dataset have a resolution of 960 pixels in width and 720 pixels in height. These are relatively high-resolution images. The images are in the RGB color space. This means that each pixel in the images is represented as a combination of red (R), green (G), and blue (B) color channels. There is a corresponding mask for each image in the dataset. These masks are also of the same size (960×720) as the original images. In semantic segmentation, masks are typically grayscale images where each pixel corresponds to a class label. Each pixel in a mask indicates the class of the corresponding pixel in the original image.

This study compares the query's image feature vector and the indexed feature vector datasets using the Euclidean distance metric. Finding the images that are most pertinent to the query image is the goal of query matching. As a consequence of the retrieval, images of feature vectors with the smallest euclidean distance to the query feature vector are presented. This is how the Euclidean distance is calculated:

Algorithm 1. *Image Retrieval*

Require: *Image source folder and Q_{image}*
Ensure: N no. of image relevant to Q_{image}
1: Total number of image N in DB_{images}.
2: net = Load pre-trained network.
3: **for** $i = 1$ to N **do**
4:      image $\leftarrow I \in \mathbb{R}^{N \times M \times 3}$
5:      C = semanticseg(I, *net*);
6:      Image_features $\leftarrow$ layer activation(C).
7: **end for**
8: DB_F = [Image_features label]
9: $Q_{image} \leftarrow I \in \mathbb{R}^{N \times M \times 3}$
10: C = semanticseg(Q_{image}, *net*);
11: Image_features $\leftarrow$ layer activation(C).
12: $Q_F \leftarrow$ [Image_features label]
13: $R_i \leftarrow Euclidean(Q_F, DB_F)$
14: Display top N images $\leftarrow R_i$

$$D\left(Q, I\right) = \sqrt{\sum_{i=1}^{n} \left(F_{Q_i} - F_{I_i}\right)^2}, \tag{1}$$

In the given Eq. 1, the notations used are $D\left(Q, I\right)$ represents the Euclidean distance between the query image feature vector F_Q and the dataset image feature vector F_I. n denotes the total number of images in the dataset.

On the basis of precision Eq. 2 and recall Eq. 3 this approach is evaluated on a CamVid dataset.

$$Precision(P) = \frac{A}{B}, \tag{2}$$

$$Recall(R) = \frac{A}{C}, \tag{3}$$

where, A represent "retrieved relevant images", B represent "total number of retrieved images" and C represent "total number of relevant images in the database".

In this study, we conducted an experiment to curate a comprehensive feature dataset through both pre- and post-processing application of semantic segmentation. To facilitate our descriptive analysis, we employed all available images as queries, with a retrieval limit set at 20 images. For quantifying image similarity, we utilized the Euclidean distance metric, which gauges the resemblance between the feature vector of the query image and that of the feature vector dataset. Tables 1 and 2 provide insights into class-wise average precision and average recall, respectively. Feature extraction was conducted using AlexNet [12], VGG16 [20], and ResNet-18 [10] with and without semantic segmentation, incorporating overlay images for enhanced understanding. An overlay image is created by superimposing or combining a segmented image on top of the original image. This is often done to highlight or visually represent the regions or objects of interest that have been identified through the segmentation process. The result is an image where the segmented areas are overlaid onto the original image to provide a clear visual representation of the segmentation results.

Table 1. Class-wise Average Precision for CamVid

Dataset	Model			
	Class	AlexNet	VGG16	ResNet18
Image	1	96.89	**99.11**	**98.46**
	2	**74.85**	**78.91**	70.29
	3	**86.09**	83.01	85.90
	4	**77.88**	**76.91**	80.08
Segmented image	1	88.42	84.19	86.81
	2	60.09	57.12	59.95
	3	80.24	**85.18**	**86.24**
	4	61.67	56.50	61.41
Overlay image	1	**97.50**	97.17	95.88
	2	59.10	68.61	**74.80**
	3	81.59	82.54	83.57
	4	77.58	72.73	77.26

Table 2. Class-wise Average Recall for CamVid

Dataset	Model			
	Class	AlexNet	VGG16	ResNet18
Image	1	15.62	15.98	15.88
	2	14.82	15.62	13.92
	3	5.64	5.44	5.63
	4	9.16	9.04	9.42
Segmented image	1	14.26	13.57	14.00
	2	11.90	11.31	11.87
	3	5.26	5.58	5.65
	4	7.25	6.65	7.22
Overlay image	1	15.72	15.62	15.46
	2	11.70	13.58	14.81
	3	5.35	5.41	83.57
	4	9.12	8.55	9.08

6   Conclusion

In this study, we have presented a methodology for CBIR utilizing semantic segmentation for image representation and facilitate image retrieval based on specific objects or regions of interest within images. The core of our approach lies in the construction of image signatures, which are compact, informative representations of raw image data. By leveraging pixel-level semantic segmentation masks generated through a pretrained model DeepLab v3+, we achieve precise segmentation of objects or regions within images. Subsequent feature extraction using a pretrained Deep Convolutional Neural Network (DCNN) enables us to capture high-level features specific to each segmented region. These features, organized alongside image filenames and class labels, compose the image signature, serving as a fundamental tool for efficient retrieval and comparison in a CBIR system.

During the retrieval phase, user queries are processed by segmenting the query image using the pre-trained semantic segmentation model and extracting features from the segmented regions. This proposed methodology emphasizes the significance of a well-designed image signature in achieving image retrieval within CBIR systems. Based on the information presented in Table 1, it can be inferred that prior to semantic segmentation, the performance of three classes is superior in AlexNet and VGG16 models, whereas ResNet-18 performs better for two classes. However, after applying semantic segmentation, there are notable shifts in performance, indicating some changes for the respective classes. Future research focus on optimizing the methodology further, exploring feature extraction techniques, and evaluating the approach on diverse datasets to enhance its applicability and efficiency in real-world scenarios.

Acknowledgment. This work is supported under Institute of Eminence(IoE) grant of Banaras Hindu University (B.H.U), Varanasi, India.

References

1. Camvid - the Cambridge-driving labeled video database. http://mi.eng.cam.ac.uk/research/projects/VideoRec/CamVid/
2. Alzu'bi, A., Amira, A., Ramzan, N.: Semantic content-based image retrieval: a comprehensive study. J. Vis. Commun. Image Represent. **32**, 20–54 (2015)
3. Babenko, A., Slesarev, A., Chigorin, A., Lempitsky, V.: Neural codes for image retrieval. In: Fleet, D., Pajdla, T., Schiele, B., Tuytelaars, T. (eds.) Computer Vision – ECCV 2014. ECCV 2014. LNCS, vol. 8689, pp. 584–599. Springer, Cham (2014). https://doi.org/10.1007/978-3-319-10590-1_38
4. Bronstein, A.M., Bronstein, M.M., Guibas, L.J., Ovsjanikov, M.: Shape google: geometric words and expressions for invariant shape retrieval. ACM Trans. Graph. (TOG) **30**(1), 1–20 (2011)
5. Carneiro, G., Chan, A.B., Moreno, P.J., Vasconcelos, N.: Supervised learning of semantic classes for image annotation and retrieval. IEEE Trans. Pattern Anal. Mach. Intell. **29**(3), 394–410 (2007)

6. Chen, L.C., Zhu, Y., Papandreou, G., Schroff, F., Adam, H.: Encoder-decoder with atrous separable convolution for semantic image segmentation. In: Proceedings of the European Conference on Computer Vision (ECCV), pp. 801–818 (2018)
7. Flickner, M., et al.: Query by image and video content: the QBIC system. Computer **28**(9), 23–32 (1995)
8. Gurrin, C., Smeaton, A.F., Doherty, A.R., et al.: Lifelogging: personal big data. Found. Trends Inf. Retr. **8**(1), 1–125 (2014)
9. Haralick, R.M., Shanmugam, K., Dinstein, I.H.: Textural features for image classification. IEEE Trans. Syst. Man Cybern. **6**, 610–621 (1973)
10. He, K., Zhang, X., Ren, S., Sun, J.: Deep residual learning for image recognition. In: Proceedings of the IEEE Conference on Computer Vision and Pattern Recognition, pp. 770–778 (2016)
11. Huang, J., Kumar, S.R., Mitra, M., Zhu, W.J., Zabih, R.: Image indexing using color correlograms. In: Proceedings of IEEE Computer Society Conference on Computer Vision and Pattern Recognition, pp. 762–768. IEEE (1997)
12. Krizhevsky, A., Sutskever, I., Hinton, G.E.: Imagenet classification with deep convolutional neural networks. Adv. Neural Inf. Process. Syst. **25** (2012)
13. Kumar, S., Singh, M.K., Mishra, M.: Efficient deep feature based semantic image retrieval. Neural Process. Lett. 1–24 (2023)
14. Liu, W., Wang, Z., Liu, X., Zeng, N., Liu, Y., Alsaadi, F.E.: A survey of deep neural network architectures and their applications. Neurocomputing **234**, 11–26 (2017)
15. Mezzoudj, S., Behloul, A., Seghir, R., Saadna, Y.: A parallel content-based image retrieval system using spark and tachyon frameworks. J. King Saud Univ.-Comput. Inf. Sci. **33**(2), 141–149 (2021)
16. Ojala, T., Pietikainen, M., Maenpaa, T.: Multiresolution gray-scale and rotation invariant texture classification with local binary patterns. IEEE Trans. Pattern Anal. Mach. Intell. **24**(7), 971–987 (2002)
17. Pentland, A., Picard, R.W., Sclaroff, S.: Photobook: content-based manipulation of image databases. Int. J. Comput. Vis. **18**(3), 233–254 (1996)
18. Rui, Y., Huang, T.S., Chang, S.F.: Image retrieval: current techniques, promising directions, and open issues. J. Vis. Commun. Image Represent. **10**(1), 39–62 (1999)
19. Shakarami, A., Tarrah, H.: An efficient image descriptor for image classification and CBIR. Optik **214**, 164833 (2020)
20. Simonyan, K., Zisserman, A.: Very deep convolutional networks for large-scale image recognition. arXiv preprint arXiv:1409.1556 (2014)
21. Singha, M., Hemachandran, K.: Content based image retrieval using color and texture. Signal Image Process. **3**(1), 39 (2012)
22. Smeulders, A.W., Worring, M., Santini, S., Gupta, A., Jain, R.: Content-based image retrieval at the end of the early years. IEEE Trans. Pattern Anal. Mach. Intell. **22**(12), 1349–1380 (2000)
23. Tzelepi, M., Tefas, A.: Relevance feedback in deep convolutional neural networks for content based image retrieval. In: Proceedings of the 9th Hellenic Conference on Artificial Intelligence, pp. 1–7 (2016)
24. Xu, X., Lu, H., Song, J., Yang, Y., Shen, H.T., Li, X.: Ternary adversarial networks with self-supervision for zero-shot cross-modal retrieval. IEEE Trans. Cybern. **50**(6), 2400–2413 (2019)
25. Zhu, L., Jin, H., Zheng, R., Feng, X.: Weighting scheme for image retrieval based on bag-of-visual-words. IET Image Proc. **8**(9), 509–518 (2014)

Startup Unicorn Success Prediction Using Ensemble Machine Learning Algorithm

Sattaru Harshavardhan Reddy[1]([✉]), Hemanth Bathini[1], Vamshi Nayak Ajmeera[1], Revanth Sai Marella[1], T. V. Vijay Kumar[2], and Manju Khari[2]

[1] School of Engineering, Jawaharlal Nehru University, New Delhi, India
sattaruharsha8529@gmail.com
[2] School of Computer and Systems Sciences, Jawaharlal Nehru University, New Delhi, India

Abstract. Every year, a large number of companies are created, most of them never succeed or even survive. Although many startups receive enormous investments, it is still unclear which startups will receive funding from venture capitalists. Understanding what factors contribute to corporate success, and predicting a company's performance, are of utmost importance. Recently, methods based on machine learning have been employed for accomplishing this task. This paper presents an ensemble machine learning method for predicting startup unicorns that are likely to be successful in their future ventures. This research showcases machine learning algorithms capabilities in improving the effectiveness with regard to early stage startup prediction. The results show that the proposed ensemble machine learning algorithm outperforms traditional machine learning algorithms in terms of various performance metrics thereby demonstrating its potential in predicting startup unicorns that are likely to be successful and competitive in the global market.

Keywords: Ensemble Learning · Early stage startup prediction · Performance metrics

1 Introduction

The success of startups is a vital aspect of any economy, as they can create significant employment opportunities and contribute to economic growth. However, predicting the success of startups remains a challenging task for investors and entrepreneurs. Machine learning has emerged as a powerful tool in the field of startup prediction, with various algorithms and techniques being developed for this purpose. Therefore, there is a need for more robust and accurate methods that can generalize well on the new data. This paper aims to develop an ensemble learning based method for predicting successful startup unicorns. The paper highlights the importance of predicting a startup's success, given that many startups fail to survive or succeed despite receiving enormous investments. The proposed method combines the predictions of multiple machine learning models to improve the generalizability of the methodology. The proposed methodology will help the users to predict the success of the startup and assist in their business investment decision making.

© The Author(s), under exclusive license to Springer Nature Switzerland AG 2024
B. J. Choi et al. (Eds.): IHCI 2023, LNCS 14532, pp. 330–338, 2024.
https://doi.org/10.1007/978-3-031-53830-8_34

Several machine learning based techniques have focused on predicting the success of startups, businesses and early stage companies [1–6]. Though machine learning techniques such as such as Decision Trees (*DT*) [7], Logistic Regression (*LR*) [8], K-Nearest Neighbor (*KNN*) [8], Multinomial Naive Bayes (*MNB*) [9], MultiLayer Perceptron (MLP) [10], Support vector Machine (*SVM*) [11], Extremely Randomized Trees (*ERT*) [12], Random Forest (*RF*) [13] and Gradient Boosting(*GB*) [15] have been able to achieve satisfactory results, they have certain limitations such as low bias and high variance in predicting rare events such as unicorns. This necessitates a need for more robust and accurate methods that can generalize well on new data instances. In this paper, an attempt has been made to use a novel tried ensemble of *SVM*, *ERT* and *GB*, referred hereinafter as ESEG, for predicting success rate of a startup unicorns. Dataset obtained from Kaggle [14] has been used in this study. The proposed ensemble based machine learning technique is compared with the traditional machine learning techniques on performance metrics such as Accuracy, Precision, Recall and F1-score [15].

The paper is organized as follows: A literature review discussing existing works in the area of startup prediction is given in Sect. 2. Section 3 focuses on the proposed ensemble based technique *ESEG*. Experimental results are given in Sect. 4 followed by conclusion in Sect. 5.

2   Literature Review

A brief literature review in regard to existing works in the area of startup prediction is given in Table 1. This review provided a summarized glimpse of various research studies and their respective purposes, focusing on predicting the success of startups, businesses and early stage companies using machine learning and cognitive decision making models. Each study employs different methods, metrics, and algorithms to achieve its objectives.

Table 1. Literature Survey

Reference	Focus	Methods	Metrics	Findings
[1]	Predicting success of early stage startups	Hybrid intelligence method combining machine and collective intelligence; Machine learning algorithms including LR, MNB, SVM, ANN and RF	Mathew correlation coefficient	Hybrid approaches give better results than machine only or human only predictions. Combining the strengths of both machines and humans can lead to more precise predictions of early-stage startup success

(*continued*)

Table 1. (*continued*)

Reference	Focus	Methods	Metrics	Findings
[2]	Understanding the complex and uncertain decision making context of early stage ventures	–	–	The findings of the research highlight that decision-making within a given context is confronted with significant uncertainty stemming from information asymmetry and the presence of unknowable risks. Consequently, accurately predicting outcomes becomes a formidable task
[3]	Predicting success of early stage companies in venture capital investments	Multiclass approach; Machine learning algorithms including SVM, DT, RF and ERT	Accuracy, precision, recall, F1-score	SVM and ERT performed the best overall, with SVM outperforming the other algorithms in the time aware analysis
[4]	Creating classification models from textual descriptions of companies	Multi-label text classification approach; Machine learning algorithms including MNB, SVM, and Fuzzy Fingerprints	Accuracy, precision, recall, F1-score	Support Vector Machines achieved the best performance, with an accuracy above 65% using a multiclass approach. This suggests that machine learning algorithms can be effective in classifying companies based on their textual descriptions, and can provide valuable insights for investors, entrepreneurs, and researchers
[5]	Predicting business success of companies	Supervised models using machine learning algorithms including LR, RF, and KNN	F1-score	KNN gave the highest F1 score, suggesting it is the most effective algorithm for predicting the success of companies based on revenue growth and profitability

3   Proposed Algorithm *ESEG*

The proposed algorithm ESEG amalgamates the outcome of three different Machine Learning algorithms SVM, ERT, and GB on a startup success prediction dataset [14]. Figure 1 explains the structure of the ESEG (ensemble of SVM, ERT, GB) algorithm.

In the proposed algorithm *ESEG*, prior to analysis, a systematic process of feature selection and engineering is conducted. The feature selection phase involves using R regression to obtain scores for each feature. The selection of the R regression method for feature selection is based on the findings and recommendations discussed in [16]. It is suggested that R regression is suitable for scenarios where the input features and the output variable are both numerical. In this particular context, the input data consists

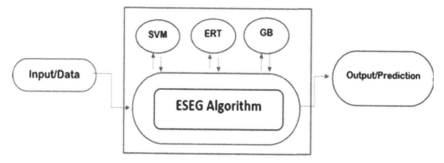

Fig. 1. Structure of the algorithm ESEG

of numerical values, and the objective is to predict a numerical output variable. These scores are determined by considering the absolute values of the feature scores from each algorithm. From these feature scores, the top features, out of 48 features in the Startup Success Prediction dataset, having feature score greater than 0.1 were identified and selected for further analysis. As a result, 13 features were selected and these are *is_otherstate, is_enterprise, avg_participants, has_roundC, has_roundB, has_roundA, funding rounds, age_first_milestone_year, age, is_top500, age_last_milestone_year, milestones* and *norm relationships*. These features have been deemed relevant and important for the subsequent steps in the algorithm.

Since the Startup Success Prediction dataset is labelled, the proposed ESEG algorithm is used to predict the startup success rate. The dataset is divided into training and testing dataset. The training data is used to train the model using ESEG algorithm. The trained model is, thereafter, tested using the testing dataset. The results produced is used to arrive at a confusion matrix using which various performance metrics such as Accuracy, Precision, Recall and F1-score are evaluated.

The pseudocode for the ensemble prediction methodology using three different machine learning algorithms: ERT, SVM and GB is given as under:

Step 1: Send the train data to ERT, SVM, and GB Algorithms
Train the Extremely Randomised Tree model
Train the Support Vector Machine model
Train the Gradient Boosting model
Step 2: Send the test data to all three algorithms
Make predictions using the Extremely Randomised Tree model
Make predictions using the Support Vector Machine model
Make predictions using the Gradient Boosting model
Step 3: Take the predictions from all three algorithms and store them in dictionary
Step 4: Iterate over the dictionary
Count of 1's in the predictions for
model, pred in predictions.items(): if pred
is 1 then ones_count increases by 1
Step 5: If there are two or more 1's, then give final prediction as 1 if
ones_count is greater than or equals to 2: then final_prediction = 1
else:

Step 6: Else give final prediction as 0 final_prediction
= 0

Experimental results are discussed next.

4 Experimental Results

As part of experimentation, the performance of the proposed algorithm ESEG is evaluated on various performance metrics such as accuracy, precision, recall and F1score.
The proposed model ESEG along with traditional machine learning models such as DT,

DT		Predicted	
		False	True
Actual	False	189	21
	True	48	55

LR		Predicted	
		False	True
Actual	False	192	18
	True	44	59

KNN		Predicted	
		False	True
Actual	False	192	18
	True	45	58

MNB		Predicted	
		False	True
Actual	False	173	37
	True	67	36

MLP		Predicted	
		False	True
Actual	False	188	22
	True	55	48

SVM		Predicted	
		False	True
Actual	False	191	19
	True	49	54

ERT		Predicted	
		False	True
Actual	False	190	20
	True	46	57

RF		Predicted	
		False	True
Actual	False	189	21
	True	59	44

GB		Predicted	
		False	True
Actual	False	208	2
	True	95	8

ESEG		Predicted	
		False	True
Actual	False	198	12
	True	43	60

Fig. 2. Confusion Matric for Machine Learning Models under Consideration

LR, KNN, MNB, MLP, SVM, ERT, RF and GB were implemented and applied on the startup success prediction dataset. Based on the results, a confusion matrix is arrived at and is shown in Fig. 2.

Based on the confusion matrix, the Accuracy, Precision, Recall and F1-score values are computed and are shown in Table 2.

Table 2. Accuracy, Precision, Recall and F1-Score values of the Machine Learning Models under Consideration

	Accuracy	Precision	Recall	F1-Score
DT	0.779553	0.723684	0.533981	0.614525
LR	0.801917	0.766234	0.572816	0.655556
KNN	0.798722	0.763158	0.563107	0.648045
MNB	0.667732	0.493151	0.349515	0.409091
MLP	0.753994	0.685714	0.466019	0.554913
SVM	0.782748	0.739726	0.524272	0.613636
ETC	0.789137	0.740259	0.553398	0.633333
RF	0.744409	0.676923	0.427184	0.523810
GB	0.690090	0.80000	0.077660	0.141590
ESEG	0.824281	0.833333	0.582524	0.685714

The graphs, based on Table 2, are plotted for Accuracy, Precision, Recall and F1-score and are shown in Fig. 3, Fig. 4, Fig. 5 and Fig. 6 respectively.

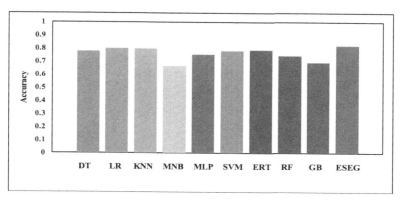

Fig. 3. Accuracy based comparison of Machine Learning Models under Consideration

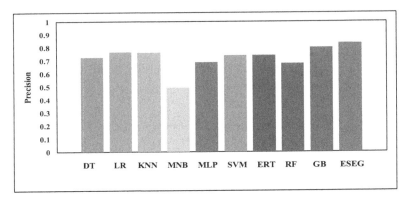

Fig. 4. Precision based comparison of Machine Learning Models under Consideration

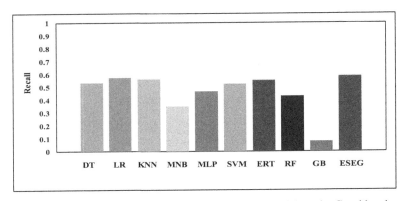

Fig. 5. Recall based comparison of Machine Learning Models under Consideration

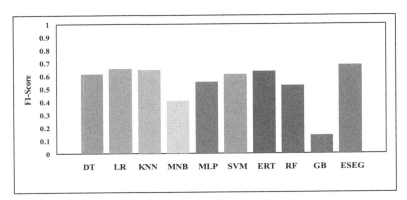

Fig. 6. F1-Score based comparison of Machine Learning Models under Consideration

It can be inferred from the bar graphs shown in Fig. 3, Fig. 4, Fig. 5 and Fig. 6 that the proposed ensemble algorithm ESEG performs comparatively better than the traditional algorithms DT, LR, KNN, MNB, MLP, SVM, ERT, RF and GB on all performance metrics. ESEG achieves an Accuracy as 0.824281, Precision as 0.833333, Recall as 0.582524 and F1-SCORE as 0.685714. Thus, it can be stated that the use of ensemble technique ESEG has led to improvement in the prediction ability and thereby has capability of predicting the success of a startup at an early stage.

5 Conclusion

Predicting startup success is a challenging task, and the proposed ESEG Algorithm aimed to combine the strengths of SVM, ERT and GB in order to improve the predictive capability in terms of predicting the success rate of startups so that better funding decisions can be taken at early stages. The proposed model first identifies relevant and important features in the startup success prediction dataset and then use the data instances, with these features, to train and test the model. Experiment based comparison of the proposed model ESEG with the traditional machine learning models DT, LR, KNN, MNB, MLP, SVM, ERT, RF and GB show that the ESEG outperforms the traditional models on all performance metrics such as accuracy, precision, recall and F1-score.

Though machine learning is a useful solution, techniques for handling extreme uncertainties should also be considered. This paper lays the foundation for future studies to improve the performance with respect to various metrics. Overall, the proposed solution may assist angel investors in making better investment decisions while reducing the instances of bad investment decisions.

The primary concern that needs to be addressed revolves around the availability and collection of data pertaining to startups. Conducting individual interviews and surveys with startups consumes substantial time and resources, while also lacking reproducibility. Moreover, this approach can introduce a response bias. This study has demonstrated that utilizing reproducible models, that train on readily available data, without detailed information about the entrepreneur's personality or the management team's characteristics. However, it is important to note that the dataset used in this research lacks crucial information concerning the personality traits of the entrepreneur and the management team. Incorporating these essential features has the potential to further enhance the performance of the model.

There are several potential enhancements that can be made to improve the accuracy of startup success prediction models. Firstly, establishing a standardized framework for gathering data that is rich in information and, therefore, would be crucial for building robust prediction models. Furthermore, incorporating widely recognized features such as the personality traits of the entrepreneur and the characteristics of the management team can significantly contribute to improved performance of the model. Secondly, the utilization of longitudinal metrics such as growth rate and changes in the number of employees can provide valuable insights and enhance the accuracy of the predictions. Thirdly, by defining success and failure in a much more specific and nuanced manner, it is possible to mitigate class imbalance issues and improve the overall reliability of the predictions. Fourthly, exploring model selection techniques through the implementation of cost functions or matrices could open a fruitful pathway for further investigation. Lastly, narrowing the research focus to a particular industry or subcategory within industries can offer valuable insights into the key drivers of its specific success and facilitate more accurate predictions of business success. These proposed improvements have the potential to augment the predictive capability of quantitative models in predicting startup success and in identifying future high performing ventures.

References

1. Dellermann, D., Lipusch, N., Ebel, P., Popp, K.M., Leimeister, J.M.: Finding the Unicorn: Predicting Early Stage Startup Success through a Hybrid Intelligence Method (2016)
2. Varma, S.: Machine Learning Based Outcome Prediction of New Ventures (2019)
3. Arroyo, J., Corea, F., Jiménez-Díaz, G., Recio-García, J.: Assessment of Machine Learning Performance for Decision Support in Venture Capital Investments (2019)
4. Felgueiras, M., Batista, F., Paulo, J.: Creating Classification Models from Textual Descriptions of Companies Using Crunchbase (2020)
5. Pan, C., Gao, Y., Luo, Y.: Machine Learning Prediction of Companies' Business Success (2018)
6. Peterson, J.C., Bourgin, D.D., Agrawal, M., Reichman, D., Griffiths, T.L.: Using large-scale experiments and machine learning to discover theories of human decision-making. Science 372(6547), 1209–1214 (2021)
7. Rokach, L., Maimon, O.: Decision trees. In: Maimon, O., Rokach, L. (eds.) Data Mining and Knowledge Discovery Handbook, pp. 165–192. Springer, Boston (2005). https://doi.org/10.1007/0-387-25465-X_9
8. Peng, C.J., Lee, K.L., Ingersoll, G.M.: An introduction to logistic regression analysis and reporting. J. Educ. Res. 96(1), 3–14 (2002)
9. Guo, G., Wang, H., Bell, D., Bi, Y., Greer, K.: KNN model-based approach in classification. In: Meersman, R., Tari, Z., Schmidt, D.C. (eds.) OTM 2003. LNCS, vol. 2888, pp. 986–996. Springer, Heidelberg (2003). https://doi.org/10.1007/978-3-540-39964-3_62
10. Rish, I.: An empirical study of the Naive Bayes classifier, vol. 3, no. 22, pp.4863–4869 (2001). https://doi.org/10.1039/b104835j
11. Gurney, K.: An Introduction to Neural Networks, vol. 346. UCL Press Limited (1997). https://doi.org/10.1016/S0140-6736(95)91746-2
12. Evgeniou, T., Pontil, M.: Support vector machines: theory and applications. In: Paliouras, G., Karkaletsis, V., Spyropoulos, C.D. (eds.) ACAI 1999. LNCS (LNAI), vol. 2049, pp. 249–257. Springer, Heidelberg (2001). https://doi.org/10.1007/3-540-44673-7_12
13. Geurts, P., Ernst, D., Wehenkel, L.: Extremely randomized trees. Mach. Learn. 63, 3–42 (2006). https://doi.org/10.1007/s10994-006-6226-1
14. Biau, G.: Analysis of random forests model. J. Mach. Learn. Res. 13, 1063–1095 (2012). https://doi.org/10.5555/2188385.2343682
15. Zhang, C., Zhang, Y., Shi, X., Almpanidis, G., Fan, G., Shen, X.: On incremental learning for gradient boosting decision trees. Neural. Process. Lett. 50(1), 957–987 (2019). https://doi.org/10.1007/s11063-019-09999-3
16. Startup Success Prediction. Kaggle (2021). https://www.kaggle.com/datasets/manishkc06/startup-success-prediction. Accessed 29 May 2023
17. Liu, Y., Zhou, Y., Wen, S., Tang, C.: A strategy on selecting performance metrics for classifier evaluation. Int. Mob. Comput. Multimed. Commun. 6(4), 20–35 (2014)
18. Java point (n.d.): Feature selection techniques in machine learning. Java Point. https://www.javatpoint.com/feature-selection-techniques-in-machine-learning. Accessed 29 May 2023

Author Index